The Psychology of
Workplace
Safety

The Psychology of Workplace Safety

Edited by Julian Barling and Michael R. Frone

American Psychological Association • Washington, DC

Published by
American Psychological Association
750 First Street, NE
Washington, DC 20002
www.apa.org

To order
APA Order Department
P. O. Box 92984
Washington, DC 20090-2984

Tel: (800) 374-2721; Direct: (202) 336-5510
Fax: (202) 336-5502; TDD/TTY: (202) 336-6123
Online: www.apa.org/books
E-mail: order@apa.org

In the U.K., Europe, Africa, and the Middle East, copies may be ordered from
American Psychological Association
3 Henrietta Street
Covent Garden, London
WC2E 8LU England

Typeset in Goudy by Argosy, Waltham, MA

Printer: Data Reproductions, Inc., Auburn Hills, MI
Cover Designer: Mercury Publishing Services, Rockville, MD
Project Manager: Argosy, Waltham, MA

Library of Congress Cataloging-in-Publication Data

The psychology of workplace safety / edited by Julian Barling and Michael R. Frone—1st ed.
p. cm.
Includes bibliographical references and indexes.
ISBN 1-59147-068-4 (alk. paper)
1. Industrial safety. I. Barling, Julian. II. Frone, Michael Robert.

T55.P79 2004
613.6'2—dc21

2003051920

British Library Cataloguing-in-Publication Data

A CIP record is available from the British Library.

Printed in the United States of America
First Edition

To Annette Lilly, who provided me with more support, motivation, and fun at work than I could ever have expected.

—Julian Barling

To my parents, Henry and Mary Ann, and to my wife, Joan, for all of their encouragement and support.

—Michael R. Frone

CONTENTS

CONTRIBUTORS

Julian Barling, School of Business, Queen's University, Kingston, Ontario, Canada

Philip Bohle, University of New South Wales, Sydney, Australia

Alexander Cohen, Occupational Human Factors Consultant, Cincinnati, OH

Michael J. Colligan, National Institute for Occupational Safety and Health, Cincinnati, OH

Michael R. Frone, Research Institute on Addictions, State University of New York at Buffalo

Mark A. Griffin, Australian Centre in Strategic Management, School of Management, Queensland University of Technology, Brisbane

David A. Hofmann, Kenan-Flagler Business School, University of North Carolina at Chapel Hill

E. Kevin Kelloway, Department of Management, Saint Mary's University, Halifax, Nova Scotia, Canada

Niklas Krause, University of California, San Francisco

Catherine Loughlin, Joseph L. Rotman School of Management, University of Toronto, Ontario, Canada

Thomas Lund, National Institute for Occupational Health, Copenhagen, Denmark

Frederick P. Morgeson, Eli Broad Graduate School of Management, Michigan State University, East Lansing

Andrew Neal, School of Psychology and Key Centre for Human Factors, The University of Queensland, Brisbane, Australia

Sharon K. Parker, Australian Graduate School of Management, University of New South Wales, Sydney

Tahira M. Probst, Department of Psychology, Washington State University at Vancouver
Michael Quinlan, School of Industrial Relations and Organisational Behaviour, University of New South Wales, Sydney, Australia
Robert R. Sinclair, Department of Psychology, Portland State University, Portland, OR
Lois E. Tetrick, Department of Psychology, University of Houston, TX
Nick Turner, Institute of Work Psychology, The University of Sheffield, England
Anthea Zacharatos, School of Business, Queen's University, Kingston, Ontario, Canada

ACKNOWLEDGMENTS

This book could not have happened without the cooperation of a number of people. We thank the contributors for being prompt and responsive and for their investment of time and energy into this project. Judy Nemes at the American Psychological Association provided invaluable help as this project moved through the publication process. The anonymous reviewers provided helpful and detailed commentary that contributed to this project. Finally, we would like to acknowledge the editorial assistance of Alex Bilsky at Argosy.

I

INTRODUCTION

1

OCCUPATIONAL INJURIES: SETTING THE STAGE

JULIAN BARLING AND MICHAEL R. FRONE

Occupational safety is an important concern for all working individuals. Worldwide, economically active individuals constitute 60–70% of the adult male population, 30–60% of the adult female population, and 23% of the population of children and adolescents who are 5 to 17 years old (International Labor Organization, 2002; World Health Organization, 1995). Over the course of the 20th century, substantial reductions were seen in the numbers of occupational deaths and injuries, at least in industrialized nations (e.g., Institute of Medicine, 2000; Stout & Linn, 2002; World Health Organization, 1995). Nonetheless, despite this progress, safety concerns in the workplace are still paramount. An examination of the available data shows that work-related injuries and fatalities occur at an alarming and unacceptably high rate throughout industrialized and developing nations. For example, Leigh, Macaskill, Kuosma, and Mandryk (1999) estimated the number of new cases of occupational injuries (fatal and nonfatal) that occur every year in the eight World Bank regions: Established Market Economies, Former Socialist Economies of Europe, India, China, Other Asia and Islands, Sub-Saharan Africa, Latin American and the Caribbean, and

Middle Eastern Crescent. The estimates of yearly occupational injuries ranged from 4.85 million in the Former Socialist Economies of Europe to 24.42 million in China, with an overall estimate of 100.69 million for the world. Approximately 100,000 of these occupational injuries worldwide were fatal. Leigh et al. (1999) argued that, because of problems in obtaining accurate data, their estimates are very conservative. More recent projections by the World Health Organization for the year 2002, using data from 1998, suggested that annually there are 270.7 million occupational injuries and 354,753 occupational fatalities worldwide (Takala, 2002).

Not surprisingly, the economic cost of occupational injuries is staggering. For example, in the United States, Leigh, Markowitz, Fahs, Shin, and Landrigan (1997) estimated that 6,529 people died of work-related injuries and 13.25 million others experienced nonfatal occupational injuries in 1992. Moreover, they estimated that the total cost of occupational injuries in 1992 was $145 billion. In Canada, there were 758 work-related fatalities in 1993 (Statistics Canada, 1994). In the same year, 423,184 Canadian workers experienced injuries serious enough to be compensated either for wages lost due to time off work or for a permanent disability (Statistics Canada, 1994). It has been estimated that the average cost of each workplace injury in Canada is $6,000 and the average cost of each workplace fatality is $492,000 (Marshall, 1996). In the United Kingdom, approximately 1.1 million people were injured at work each year between 1993 and 1996 (Health and Safety Executive, 1997), with the corresponding annual financial cost to the British economy of £4 to £9 billion.

These data become all the more compelling when it is noted that reports of occupational injuries may well be underestimated. For example, individuals in the military are excluded from such data, and epidemiologic studies have shown that many injuries are either not reported or undocumented (Conway & Svenson, 1998; Donald, 1995; Parker, Carl, French, & Martin, 1994; Veazie, Landen, Bender, & Amandus, 1994). Furthermore, occupational injuries have a broad social cost (e.g., Dembe, 2001; Keller, 2001). The official estimates of the number of occupational injuries and their associated economic costs fail to provide an adequate picture of the overall human suffering that results from work-related injuries; those who suffer include not only the injured worker but his or her family and friends as well. Also, the impact of work injuries on human productivity reaches well beyond the workplace and includes a worker's ability to contribute to family and community.

The statistics given in the preceding paragraphs clearly suggest that workplace safety is a critical issue in need of further research. In the rest of this introduction, we will (a) more precisely define and explain our focus in this book, (b) briefly discuss the role of psychology in the study of unintentional occupational injuries, and (c) present an overview of the chapters.

FOCUS ON UNINTENTIONAL OCCUPATIONAL INJURIES

Occupational illness (or health) and occupational injury (or safety) are often discussed together, as can be seen in discussions of occupational health and safety. This clearly makes sense for a number of purposes. However, there are important differences between occupational injury and occupational illness:

> *Occupational injury* represents a wound or damage to the body resulting from unintentional or intentional acute exposure to energy (kinetic, chemical, thermal, electrical, and radiation) or from the acute absence of essential elements (e.g., heat, oxygen) caused by a specific event, incident, or series of events within in a single workday or shift (Baker, O'Neill, Ginsburg, & Li, 1994; Bureau of Labor Statistics, 1997, 2002; Institute of Medicine, 1999). An occupational injury can be fatal or nonfatal. Common forms of occupational injuries include cuts and lacerations, bruises and contusions, burns, strains and sprains, fractured bones, and dislocated joints.
>
> *Occupational illness* represents any abnormal condition or disorder, other than one resulting from an occupational injury, caused by exposure to factors associated with employment. It includes acute and chronic illnesses or diseases that may be caused by inhalation, absorption, ingestion, or direct contact. Occupational illnesses may be fatal or nonfatal. Occupational illnesses may be classified into one of seven categories: occupational skin diseases or disorders (e.g., contact dermatitis, eczema, rash), dust diseases of the lungs (e.g., silicosis, asbestosis), respiratory conditions due to toxic agents (e.g., pharyngitis, pneumonitis), poisoning due to the systematic effects of toxic agents (e.g., lead poisoning, arsenic poisoning, carbon monoxide poisoning), disorders due to physical agents other than toxic materials (e.g., sunstroke, frostbite, welding flash), disorders associated with repeated trauma (e.g., carpal tunnel syndrome, noise-induced hearing loss, bursitis), and all other occupational illnesses (e.g., anthrax, infectious hepatitis, malignant and benign tumors, food poisoning). (Bureau of Labor Statistics, 1997, 2002)

Thus, the study of the etiology and prevention of occupational injury comprises the occupational safety literature, whereas the study of the etiology and prevention of occupational illness comprises the occupational health literature. As noted by Baker et al. (1994), there may not always be a sharp scientific distinction between occupational injury and illness. Nonetheless, a central defining feature of an occupational injury is that it occurs immediately or very close in time to an acute occupational exposure. Furthermore, as noted in the preceding definition of *occupational injury*, one can distinguish between unintentional and intentional injuries (Institute of Medicine, 1999). Unintentional occupational injuries are unplanned, whereas intentional occupational injuries occur as the result of purposeful human action directed either at oneself (self-inflicted injury or suicide) or at others (assault or homicide) (Institute of Medicine, 1999).

The collective focus of the chapters in the present book is on the etiology and underlying processes leading to unintentional occupational injuries, as well their prevention and management. Nonetheless, reference to occupational health is made in several chapters because the extant literature often does not make a clear distinction between occupational injury and health. The present focus on unintentional occupational injuries does not imply that occupational illnesses and intentional occupational injuries are unimportant. However, unintentional occupational injuries have a far greater rate of occurrence than occupational illnesses or intentional occupational injuries. The Bureau of Labor Statistics (2002) reported, for example, that of the 5.2 million occupational injuries and illnesses recorded in the United States during 2001, 4.9 million (94%) were injuries. Leigh et al. (1999) estimated that of the 111.37 million occupational injuries and illness that occur in a given year worldwide, 100.69 million (90%) are injuries. Likewise, of the approximately 6,500 occupational fatalities in the United States in 1992 (Leigh et al., 1997), only approximately 1,000 (15%) were intentional (i.e., workplace homicide); and though there were 13.25 million nonfatal occupational injuries that same year (Leigh et al., 1997), there were only 668,000 (5%) incidents of intentional workplace violence that could lead to injury (Bulatao & VandenBos, 1996). Another, and perhaps more important, reason for this book's focus on unintentional occupational injuries is that the etiology and prevention of unintentional occupational injuries, intentional occupational injuries, and occupational illnesses are sufficiently diverse to preclude covering all three dimensions of workplace health and safety in a single volume.

PSYCHOLOGY AND UNINTENTIONAL OCCUPATIONAL INJURIES: A RELATIONSHIP OF NEGLECT?

Before examining the role played by industrial/organizational psychology in the study of unintentional occupational injuries, we explore briefly why unintentional occupational injuries remain a critical issue in the workplace. First, the prevention of unintentional occupational injury has been largely a public policy issue to which governments have historically responded by enacting legislation that prescribes minimum safety standards, with penalties meted out for deviations from these minimum standards. What is notable about this approach is the variation (a) in the standards and penalties across jurisdictions both within and across countries, and (b) in the extent to which standards are enforced and penalties are imposed for deviations from standards. Second, where they exist, unions have traditionally made occupational safety one of their primary goals, and what unions could not achieve through influence over legislation has typically been sought at the bargaining table (Barling, Fullagar, & Kelloway, 1992). Finally, management has typically tried to limit occupational injuries in two different ways. One approach has been the introduction of greater attention to the principles of ergonomics in the design and use of equipment. A second approach has been to rely more on a control orientation (Barling & Hutchinson, 2000), whereby management sets minimum safety standards (as opposed to such standards being imposed externally by government legislation) and imposes sanctions on employees for any deviations. Thus, unintentional injuries may represent a continuing concern, in part because appropriate safety standards, whether developed through government legislation, union action, or management control, may not exist or may not be enforced.

Even if appropriate standards did exist and were consistently enforced, unintentional occupational injuries would likely remain an issue because the workplace and labor force are not static entities (e.g., National Institute for Occupational Safety and Health, 2002; National Research Council, 1999; Offermann & Gowing, 1990; World Health Organization, 1995). For example, just within the last decade or so, companies have restructured, downsized, increased reliance on contingent and part-time workers, and adopted more flexible and lean production technologies. In many parts of the world, the manufacturing sector is shrinking and the service sector is expanding. Finally, although trends may differ somewhat across various regions of the world, the demographic composition of the labor force is changing. Diversity is increasing along the lines of gender and racial/ethnic background, a growing number of workers from single-parent or dual-earner families are struggling with the simultaneous demands of work and family life, and there are growing concerns regarding the basic skills of many entry-level workers. All of these changes are important because they represent costs and opportunities. That is, some of these changes may undermine workplace safety and others may create new avenues to manage workplace safety. In other words, attention to occupational injury prevention needs to be ongoing.

The study of occupational safety is an interdisciplinary endeavor involving engineering, toxicology, epidemiology, medicine, sociology, economics, and psychology (e.g., Smith, Karsh, Carayon, & Conway, 2003). But what has been the role of psychology in addressing this ongoing need for research on the etiology and prevention of unintentional occupational injuries? Initial interest in this issue can be found as far back as the turn of the 20th century. Hugo Munsterberg, who served as president of the American Psychological Association in 1898 and who has been described as one of the four leading figures in the development of industrial/organizational psychology (Landy, 1997), was interested in industrial injuries and accidents (Quick, 1999). Moreover, chapters on accidents and safety can be found in the early textbooks (1930s–1940s) on industrial/organizational psychology (Ilgen, 1990).

Over a decade ago, in a special issue of the *American Psychologist* devoted to organizational psychology, Ilgen (1990) pointed out that workplace safety represented an important element in the broader spectrum of issues related to employee health and well-being. But Ilgen also suggested that, despite the importance of workplace safety, psychologists in general, and industrial/organizational psychologists in particular, had not played a major role in the study of this issue. Since Ilgen's article was published, it is clear that the study of workplace safety and the publication of such research in mainstream outlets for industrial and organizational psychology and organizational behavior have remained underdeveloped.

One indication of this underdevelopment comes from Campbell, Daft, and Hulin's (1982) survey of topics covered by management scholars. On the basis of a content analysis of several mainstream management and organizational psychology journals, they found that less than 1% of management research focused on occupational safety. More recently, Zacharatos (2001) updated their analysis and reported similar results. Zacharatos examined major scholarly management journals (*Academy of Management Journal, Administrative Science Quarterly*) and organizational psychology journals (*Journal of Applied Psychology, Journal of Organizational Behavior, Journal of Occupational and Organizational Psychology*) published between 1990 and 1999. Articles published during this 10-year period that dealt with any aspect of employee safety, or included the term *employee* (or *occupational*) *safety* as a key or descriptive term, were counted in proportion to those that did not. Zacharatos's results showed clearly that the situation had not improved. Overall, of the empirical articles published in the three mainstream outlets for industrial/organizational psychologists that were examined (*Journal of Applied Psychology, Journal of Organizational Behavior*, and *Journal of Occupational and Organizational Psychology*), substantially less than 1% focused on employee safety.

OVERVIEW OF THE CHAPTERS

We believe strongly, and this book is based on the fundamental premise, that the theoretical and methodological traditions of psychology have much to offer to the understanding of the causes and prevention of occupational injuries. Because of the unacceptably high prevalence rates and costs of occupational injuries, the goal of this book is to provide an overview of key topics in the etiology and management of unintentional occupational injuries in order to promote further the study of this issue by psychologists. Toward this end, we have invited a number of behavioral researchers who have been actively involved in studying different aspects of the etiology, prevention, and management of occupational injuries to showcase their areas of research, to summarize what is known and what is not known, and to raise interesting questions for research and implications for practice.

Among the main contributing factors to unintentional occupational injuries identified by the World Health Organization (Takala, 2002) are poor worker–employer collaborative mechanisms, lack of safety management systems, poor safety culture, poor training and lack of knowledge, and lack of incentive-based compensation systems. Each of these contributing factors falls within the traditional purview of industrial/organizational psychology. The chapters in this book address these key issues and several others.

Although the distinction between risks that cause occupational injuries and factors relevant to the prevention or management of occupational injuries is sometimes arbitrary, chapters 2 through 13 are divided thematically into two parts. Part II, "Climate, Working Conditions, and Culture," contains chapters that outline major workplace factors and psychological issues primarily involved in the etiology of occupational injuries. Chapter 2, by Andrew Neal and Mark A. Griffin, reviews the nature and consequences of workplace safety climate. In this chapter, *safety climate* is defined and a process model is presented that draws important distinctions regarding the key dimensions of safety behavior: determinants of safety performance (i.e., safety knowledge and safety motivation); safety performance (i.e., following safety protocols); and safety outcomes (i.e., occurrence or nonoccurrence of injuries). In chapter 3, Nick Turner and Sharon K. Parker explore the link between work teams and workplace safety. Their chapter explores group processes and the role of health and safety committees, autonomous work teams, and cockpit crews in creating work environments that reduce or increase the potential for occupational injury. Tahira M. Probst, in chapter 4, turns our attention to job insecurity, one of the major issues that have confronted employees since the late 1980s and early 1990s. Probst reviews recent studies and presents a process model to help outline how job insecurity has a negative impact on occupational injuries. Perhaps related to the underlying structural factors that led to an increase in job insecurity in the last 15 years was a move toward greater use of contingent workers. In chapter 5, Michael Quinlan and Philip Bohle discuss occupational safety issues confronting employees involved in contingent work and precarious employment. In chapter 6, Catherine Loughlin and Michael R. Frone discuss the risk of injuries and safety issues for young workers who are just entering the labor force. In chapter 7, Michael R. Frone focuses on the link between employee alcohol and drug use and occupational injuries. A detailed process model is provided to explain inconsistent findings in past research. As part of this explanation, the relation of employee substance use to occupational injuries is placed in the broader context of workplace productivity.

Part III, "Managing Safety and the Return to Work," contains chapters that primarily focus on the management of occupational safety. Collectively, these chapters show the variety of ways in which industrial/organizational psychology can play a direct role in reducing unintentional occupational injuries. In chapter 8, David A. Hofmann and Frederick P. Morgeson begin by discussing the potentially beneficial effects of leadership on occupational safety, an approach that contrasts with the control orientation currently witnessed in many organizations. In chapter 9, Robert R. Sinclair and Lois E. Tetrick turn attention to the link between compensation systems (i.e., pay and benefits) and occupational safety—a link that has attracted considerable interest from management and unions. Another issue that has attracted considerable attention during the past decade in the popular press and in the scientific community is high-performance work systems. Anthea Zacharatos and Julian Barling, in chapter 10, investigate the extent to which high-performance work systems might be used to reduce occupational injuries. Training is undoubtedly the intervention that has been used most frequently in organizations to improve occupational safety. In chapter 11, Michael J. Colligan and Alexander Cohen provide a detailed review of the literature on safety training. The role of unions in the occupational safety process has attracted considerable debate for many decades. E. Kevin Kelloway, in chapter 12, explores the relationship between unions and workplace safety. After workers have been injured and thereby kept away from work, their reintegration into the workplace is a critical issue that can present a significant challenge. Therefore, in chapter 13, Niklas Krause and Thomas Lund review the literature on the organizational factors that facilitate or impede returning to work after a disabling occupational injury.

Each of the chapters in parts II and III demonstrate explicitly and implicitly how industrial/organizational psychology can help provide an understanding of the etiology of unintentional workplace injuries and their prevention. In part IV, chapter 14, we revisit the issue of unintentional occupational injuries, point to common themes across chapters, and raise questions for the next generation of research on the psychology of workplace safety.

REFERENCES

Baker, S. P., O'Neill, B., Ginsburg, M. J., & Li, G. (1994). *The injury fact book* (2nd ed.). New York: Oxford University Press.

Barling, J., Fullagar, C., & Kelloway, E. K. (1992). *The union and its members: A psychological approach.* New York: Oxford University Press.

Barling, J., & Hutchinson, I. (2000). Commitment vs. control-oriented safety practices, safety reputation, and perceived safety climate. *Canadian Journal of Administrative Science, 17*, 76–84.

Bulatao, E. Q., & VandenBos, G. R. (1996). Workplace violence: Its scope and the issues. In G. R. VandenBos & E. Q. Bulatao (Eds.), *Violence on the job: Identifying risks and developing solutions* (pp. 1–23). Washington, DC: American Psychological Association.

Bureau of Labor Statistics. (1997). *BLS handbook of methods*. Washington, DC: Author. Retrieved January 6, 2003, from http://www.bls.gov/opub/hom/homtoc.htm

Bureau of Labor Statistics. (2002). *Workplace injuries and illnesses in 2001*. (USDL Publication No. 02-687). Washington, DC: Author.

Campbell, J. P., Daft, R. L., & Hulin, C. L. (1982). *What to study: Generating and developing research questions*. Thousand Oaks, CA: Sage.

Conway, H., & Svenson, J. (1998). Occupational injury and illness rates, 1992–1996: Why they fell. *Monthly Labor Review, 121*, 36–58.

Dembe, A. E. (2001). The social consequences of occupational illnesses and injuries. *American Journal of Industrial Medicine, 40*, 403–417.

Donald, I. (1995). Psychological insights into managerial responsibility for public and employee safety. In R. Bull & D. Carson (Eds.), *Handbook of psychology in legal contexts* (pp. 625–642). New York: Wiley.

Health and Safety Executive. (1997). *The cost of accidents at work*. Sudbury, UK: HSE Books.

Ilgen, D. R. (1990). Health issues at work: Opportunities for industrial/organizational psychology. *American Psychologist, 45*, 273–283.

Institute of Medicine. (1999). *Reducing the burden of injury: Advancing prevention and treatment*. Washington, DC: National Academy Press.

Institute of Medicine. (2000). *Safe work in the 21st century*. Washington, DC: National Academy Press.

International Labor Organization. (2002). *IPEC action against child labour: Highlights 2002*. Geneva, Switzerland: Author.

Keller, S. D. (2001). Quantifying social consequences of occupational injuries and illnesses: State of the art and research agenda. *American Journal of Industrial Medicine, 40*, 438–451.

Landy, F. J. (1997). Early influences on the development of industrial and organizational psychology. *Journal of Applied Psychology, 82*, 467–477.

Leigh, J. P., Macaskill, P., Kuosma, E., & Mandryk, J. (1999). Global burden of disease and injury due to occupational factors. *Epidemiology, 10*, 626–631.

Leigh, J. P., Markowitz, S. B., Fahs, M., Shin, C., & Landrigan, P. J. (1997). Occupational injury and illness in the US: Estimates of costs, morbidity and mortality. *Archives of Internal Medicine, 157*, 1557–1568.

Marshall, K. (1996, summer). A job to die for. *Perspectives on Labor and Income*, 26–31.

National Institute for Occupational Health and Safety. (2002). *The changing organization of work and the safety and health of working people.* (NIOSH Publication No. 2002-116). Cincinnati, OH: Author.

National Research Council. (1999). *The changing nature of work: Implication for occupational analysis.* Washington, DC: National Academy Press.

Offermann, L. R., & Gowing, M. K. (1990). Organizations of the future: Changes and challenges. *American Psychologist, 45*, 95–108.

Parker, D., Carl, W. R., French, L. R., & Martin, F. B. (1994). Characteristics of adolescent work injuries reported to the Minnesota Department of Labor and Industry. *American Journal of Public Health, 84*, 606–611.

Quick, J. C. (1999). Occupational health psychology: Historical roots and future directions. *Health Psychology, 18*, 82–88.

Smith, M. J., Karsh, B. T., Carayon, P., & Conway, F. T. (2003). Controlling occupational safety and health hazards. In J. C. Quick & L. E. Tetrick (Eds.), *Handbook of occupational health psychology* (pp. 35–68). Washington, DC: American Psychological Association.

Statistics Canada. (1994). *Work injuries 1991–1993* (Cat. No. 72-208). Ottawa: Author.

Stout, N. A., & Linn, H. I. (2002). Occupational injury prevention: Progress and priorities. *Injury Prevention, 8(Suppl. IV)*, iv9–iv14.

Takala, J. (2002, May). *Introductory report: Decent work—safe work.* Paper presented at the XVIth World Congress on Safety and Health at Work, Vienna.

Veazie, M. A., Landen, D. D., Bender, T. R., & Amandus, H. E. (1994). Epidemiologic research on the etiology of injuries at work. *Annual Review of Public Health, 15*, 203–221.

World Health Organization. (1995). *Global strategy on occupational health for all.* Geneva, Switzerland: Author.

Zacharatos, A. (2001). *An organization and employee level investigation of the relationship between high performance work systems and workplace safety.* Unpublished doctoral dissertation, Queen's University, Kingston, Ontario, Canada.

II

CLIMATE, WORKING CONDITIONS, AND CULTURE

2

SAFETY CLIMATE AND SAFETY AT WORK

ANDREW NEAL AND MARK A. GRIFFIN

This chapter explores the nature and importance of safety climate in organizations. Perceptions of safety climate can influence employees' attitudes toward safety, the way employees perform their work, and they way employees interact with each other with regard to safety issues. Each of these factors can have a direct impact on safety outcomes such as accidents. The chapter presents a systematic model of safety climate in organizations and examines the antecedents and consequences of safety climate at the individual and organizational level. The practical relevance of safety climate is addressed through issues such as the measurement of climate at work and strategies for the management of safety climate in the workplace.

A FRAMEWORK FOR LINKING SAFETY CLIMATE AND SAFETY BEHAVIOR

We first present a framework for conceptualizing safety climate and safety behavior, and for integrating research in these areas. We derive this framework from research on the more general construct of psychological climate (e.g., James & McIntyre, 1996) in organizations and from research concerning the nature of performance in organizations (e.g., Campbell, McCloy, Oppler, & Sager, 1993). The integration of these two research areas provides a theoretical foundation for reviewing safety climate research from diverse literatures. The framework also provides a basis for enhancing our practical knowledge about the causes and consequences of safety in organizations.

The framework is depicted graphically in Figure 2.1 and centers on the safety-related behavior of individuals. The behaviors that comprise individual safety are termed the *components of safety behavior.* These components are the different types of behaviors that contribute to safety in the workplace. Using the correct safety equipment and participating in safety meetings are examples of specific behaviors related to safety.

The framework identifies the key drivers and outcomes of the components of behavior. The proximal drivers of safety are termed the *determinants* of safety behavior. The determinants represent factors that are directly responsible for individual differences in safety behavior. Campbell et al. (1993) have argued that there are three determinants of individual differences in behavior: knowledge, skill, and motivation. Safety behaviors are most directly determined by the knowledge and skills necessary to enact the behaviors and by the motivation of individuals to actually perform the behaviors.

The antecedents of performance represent more distal factors that affect behavior through their effects on knowledge, skill, and motivation. Many factors are potential antecedents of safety. Figure 2.1 provides examples of individual- and organization-level antecedents. Our focus is the role of safety climate as an organizational antecedent to individual safety behavior.

SAFETY CLIMATE AS AN ANTECEDENT OF INDIVIDUAL BEHAVIOR

We differentiate between two components of safety behavior. The first type of behavior is termed *compliance.* Compliance behaviors include obeying safety regulations, following the correct procedures, and using appropriate equipment. The second type of behavior is termed *participation.* Drawing on work in the area of contextual performance (e.g., Borman & Motowidlo, 1993), we define *participation* as behavior that does not directly contribute to an individual's personal safety but that does support safety in the wider organizational context. For example, volunteering to attend safety meetings may not directly support an individual's own safety but can contribute to the overall safety of the organization, as can helping coworkers and making

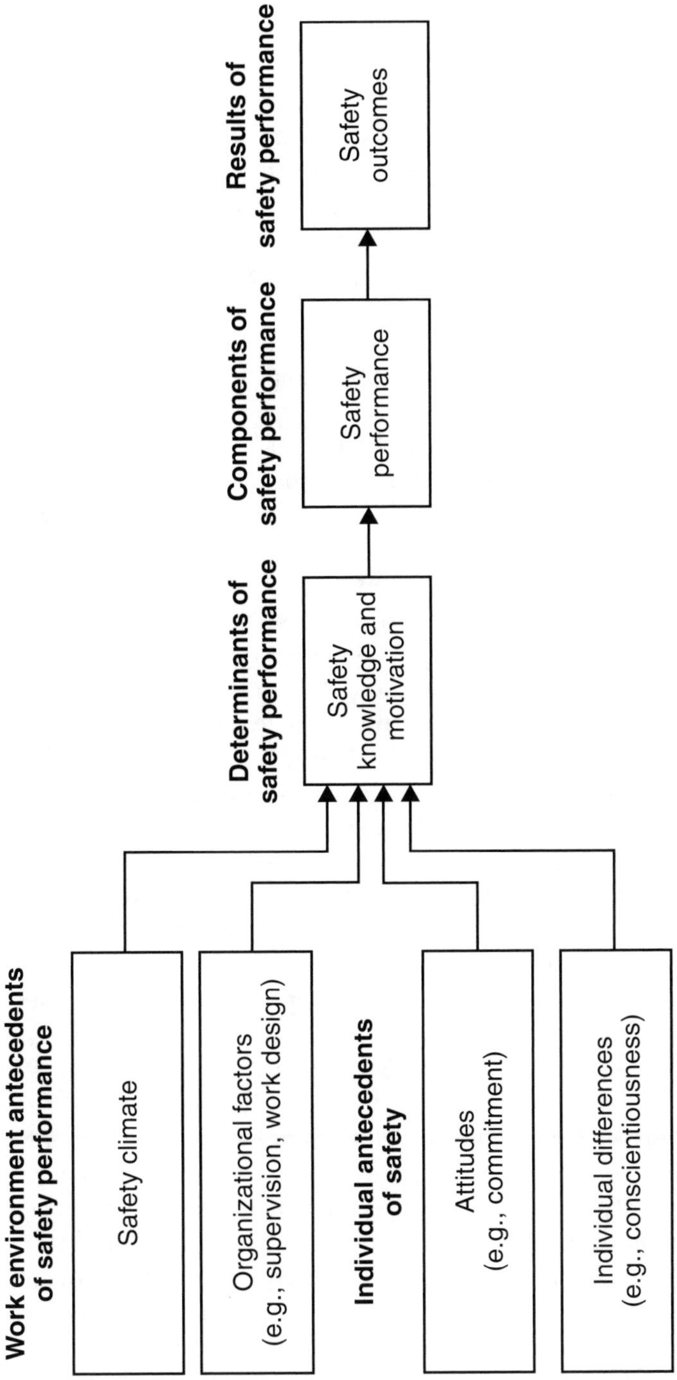

Work environment antecedents of safety performance

Safety climate

Organizational factors (e.g., supervision, work design)

Individual antecedents of safety

Attitudes (e.g., commitment)

Individual differences (e.g., conscientiousness)

Determinants of safety performance

Safety knowledge and motivation

Components of safety performance

Safety performance

Results of safety performance

Safety outcomes

Figure 2.1. A framework for conceptualizing safety climate and safety behavior.

suggestions to improve safety. The model emphasizes the importance of this wider range of safety behaviors and provides a basis for managing and promoting these behaviors.

The framework identifies safety climate as an antecedent of these safety behaviors. The general construct of organizational climate is typically defined by perceptions of the policies, procedures, and practices that operate in the work environment. The term *safety climate*, more specifically, refers to perceptions of the policies, procedures, and practices relating to safety (Barling, Loughlin & Kelloway, 2002; Griffin & Neal, 2000; Zohar, 2003). Further, these aspects of safety in the work environment provide clues to the overall importance of safety in a particular workplace. At its broadest level, safety climate describes employee perceptions about the value of safety in an organization.

It is sometimes easy to judge whether safety is taken seriously in a particular workplace. Employees might judge the value that is placed on safety by overt statements and actions of managers and coworkers that promote safety and sanction unsafe behavior. Employees also perceive implicit messages about the relative status of safety compared with other priorities such as productivity, the pace of work, and teamwork. These perceptions about the importance of safety form the basis of the construct of safety climate.

A critical feature of psychological climate is the degree to which these perceptions of the work environment are shared among individuals. When individuals share similar perceptions of safety in a particular work environment, it is possible to define a group safety climate or an organizational safety climate (James, James, & Ashe, 1990).

Of course, sometimes it is not easy to judge the overall importance of safety in a particular workplace. Considerable attention has been paid to the conceptual and empirical problems that arise when psychological constructs are defined on the basis of shared perceptions. We will not review these issues in detail, but we note that recent research provides some good ideas about how to make use of constructs based on shared perceptions (e.g., Klein, Danserau, & Hall, 1994; Morgeson & Hofmann, 1999; Schneider, White, & Paul, 1998). Our review focuses on the shared perceptions of safety in the workplace and develops a framework for describing these perceptions.

The description in the preceding paragraphs suggests that a general climate for safety is derived from specific perceptions about policies, practices, and procedures. Perceptions of safety should therefore be hierarchically structured. At the broadest level, all perceptions of safety should load onto a general higher-order factor, which we term *safety climate–general* (SC_g), reflecting the extent to which individuals believe that safety is valued in the workplace. This overall construct is derived from perceptions of more specific procedures, practices, and policies. Perceptions of specific aspects of safety should therefore load onto different first-order factors. Evidence supporting this hypothesis was provided by Griffin and Neal (2000), who found that perceptions of management values, safety communication, safety practices, safety training, and safety equipment loaded onto five separate first-order factors and that these first-order factors, in turn, loaded onto a common higher-order factor.

Our framework shows that safety climate is one of many antecedents that could influence safety behavior. Other organizational antecedents of safety behavior include factors such as leadership, training, and work design. These features are not unrelated to safety climate. Indeed, later in this chapter we discuss the central importance of practices such as leadership for enhancing safety climate (see also chapter 8 of this book). However, it is important to differentiate safety climate—perceptions about the value of safety in an organization—from other aspects of the work environment. Differentiating these constructs allows us to examine the relationships among them and the different mechanisms by which they affect safety behavior.

At the individual level, employees bring a variety of attitudes, experiences, and dispositions to their work roles. These individual antecedents may also influence motivation, knowledge, skill, and behavior at work. These may include attitudes toward safety, attitudes toward the organization (e.g., organizational commitment), and dispositions such as conscientiousness. Again, these examples of individual antecedents are not unrelated to safety climate perceptions and may influence the way individuals perceive the value of safety in their organization. Nevertheless, attitudes such as commitment are conceptually distinct from safety climate and can usefully be differentiated from other constructs that influence individual safety.

In summary, the framework indicates that safety climate reflects a psychological environment that provides a motivational antecedent for safety behaviors. The psychological environment is related to other organizational and individual antecedents but is conceptually distinct from them.

PRIOR RESEARCH

In this section we present a selective review of prior research examining safety climate, safety attitudes, individual differences, and safety behavior. Safety research has examined a wide range of perceptions, attitudes, and behaviors relating to safety. This literature has made use of a wide number of different measures. These measures have different developmental histories and vary widely in terms of content, format, industry, and country of origin (Flin, Mearns, O'Conner, & Bryden, 2000). Furthermore, many of the studies in this area have not differentiated between perceptions of safety climate and attitudes toward safety. The variety of constructs and measures that have been used makes it difficult to trace the development of a cumulative body of knowledge in this area. We first review studies that have focused predominantly on perceptions of safety climate and then consider studies that have evaluated attitudes toward safety and individual differences that may affect safety.

Perceptions of Safety Climate

Zohar (1980), the first person to introduce the concept of safety climate, identified eight dimensions of it: perceived importance of safety training programs, perceived management attitudes toward safety, perceived effects of safe conduct on promotion, perceived level of risk at the workplace, perceived effects of workpace on safety, perceived status of the safety officer, perceived effects of safe conduct on social status, and perceived status of the safety committee. There was agreement between employees in their perceptions of safety climate, demonstrating that it was meaningful to aggregate individual perceptions to the organizational level. At the organizational level, ratings of safety climate were found to correlate with ratings of the effectiveness of safety programs made by safety inspectors.

Brown and Holmes (1986) examined the factor structure of a shortened version of Zohar's (1980) measure, using a U.S. sample. Brown and Holmes identified only three factors: management concern, management action, and physical risk. Perceptions of safety climate on these dimensions were found to be associated with accidents at the individual level. Employees who reported that (a) managers were concerned for their welfare, (b) managers took action to maintain a safe working environment, and (c) the work environment had few risks had fewer accidents than those who reported otherwise. Dedobbeleer and Beland (1991) attempted to replicate Brown and Holmes's findings but were able to identify only two dimensions of safety climate: management commitment to safety and worker involvement in safety activities.

Hayes, Peranda, Smecko, and Trask (1998) argued that the reason why previous studies had managed to find only a relatively small number of dimensions of safety climate was that they had sampled only a relatively limited component of the work safety domain. Hayes et al. developed a 50-item measure to assess perceptions of safety climate. This measure produced five factors: global perceptions of workplace safety (e.g., the extent to which the job is dangerous, hazardous, and unsafe); coworker safety (e.g., the extent to which coworkers ignore safety rules, look out for others, and encourage others to be safe); management safety practices (e.g., the extent to which management provides safe equipment and responds quickly to hazards); supervisor safety (e.g., the extent to which supervisors reward safe behaviors); and satisfaction with the safety program (e.g., the extent to which the safety program is worthwhile and effective). Coworker safety, supervisor safety, and management safety practices were found to predict safety compliance, whereas supervisor safety and management safety practices were found to predict self-reported accidents. Perceptions of safety climate were also found to be negatively associated with job stress and physical complaints. All of these analyses were carried out at the individual level.

Recently, Cox and Cheyne (2000) have developed a measure of safety climate for use in offshore oil- and gas-processing organizations. The items for this measure were derived from instruments used in a range of different industries, including nuclear power, energy supply, and manufacturing (Cox & Cox, 1991; Cox, Tomas, Cheyne, & Oliver, 1998). These items loaded onto nine factors: management commitment (e.g., "Management acts decisively when a safety concern is raised"); priority of safety (e.g., "Management clearly considers the safety of employees of great importance"); communication (e.g., "There is good communication here about safety issues which affect me"); safety rules (e.g., "Some health and safety rules and procedures are not really practical"); supportive environment (e.g., "I am strongly encouraged to report unsafe conditions"); involvement (e.g., "I am involved with safety issues at work"), personal priorities and need for safety (e.g., "Safety is the number one priority when completing my job"); personal appreciation of risk (e.g., "I am sure it is only a matter of time before I am involved in an accident"); and work environment (e.g., "Sometimes I am not given enough time to get the job done safely").

Varonen and Mattila (2000) examined safety climate in eight Finnish wood-processing companies. Their measure of safety climate included four factors, three of which we would classify as climate: organizational responsibility, safety supervision, and company safety precautions. The fourth factor assessed employee attitudes toward safety. Employee perceptions of organizational responsibility and company safety precautions were negatively associated with expert ratings of workplace hazards and with accidents at the organizational level of analysis.

One of the factors frequently identified in assessments of safety climate is perceived risk. Rundmo has carried out a program of research examining perceptions of risk in the workplace and the relationship between perceptions of risk and other aspects of safety. Rundmo (1992a) identified three dimensions of risk perception for personnel working on offshore oil platforms: disasters and major accidents (e.g., blowouts, explosions, and fires); ordinary work injuries; and postaccident measures. Risk perception and work stress were positively correlated with injuries and human errors at the individual level of analysis (Rundmo, 1992b). Employee perceptions of safety and contingency measures, and of management and employee commitment and involvement in safety, were found to be important predictors of perceived risk and work stress. In a follow-up study, Rundmo (1994) found that, compared with other employees, employees working on oil platforms with high injury rates reported lower levels of satisfaction with safety and contingency measures, even if they had not been injured themselves. The injury rate on the oil platform was found to be a stronger predictor of safety perceptions than prior injury experience, suggesting that safety perceptions are not merely a reflection of whether an individual has been injured or not. These findings provide some support for the claim that perceptions of safety influence the likelihood of injury, rather than vice versa.

Recently, a number of papers have examined perceptions of safety climate at different levels of analysis. Hofmann and Stetzer (1996) examined the effects of factors at both the individual and group levels on safety behavior and accidents. Role overload was argued to be an individual-level construct because it varies between individuals within the same group and organization. However, Hofmann and Stetzer argued that group processes (e.g., planning, decision making, and sharing of information), management commitment and employee involvement in safety (Dedobbeleer & Beland, 1991), and behavioral intentions (intention to approach team members engaged in unsafe acts) can be conceptualized at the group level. As expected, there was sufficient homogeneity within groups on these factors to justify aggregation to the group level. Perceptions of role overload were found to be significantly associated with safety behavior at the individual level, whereas group processes, management commitment and employee involvement in safety, and approach intentions were found to exert significant cross-level effects on safety behavior. Furthermore, there was some evidence to suggest that approach intentions mediated the relationship between group processes and unsafe behaviors. Finally, all factors were found to be significantly correlated with accidents at the group level of analysis.

Zohar (2000) also argued that different facets of safety climate can be conceptualized at different levels of analysis and that the majority of previous research into safety climate has examined perceptions of organizational attributes. These perceptions predominantly concern organizational policies and procedures that are consistent across groups within organizations. Organizational policies and procedures, however, have to be implemented at the group level. The implementation of these policies and procedures can vary between groups, producing differences in practices within the same organization. Zohar (2000) found that perceptions of supervisory practices were shared among individuals within groups and that there was significant variance between groups in supervisory practice within a single organization. Perceptions of supervisory practices were therefore aggregated to the group level. Aggregate perceptions of supervisory practices were found to predict subsequent accidents at the individual level, after controlling for individual differences in role overload. Furthermore, there was a significant cross-level interaction effect, demonstrating that the effects of overload on personal injury were greater in groups where supervisors provided little support.

Attitudes Toward Safety

A number of studies have examined attitudes toward safety as well as perceptions of safety climate. *Safety climate* describes the shared perception of the value of safety in the work environment and can be differentiated from *attitudes*, which are individual beliefs and feelings about specific objects or activities. Cox and Cox (1991) examined safety climate and safety attitudes in a European gas company. They identified two factors that we would classify as climate (perceived safeness of the work environment and effectiveness of arrangements for safety) and three factors that we would classify as attitudes (personal skepticism, individual responsibility, and personal immunity). Williamson, Feyer, Cairns, and Biancotti (1997) developed a measure that included both perceptual and attitudinal questions. Five factors were extracted from these items. The first factor was personal motivation for safe behavior, which incorporated both perceptual and attitudinal elements (e.g., "It would help me to work more safely if management listened to my recommendations"). The second factor was positive safety practice (e.g., "There is adequate safety training in my workplace"), which is a perceptual measure. The final three factors were attitudinal: risk justification (e.g., "When I have worked unsafely, it has been because I was not trained properly"); fatalism (e.g., "If I worried about safety all the time, I would not get my job done"); and optimism (e.g., "It is not likely that I will have an accident, because I am a careful person").

Mearns, Flin, Gordon, and Fleming (1998) examined perceptions, behaviors, and attitudes relating to safety on offshore oil platforms. A 52-item measure was developed to assess attitudes toward safety. Illustrative examples of items were not provided, so it is difficult to assess whether all of the items are truly attitudinal or whether some of them include perceptual or behavioral elements. The items loaded onto nine factors: speaking up about safety, attitudes toward violations, supervisor commitment, attitudes toward rules and regulations, management commitment, safety regulation, cost versus safety, personal responsibility for safety, and overconfidence in safety. Responses on seven of these scales were significantly correlated with accident history. For example, employees who had been injured reported feeling under more pressure to violate rules and regulations, were less prepared to take personal responsibility for their own safety, and were less positive about speaking up about safety. The scales not correlated with accident history were management commitment, safety regulation, and overconfidence in safety. A number of other perceptual and behavioral measures were also used in this study. Perceptual and behavioral measures that were correlated with accidents included communication, safety behavior, perceptions of work task hazards, and satisfaction with safety measures. Interestingly, Mearns et al. found that there was greater variability between respondents on the attitudinal measures than on the perceptual measures. Mearns et al. argued that these data suggest that there are fragmented "safety subcultures" on the oil rigs that vary across different types of employees. From our perspective, the finding that there is greater variability in attitudinal measures than perceptual measures supports the claim that attitudes toward safety should be clearly differentiated from perceptions of safety climate. Attitudes are influenced by individual differences as well as by environmental factors. As a result, there should be less agreement between individuals with respect to attitudes than in relation to perceptions of the work environment.

Individual Differences

There is a long tradition of research examining the role of individual differences in safety. In a pioneering study, Greenwood and Woods (1919) found that the distribution of accidents across individuals was not equal, suggesting that some individuals were more prone to accidents than others. Subsequent research has attempted to identify the factors that might be responsible for individual differences in safety. A brief review of the personality and cognitive variables that have been examined within this literature is provided in the following paragraphs.

A number of studies have found that neuroticism is positively associated with accident involvement or risk taking (Frone, 1998; Hansen, 1989; Iverson & Erwin, 1997; Salminen, Klen & Ojanen, 1999; Shaw & Sichel, 1971; Sutherland & Cooper, 1991). High levels of neuroticism are characterized by negative emotional states such as anxiety, stress, and depression. Hansen argued that individuals with high levels of neuroticism are more distractible than those with low levels and are likely to suffer from lapses of attention, which predisposes them to making mistakes.

There is also evidence to suggest that Type A behavior, sensation seeking, and extremely high levels of extraversion are positively associated with risk taking or accident involvement (Beirness & Simpson, 1988; Cooper & Sutherland, 1987; Dahlback, 1991; Hansen, 1989; Shaw & Sichel, 1971; Sutherland & Cooper, 1991). Frone (1998) argued that Type A individuals may be more likely than other people to be involved in accidents because they have an elevated sense of haste and time urgency. Iverson and Erwin (1997) argued that individuals with extremely high levels of extraversion take more risks than other people because they are overconfident, intolerant, and aggressive. However, Iverson and Erwin also found that positive affectivity, a trait reflecting the more socially adjusted components of extraversion such as enthusiasm and self-efficacy, was negatively associated with subsequent accident involvement. They argued that individuals with high levels of positive affectivity are more likely than others to actively engage with and control their environment.

A number of studies have also examined the relationship between locus of control and accident involvement, although the results are mixed (Lawton & Parker, 1998). Several studies have found that individuals with an internal locus of control are less likely to have accidents than individuals with an external locus of control (e.g., Jones & Wuebker, 1985). However, other studies have found that this relationship holds only for certain groups of individuals (Clement & Jonah, 1984; Foreman, Ellis, & Beavan, 1983) or have found no relationship at all (Sims, Graves, & Simpson, 1984).

Most of the research examining the relationship between cognitive abilities and safety has focused on road accidents (Lawton & Parker, 1998). In a meta-analysis of available studies, Arthur, Barrett, and Alexander (1991) found that selective attention and cognitive ability were significant predictors of involvement in vehicular accidents, together with locus of control and regard for authority. Within the safety literature more generally, research suggests that up to 80% or 90% of industrial accidents are caused by errors and violations committed by individuals (e.g., Reason, 1990). Errors represent instances of noncompliance that are not intentional, whereas violations represent instances of intentional noncompliance. Common types of errors include slips of action, lapses in memory, and rule-based or knowledge-based mistakes. Broadbent, Cooper, Fitzgerald, and Parkes (1982) argued that some individuals are more prone than others to slips of action and lapses of memory, and that these individuals are more vulnerable to subsequent stressors. A number of authors (e.g., Lawton & Parker, 1998; Wickens, 1996) have argued that negative emotional states, such as stress, may (a) alter the way in which individuals allocate attention to tasks (e.g., via attentional tunneling); (b) interfere with the representation of information in working memory; and (c) change the strategies individuals use to perform tasks (e.g., by causing a shift from accuracy to speed).

In summary, relatively little research has examined the mechanisms responsible for the relationship between individual difference variables and accident involvement. However, a number of different mechanisms have been proposed. Lawton and Parker (1998) identified two broad mechanisms from this body of research: cognitive and motivational. The cognitive mechanism involves failures in information processing or skill, and results in reduced safety compliance, because people make slips, lapses, or mistakes. Personality traits, such as neuroticism and Type A behavior, may increase individuals' vulnerability to these kinds of cognitive failures when the individuals are exposed to stressors. The motivational mechanism results in reduced safety compliance because people deliberately violate safety procedures.

Personality traits such as extreme extraversion and sensation seeking may act via this motivational pathway.

Dimensions of Safety

The review in the preceding paragraphs describes a number of dimensions of safety climate and attitudes toward safety that have emerged from the literature. In a review of 18 studies published between 1980 and 1998, Flin et al. (2000) identified five common themes that had emerged from the literature: perceptions of managers' and supervisors' attitudes and behavior; the perceived effectiveness of safety systems; perceptions of, and attitudes toward, risk; work pressure; and perceptions of the knowledge, skill, and competence of employees. Flin et al. also argued that perceptions of, and attitudes toward, procedures and rules are important, but that this issue has received less attention in the literature. Guldenmund (2000), in a review of 15 studies, also identified six common dimensions from the literature: management, risk, safety arrangements, procedures, training, and work pressure.

In Table 2.1, we present a model of safety climate derived from this body of prior research. Following Griffin and Neal (2000), we assume that perceptions of specific aspects of safety will load onto first-order factors and that these first-order factors will load onto a common higher-order factor, reflecting global perceptions of safety. In conceptualizing the first-order factors, we distinguish between perceptions of organizational policies and procedures, and perceptions of local work conditions and practices (Zohar, 2000). At the organizational level, we distinguish among perceptions of management commitment, perceptions of human resource management practices, and perceptions of the quality/effectiveness of safety systems. At the local work-group level, we distinguish among supervisor support, internal group processes, boundary management, risk, and work pressure.

It is possible to formulate attitudinal questions to assess a wide range of variables. For example, attitudinal questions can be asked in relation to both organizational policies/procedures and local work conditions/practices. However, we believe that there is little practical value in asking attitudinal questions about environmental factors, because (a) attitudes toward environmental factors are likely to be highly correlated with perceptions of environmental factors and (b) attitudinal questions are more likely to be confounded by individual differences than perceptual questions. If the aim of a measure is to assess the effect of different environmental factors on the safety of employees, then we believe that it is better to frame the questions perceptually rather than attitudinally. The major area in which attitudes are likely to be important is that of behavior. The types of behavior-oriented attitudes that have been identified in the literature include general attitudes

TABLE 2.1
Proposed First-Order Dimensions of Safety Climate at the Organizational and Group Levels of Analysis

Dimensions	Elements
Organizational policies and procedures	
Management commitment	The extent to which management is perceived to place a high priority on safety, and communicate and act on safety issues effectively.
Human resource management practices	The extent to which recruitment, selection, training, performance management, and compensation practices are perceived to enhance safety.
Safety systems	The perceived quality and effectiveness of hazard management systems, incident investigations, and safety policies and procedures.
Local work conditions and practices	
Supervisor support	The extent to which supervisors are perceived to place a high priority on safety, respond to safety concerns, and provide support and encouragement for subordinates who comply with safety procedures and participate in safety activities.
Internal group processes	The extent to which employees perceive that there is adequate communication, support, and backup in relation to safety issues within the group.
Boundary management	The perceived quality of communication between the work group and other relevant stakeholders regarding safety issues.
Risk	The extent to which work tasks are perceived to be dangerous, hazardous, or unsafe.
Work pressure	The extent to which workload is perceived to exceed employees' capacity to perform their tasks safely.

(e.g., personal responsibility, skepticism, immunity, and fatalism) and more specific attitudes (e.g., self-efficacy).

Linkages Among Perceptions, Attitudes, and Behavior

In the following paragraphs we review available evidence regarding the linkages among perceptions of safety, attitudes toward work and safety, and safety behavior. In particular, we focus on the mechanisms by which perceptions and attitudes affect safety behavior.

Rundmo (1997) examined the role of satisfaction and stress as mediators of the relationships between (a) perceptions of safety and compliance, and (b) attitudes toward safety and compliance. Perceptions of management and employee commitment to safety, perceptions of physical working conditions, and attitudes toward safety were found to predict satisfaction with safety and contingency measures, job stress, perceived risk, job strain, and safety behavior. Furthermore, the effects of management/employee commitment, physical working conditions, and safety attitudes on safety behavior were partially mediated by satisfaction with safety and contingency measures, job stress, perceived risk, and job strain. Employees who (a) believed that management was not committed to safety, (b) rated physical working conditions poorly, and (c) had poor attitudes to safety had lower levels of satisfaction with safety procedures, felt unsafe, and had higher levels of stress and strain than those who held other beliefs and attitudes. These employees were less likely than others to comply with safety procedures. Safety behavior, in turn, was found to predict accidents and near misses. Similar findings have been reported in health care settings. Dejoy, Searcy, Murphy, and Gershon (2000) identified safety climate as a reinforcing factor in the work environment that increased both general safety compliance and compliance with personal protective equipment. McGovern et al. (2000) found that safety climate, attitudes toward risk, job tenure, knowledge, and training predicted self-reported compliance with safety procedures.

Research has also begun to examine the role of other organizational factors in relation to safety. Much of this work has been influenced by the concept of high-performance work systems (Pfeffer, 1998) and has focused on the concepts of leadership and commitment. Barling and Zacharatos (1999) argued that work practices generating high levels of commitment should also enhance safety behavior. Evidence to support this hypothesis was provided by Hofmann and Morgeson (1999), who found that leader-member exchange (LMX) and perceived organization support influenced accidents and that this relationship was mediated by self-reported safety communication and supervisor ratings of safety commitment. These results suggest that individuals are increasingly likely to be committed to safety and to engage in open communication regarding safety when they perceive the organization to be supportive and they have high-quality relationships with their leaders. Barling et al. (2002) found that transformational leadership behaviors predicted safety outcomes and that this relationship was mediated by safety climate. Other factors that have been found to be associated with safety climate, safety behavior, or safety outcomes include (a) general organizational climate (Neal, Griffin, & Hart, 2000); (b) job satisfaction, job involvement, organizational commitment, and stress (Morrow & Crum, 1998); (c) job insecurity (Probst & Brubaker, 2001); and (d) pay systems, training, working hours, and work design (Kaminski, 2001).

Recently, a number of longitudinal studies have examined the linkages among leadership, safety climate, and safety behavior. Griffin, Burley, and Neal (2000) examined the effects of leadership and safety climate on safety behavior over time; they found that changes in leadership exerted a lagged effect on safety climate and that changes in leadership and safety climate predicted changes in safety compliance and participation in safety activities. Furthermore, conscientiousness was found to have a direct effect on safety motivation and safety behavior, independent of any effects of leadership and safety climate. Zohar (2002) examined the impact of supervisory training on safety climate and safety outcomes over time. This study demonstrated that a training program that taught supervisors to provide feedback about safety and to make safety an explicit performance goal produced a marked reduction in safety-related episodes and produced an improvement in safety climate.

CAN WE USE HRM PRACTICES TO CREATE AND MAINTAIN A SAFE WORKING CLIMATE?

A limitation of safety climate research to date is the lack of specific guidelines for improving safety climate and, subsequently, safety. The research reviewed in the preceding sections suggests that factors such as leadership and organizational commitment are likely to have complex relationships with safety climate in organizations. A research priority for the future is to develop a better understanding of these antecedents and their relationships with each other so that we can identify ways to improve safety climate.

The strategic human resource management (HRM) literature provides some direct information about ways to enhance safety climate in organizations. Research in this area suggests that organizations can use HRM practices to achieve specific outcomes, such as enhanced productivity and profitability (e.g., Huselid, 1995). So-called progressive HRM practices are thought to enhance productivity and profitability by maximizing the knowledge, skill, and motivation of employees (Neal & Griffin, 1999). Progressive HRM practices include selective hiring, extensive training, systematic performance appraisals, nonmonetary benefits, incentives, job enrichment, teamworking, supportive leadership, and participation in decision making. Organizations may also be able to use these practices to support safety in organizations. Guthrie and Olian (1990) reviewed HRM practices in terms of the psychological processes through which these practices might influence safety outcomes such as disease and injury. They identified affective states, cognitive structures, and individual and group behaviors that could be influenced by HRM practices such as staffing, training, and job design. They noted that HRM practices have traditionally focused on capitalizing on individual differences and encouraging behaviors that support organizational goals. From a safety management perspective, the main task is to incorporate health and safety as important organizational goals along with others such as productivity. Many of the constructs and processes linking HRM to individual behavior are familiar to researchers and practitioners.

However, the incorporation of safety as a core organizational goal is less well integrated.

A focus on HRM and safety can translate directly into safety benefits for organizations. For example, training programs that identify appropriate safety procedures can provide employees with the knowledge and skill required to comply with safety procedures and participate effectively in safety programs. Performance management practices (e.g., leadership, performance appraisal, compensation) that encourage, support, and reward employees for safety may motivate them to perform their job safely. Practices such as job enrichment, teamwork, and participation in decision making are known to produce improvements in organizational commitment and job satisfaction, which may motivate employees to behave more safely (e.g., Parker, Axtell, & Turner, 2001). These changes in employee attitudes and behavior may, in turn, help create a climate in which safety is valued. More generally, HRM practices that successfully integrate safety communicate the importance of safety within the organization and provide cues to the value that is placed on safety. For example, selection procedures that use safety knowledge as a criterion not only orient selection of individuals to those participating in safety behavior but also communicate the value of safety in the organization more widely (e.g., Rynes, Bretz, & Gerhart, 1991).

Little research has addressed the most effective HRM practices for enhancing a positive safety climate. However, the recent focus on high-performance work systems suggests that HRM practices designed to enhance commitment may be most strongly related to safety climate (Barling & Zacharatos, 1999; Pfeffer, 1991). It is clear that the quality of leadership will be a critical factor for the implementation of HRM practices that enhance safety climate (e.g., Barling et al., 2002). Leaders play a role in setting group directions, enacting organizational policies and procedures, and providing support and structure within groups. In this way, leaders not only develop and support safety systems but also play a central role in creating a positive safety climate.

CONCLUSION

Safety climate is an important construct because it describes a key intersection between organizational and psychological processes and their relationship with safety. Growing evidence suggests that safety climate is an antecedent to safety-related motivation for employees and that this motivation influences both behavior of individuals and safety outcomes for the organization. An earlier focus on the measurement of safety climate is now growing to a more fully developed theoretical approach that describes the systems that develop a positive safety climate and articulates the organizational consequences of this climate.

REFERENCES

Arthur, W., Barrett, G. V., & Alexander, R. A. (1991). Prediction of vehicular accident involvement: A meta-analysis. *Human Performance, 4,* 89–105.

Barling, J., Loughlin, C., & Kelloway, E. K. (2002). Developing and testing a model of safety specific transformational leadership and occupational safety. *Journal of Applied Psychology, 87,* 488–496.

Barling, J., & Zacharatos, A. (1999). *High performance safety systems: Management practices for achieving optimal safety performance.* Paper presented at the 25th annual meeting of the Academy of Management, Toronto.

Beirness, D. J., & Simpson, H. M. (1988). Lifestyle correlates of risky driving and accident involvement among youth. *Alcohol, Drugs & Driving, 6,* 129–143.

Borman, W. C., & Motowidlo, S. J. (1993). Expanding the criterion domain to include elements of contextual performance. In N. Schmitt & W. C. Borman and Associates (Eds.), *Personnel selection in organizations* (pp. 71–98). San Francisco: Jossey-Bass.

Broadbent, D. E., Cooper, P. F., Fitzgerald, P., & Parkes, K. R. (1982). Cognitive Failures Questionnaire (CFQ) and its correlates. *British Journal of Clinical Psychology, 19,* 177–188.

Brown, R. L., & Holmes, H. (1986). The use of a factor-analytic procedure for assessing the validity of an employee safety climate model. *Accident Analysis and Prevention, 18*(6), 455–470.

Campbell, J. P., McCloy, R. A., Oppler, S. H., & Sager, C. E. (1993). A theory of performance. In N. Schmitt & W. Borman (Eds.), *Personnel selection in organizations* (pp. 35–69). San Francisco: Jossey-Bass.

Clement, R., & Jonah, B. A. (1984). Field dependency, sensation seeking and driving behavior. *Personality and Individual Differences, 5,* 87–93.

Cooper, C. L., & Sutherland, V. J. (1987). Job stress, mental health, and accidents among offshore workers in the oil and gas extraction industries. *Journal of Occupational Medicine, 29,* 119–125.

Cox, S. J., & Cheyne, A. J. T. (2000). Assessing safety culture in offshore environments. *Safety Science, 34,* 1–3.

Cox, S. J., & Cox, T. (1991). The structure of employee attitudes to safety: A European example. *Work and Stress, 5,* 93–106.

Cox, S. J., Tomas, J. M., Cheyne, A. J. T., & Oliver, A. (1998). Safety culture: The prediction of commitment to safety in the manufacturing industry. *British Journal of Management, 9,* S3–S11.

Dahlback, O. (1991). Accident-proneness and risk-taking. *Personality and Individual Differences, 12,* 79–85.

Dedobbeleer, N., & Beland, F. (1991). A safety climate measure for construction sites. *Journal of Safety Research, 22*(2), 97–103.

Dejoy, D. M., Searcy, C. A., Murphy, L. R., & Gershon, R. (2000). Behavioral-diagnostic analysis of compliance with universal precautions among nurses. *Journal of Occupational Health Psychology, 5*, 127–141.

Flin, R., Mearns, K., O'Conner, P., & Bryden, R. (2000). Measuring safety climate: Identifying the common features. *Safety Science, 34*, 177–192.

Foreman, E. I., Ellis, H. D., & Beavan, D. (1983). Mea culpa? A study of the relationships among personality traits, life events and ascribed accident causation. *British Journal of Clinical Psychology, 22*, 223–224.

Frone, M. R. (1998). Predictors of work injuries among employed adolescents. *Journal of Applied Psychology, 83*, 565–576.

Greenwood, M., & Woods, H. M. (1919). *A report on the incidence of industrial accidents with special reference to multiple accidents*. Industrial Fatigue Research Board Report No. 4. London: Her Majesty's Stationery Office.

Griffin, M. A., Burley, N., & Neal, A. (2000). *The impact of supportive leadership and conscientiousness on safety at work*. Paper presented at the 25th annual meeting of the Academy of Management, Toronto.

Griffin, M. A., & Neal, A. (2000). Perceptions of safety at work: A framework for linking safety climate to safety performance, knowledge, and motivation. *Journal of Occupational Health Psychology, 5*(3), 347–358.

Guldenmund, F. W. (2000). The nature of safety culture: A review of theory and research. *Safety Science, 34*, 215–257.

Guthrie, J. P., & Olian, J. D. (1990). Using psychological constructs to improve health and safety: The HRM niche. *Research in Personnel and Human Resource Management, 8*, 141–201.

Hansen, C. P. (1989). A causal model of the relationship among accidents, biodata, personality and cognitive factors. *Journal of Applied Psychology, 74*, 81–90.

Hayes, B. E., Peranda, J., Smecko, T., & Trask, J. (1998). Measuring perceptions of workplace safety: Development and validation of the workplace safety scale. *Journal of Safety Research, 29*(3), 145–161.

Hofmann, D. A., & Morgeson, F. P. (1999). Safety-related behavior as a social exchange: The role of perceived organizational support and leader-member exchange. *Journal of Applied Psychology, 84*, 286–296.

Hofmann, D. A., & Stetzer, A. (1996). A cross-level investigation of factors influencing unsafe behaviors and accidents. *Personnel Psychology, 49*(2), 307–339.

Huselid, M. A. (1995). The impact of human resource management practices on turnover, productivity and corporate financial performance. *Academy of Management Journal, 38*, 635–672.

Iverson, R. D., & Erwin, P. J. (1997). Predicting occupational injury: The role of affectivity. *Journal of Occupational and Organizational Psychology, 70*, 113–128.

James, L. R., James, L. A., & Ashe, D. K. (1990). The meaning of organizations: The role of cognition and values. In B. Schneider (Ed.), *Organizational climate and culture* (pp. 40–84). San Francisco: Jossey-Bass.

James, L. R., & McIntyre, M. D. (1996). Perceptions of organizational climate. In K. Murphy (Ed.), *Individual differences and behavior in organizations* (pp. 416–450). San Francisco: Jossey-Bass.

Jones, J. W., & Wuebker, L. J. (1985). Development and validation of the Safety Locus of Control (SLC) scale. *Perceptual and Motor Skills, 61*, 151–161.

Kaminski, M. (2001). Unintended consequences: Organizational practices and their impact on workplace safety and productivity. *Journal of Occupational Health Psychology*, 6, 127–138.

Klein, K. J., Dansereau, F., & Hall, R. J. (1994). Levels issues in theory development, data collection, and analysis. *Academy of Management Review, 19*, 195–229.

Lawton, R. & Parker, D. (1998). Individual differences in accident liability: A review and integrative analysis. *Human Factors, 40*, 655–671.

McGovern, P. M., Vesley, D., Kochevar, L., Gershon, R., Rhame, F. S., & Anderson, E. (2000). Factors affecting universal precautions compliance. *Journal of Business and Psychology, 15*(1), 149–161.

Mearns, K., Flin, R., Gordon, R., & Fleming, M. (1998). Measuring safety climate on offshore installations. *Work and Stress, 12*(3), 238–254.

Morgeson, F. P., & Hofmann, D. A. (1999). The structure and function of collective constructs: Implications for multilevel research and theory development. *Academy of Management Review, 24*, 249–265.

Morrow, P. C., & Crum, M. R. (1998). The effects of perceived and objective safety risk on employee outcomes. *Journal of Vocational Behavior, 53*(2), 300–313.

Neal, A., & Griffin, M. A. (1999). Developing a theory of performance for human resource management. *Asia Pacific Journal of Human Resources, 37*, 44–59.

Neal, A., Griffin, M. A., & Hart, P. M. (2000). The impact of organizational climate on safety climate and individual behavior. *Safety Science, 34*, 99–109.

Parker, S. K., Axtell, C., & Turner, N. (2001). Designing a safer workplace: Importance of job autonomy, communication quality, and supportive supervisors. *Journal of Occupational Health Psychology*, 6, 211–228.

Pfeffer, J. (1998). Seven practices of successful organizations. *California Management Review, 40*, 96–124.

Probst, T. M., & Brubaker, T. L. (2001). The effects of job insecurity on employee safety outcomes: Cross sectional and longitudinal explorations. *Journal of Occupational Health Psychology*, 6, 139–159.

Reason, J. T. (1990). *Human error*. Cambridge: Cambridge University Press.

Rundmo, T. (1992a). Risk perception and safety on offshore petroleum platforms, Part I: Perception of risk. *Safety Science, 15*, 39–52.

Rundmo, T. (1992b). Risk perception and safety on offshore petroleum platforms, Part II: Perceived risk, job stress and accidents. *Safety Science, 15*, 53–68.

Rundmo, T. (1994). Associations between organizational factors and safety and contingency measures on offshore petroleum platforms. *Scandinavian Journal of Work, Environment and Health, 20*(2), 122–127.

Rundmo, T. (1997). Associations between risk perception and safety. *Safety Science, 24*, 197–209.

Rynes, S. L, Bretz, R. D, & Gerhart, B. (1991). The importance of recruitment in job choice: A different way of looking. *Personnel Psychology, 44*, 487–521.

Salminen, S., Klen, T., & Ojanen, K. (1999). Risk taking and accident frequency among Finnish forestry workers. *Safety Science, 33*, 143–153.

Schneider, B., White, S., & Paul, M. C. (1998). Linking service climate and customer perceptions of service quality: Tests of a causal model. *Journal of Applied Psychology, 83*, 150–163.

Shaw, L., & Sichel, H. S. (1971). *Accident proneness*. Oxford: Pergamon Press.

Sims, M. T., Graves, R. J., & Simpson, G. C. (1984). Mineworkers' scores for the Rotter Internal-External Locus of Control Scale. *Journal of Occupational Psychology, 57*, 327–329.

Sutherland, V. L., & Cooper, C. L. (1991). Personality, stress and accident involvement in the offshore oil and gas industry. *Personality and Industrial Differences, 12*, 195–204.

Varonen, U., & Mattila, M. (2000). The safety climate and its relationship to safety practices, safety of the work environment and occupational accidents in eight wood-processing companies. *Accident Analysis and Prevention, 32*, 761–769.

Wickens, C. D. (1996). Designing for stress. In J. Driskell & E. Salas (Eds.), *Stress and human performance* (pp. 279–295). Mahwah, NJ: Erlbaum.

Williamson, A. M., Feyer, A. M., Cairns, D., & Biancotti, D. (1997). The development of a measure of safety climate: The role of safety perceptions and attitudes. *Safety Science, 25*, 1–3.

Zohar, D. (1980). Safety climate in industrial organizations: Theoretical and applied implications. *Journal of Applied Psychology, 65*, 96–102.

Zohar, D. (2000). A group-level model of safety climate: Testing the effect of group climate on microaccidents in manufacturing jobs. *Journal of Applied Psychology, 85*, 587–596.

Zohar, D. (2002). Modifying supervisory practices to improve sub-unit safety: A leadership-based intervention model. *Journal of Applied Psychology, 87*, 156–163.

Zohar, D. (2003). Safety climate: Conceptual and measurement issues. In J. C. Quick & L. E. Tetrick (Eds.), *Handbook of occupational health psychology* (pp. 123–142). Washington, DC: American Psychological Association.

3

THE EFFECT OF TEAMWORK ON SAFETY PROCESSES AND OUTCOMES

NICK TURNER AND SHARON K. PARKER

Stories in the media often serve to raise the awareness of workplace safety issues, bringing to life statistics that show the prevalence of work-related injuries. Take, for example, the following three stories: Top management teams lacking commitment to safety have faced "corporate manslaughter" charges (Mathiason, 2001); gangs of railside maintenance workers have reported being unclear about procedures for working on cracked rails (McVeigh, 2000); and a cockpit crew on a jetliner responded quickly and effectively to a deranged passenger bursting into the cockpit ("BA Jumbo Plunges," 2000). What may go unnoticed in such stories, however, is the fact that many employees work in teams and therefore often need to respond to hazards collectively rather than as individuals. Whether the results are positive or negative, today's workers rarely face workplace safety challenges alone.

We thank Julian Barling, Steven Fleck, Paul Goodman, Helen Williams, and an anonymous reviewer for their feedback on earlier drafts of this manuscript. Financial support from the Social Sciences and Humanities Research Council of Canada, the Center for Corporate Change, AGSM, Australia, and the Offshore Division of the UK Health and Safety Executive helped to make collaboration on this chapter possible.

Nevertheless, media stories and prevalence statistics present only a partial picture and are not particularly informative when it comes to understanding how to intervene to enhance safety at work. Rigorous research is required that analyzes the causes and contexts of workplace injuries, thereby building a knowledge base from which to design healthier workplaces. In this chapter we explore how working in teams might help or hinder occupational safety. Few of the numerous studies regarding teams in organizations address the role that working in groups has on workplace safety. We review the rather limited set of existing studies in this area and develop ideas about how to further research on this topic. To begin, we define what we mean by *teams* and *safety*.

TEAMS

Guzzo and Dickson (1996) defined *team*, or *work group*, as an entity "made up of individuals who see themselves and who are seen by others as a social entity, who are interdependent because of the tasks they perform as members of a group, who are embedded in one or more larger social systems (e.g. community, organization), and who perform tasks that affect others (such as customers or coworkers)" (pp. 308–309). This definition highlights that teams are more than a set of colocated individuals; team members are interdependent and need to coordinate their activities to achieve their goals. Although there are likely to be many processes operating within groups that also apply to individuals (e.g., a positive safety climate is likely to affect work groups and individuals), the coordination requirements of teams mean additional processes can come into play (e.g., the impact of group communication on safety). The study of work teams with regard to safety is thus likely to have some elements that are distinct from those related to the safety of individuals at work. We focus on team working, with Guzzo and Dickson's (1996) definition containing the criteria for choosing which studies to consider. For simplicity, we use the terms *team* and *group* interchangeably in this chapter.

Beyond defining basic terms, we also need to distinguish among different types of team research. Sonnentag's (1996) analysis of four types of studies relating work-group factors to employee well-being is relevant here. One type of study, a pure group-level approach, involves relating aggregated work-group scores (e.g., aggregations of individuals' perceived job autonomy scores) to aggregated group outcome scores (e.g., aggregated individual injuries). A mixed approach aggregates work-group factor scores as in the previous type but relates them to individual outcomes, with the analysis conducted at the individual level. We report on both of these approaches in this chapter. A third approach conceptualizes, measures, and analyzes all work and outcome variables at the individual level, such as when examining the relationship between work characteristics (e.g., job autonomy) and injuries under changing work designs (e.g., introduction of self-managing teams). Although this type of study represents a valid approach, we have reported on it elsewhere (see Parker, Turner, & Griffin, 2003). A final approach assesses work-group factors at the individual level (e.g., individual perceptions of team cohesion) and relates these to individual outcomes. Conceptually, this approach assumes that individual perceptions of work-group factors reflect work-group phenomena often without testing if this is in fact the case. Because this assumption might not be valid, we focus particularly on the studies that investigate group-level determinants and either group-level or individual-level safety outcomes.

SAFETY

Researchers investigating workplace safety have conceptualized and operationalized safety in several ways. The more traditional approach uses safety statistics (e.g., days away from work, injury frequencies, or near misses) as outcomes. Such measures are convenient for researchers because collaborating organizations often record such data for regulatory purposes. However, these measures also contain reporting biases and tend to capture extraordinary events, making the "objective" outcomes of safety statistics potentially invalid and unreliable.

More recently researchers have begun using conceptually broader and more valid safety measures in research and practice. One of these approaches assesses employees' safety-related behaviors. For example, Simard and Marchand (1995) asked supervisors to rate their work groups' propensity to engage in safety-related initiatives; that is, they asked the supervisors to rate how attuned the workforce was to preventing injuries by recognizing potential hazards. Minor injury indexes (e.g., Frone, 1998; Hemingway & Smith, 1999; Zohar, 2000) are also reliable indicators of frequently occurring injuries. Compared to open-ended questions about more severe injuries, indexes of minor injuries are less prone to socially desirable response sets and indeed often predicate more severe yet less frequently occurring injuries.

Our definition of *safety* for this chapter embraces these types of prevention-oriented indicators. Thus, in addition to more traditional measures of negative events (e.g., injuries, near misses), we include concepts that assess employees' safety-related behaviors, including those directed toward maintaining safety (e.g., complying with safety procedures); coping with hazards (e.g., handling errors once they have occurred); and taking a proactive approach toward safety (e.g., introducing safety-related initiatives or exhibiting citizenship behaviors).

RESEARCH ON TEAMS AND SAFETY

Research on teams in relation to safety has concerned one of three types of teams: health and safety committees (HSCs); self-managing or autonomous work teams situated in a range of organizational settings (e.g., manufacturing or service sector); and airplane cockpit crews. Although all three types may be classified as teams according to Guzzo and Dickson's (1996) criteria, they vary in important ways. HSCs tend to include members who work on their core tasks outside the team. That is, members tend to be less homogenous and less interdependent than in the case of autonomous work groups, with individuals in HSCs often representing distinct departments and levels of the organization. Participation is typically voluntary in HSCs, whereas it tends not to be voluntary in autonomous work groups and cockpit crews. Autonomous work teams usually involve relatively homogenous groups of shop-floor or service employees working together to carry out their core tasks. Although these teams operate in environments with different levels of operational uncertainty, they all tend to have stable membership, a high level of collective autonomy, and a moderate to high degree of interdependence. Finally, cockpit crews are similar to autonomous work groups in that the team members carry out their core tasks in a relatively autonomous way, yet with cockpit crews membership consists of highly trained specialists, changes frequently, and is often quite small. Additionally, cockpit crews undertake highly interdependent tasks in an environment that has the potential for a high degree of operational uncertainty.

By examining outcomes across these rather distinct types of teams, we obtain a broad perspective on the ways in which working in teams might influence safety. Our discussion of HSC effectiveness is followed by a discussion of autonomous work teams and their effect on safety. For cockpit crews, our focus is on how we can apply ideas from this much more developed research area to other organizational teams. We then draw together common threads and make recommendations for future research.

HEALTH AND SAFETY COMMITTEES

Health and safety committees (HSCs) are the keystone of official workplace safety programs in many organizations. They usually consist of representatives from the workforce and management, and there is usually one per organization or site. HSCs discuss, make recommendations, and initiate action on workplace health and safety issues with the aim of reducing the numbers of injuries and illnesses in the workplace (Eaton & Nocerino, 2000). In the United States, HSCs are the most common form of safety-specific employee participation; they exist in 75% of organizations with 50 or more employees and in 31% of organizations with fewer than 50 employees (Commission on the Future of Worker-Management Relations, 2001). Over the past decade, American policymakers have debated making HSCs a mandatory feature of organizations (Occupational Safety and Health Administration, 1991). The implications of HSC legislation, HSC membership, and the role HSCs play in setting the safety agenda in organizations are key topics of interest for researchers interested in HSC effectiveness. The next paragraphs summarize three of most recent and rigorous studies in the area.

Key Studies on HSCs

Using data from eight manufacturing plants in a U.S. building products company, O'Toole (1999) investigated how the presence of HSCs affected organizational safety outcomes and whether voluntary or mandatory implementation of HSCs in these plants moderated their effectiveness. In this quasi-experimental design, six of the plants implemented HSCs in either the last half of 1991 or the first half of 1992; the remaining two plants did not implement HSCs during this period, despite equal opportunity to do so. Additionally, three of the six plants that implemented HSCs did so to comply with regulatory requirements; the other three plants implemented HSCs voluntarily (i.e., they were not required by either regulatory or organizational pressures to do so). The safety outcome variables used were the Occupational Safety and Health Administration (OSHA) Recordable Injury Rate (i.e., rate of occupational injuries per 100 employees per year based on the employer's OSHA logs) and OSHA Severity Rate (i.e., rate of either days lost or restricted work activity per 100 employees per year based on OSHA lost time injuries) for the eight plants between 1987 and 1996.

The introduction of HSCs was associated with a lowering of injury rates and a decrease in the number of severe injuries. Furthermore, plants that had established HSCs voluntarily experienced fewer injuries than those that had introduced HSCs because they were required to do so by law. The six plants that implemented HSCs experienced fewer severe injuries overall, albeit at differing rates of change depending on whether implementation was imposed or discretional. The two plants that did not implement HSCs had no changes in either injury frequency or severity over the study period. An additional finding was that in plants that provided employees with opportunities for safety participation outside the HSC structure and that had managers who supported such participation, the injuries were less severe. Although there are limitations to this study—notably the use of a single organization (raising questions of generalizability) and the plants' self-selected rather than randomly allocated use of HSCs—the findings point to the importance of not only HSC presence but also the reasons for which HSCs are established. One could speculate, for example, that managers who voluntarily introduce HSCs are prepared to commit more resources and are more motivated to ensure that the HSC is effective than those who are required to introduce HSCs.

Whereas O'Toole's (1999) study examined HSCs in plants from a single U.S. organization, Reilly, Paci, and Holl (1995) examined how the internal composition of HSCs across multiple UK manufacturing organizations affected company injury rates. These researchers found significant differences between HSC membership and the injury rates of the associated workplace. Controlling for industry-related background variables, the researchers found that the organizations with joint consultative HSCs (i.e., those that had members from both union and management) had between 3.4 and 5.7 fewer injuries per 1,000 employees than those with nonconsultative HSCs. In particular, the greater the proportion of union-chosen members on these committees, the fewer the injuries. This study thus provided some positive evidence for the benefits of labor involvement in safety-improvement activities.

Given evidence on how HSC presence, composition, and implementation can affect safety outcomes, what is known about how internal and external factors affect HSC effectiveness? Eaton and Nocerino (2000) addressed this question by examining a stratified sample of public organizations and departments in New Jersey. Some of the organizations had HSCs (n = 180), whereas others did not (n = 247). Injury and illness data were collected from the New Jersey Department of Labor for the years 1988 and 1989 on a joint subsample (n = 251) of these two categories. Surveying one labor representative and one management representative on the HSC at each workplace, the researchers assessed (a) the breadth of the committee's responsibilities; (b) the committee's effort in addressing health and safety issues (measured by the total number of members who usually attended meetings); (c) training received by the committee members in technical, legislative, and process issues; and (d) the level of employee (nonmanagerial) involvement in setting the agenda. The study also measured the presence of senior managers on the committee, the nature of the hazards in the workplace, the presence of a comprehensive health and safety program, the level of union resources committed to health and safety, and governmental pressure exerted on the workplace (operationalized as whether the workplace had been inspected by the state).

Surprisingly, results from the analyses suggested that organizations with HSCs had more lost-work days from injuries than those without HSCs. The fact that this study used a single-year change in injuries as its outcome raises a criticism often leveled against cross-sectional studies of this sort, as well as the more general problem of statistical power. However, Eaton and

Nocerino (2000) explained their finding in two ways. First, organizations with HSCs may be more likely than other organizations to report workplace injuries. Second, in the absence of a measure of workplace hazards, it is possible that HSCs are a proxy for more dangerous workplaces.

Despite Eaton and Nocerino's (2000) counterintuitive result for HSC presence in relation to injury rates, there was nevertheless evidence that greater worker involvement with HSCs (i.e., a higher proportion of nonmanagement members), along with greater nonmanagement involvement in agenda setting, was associated with fewer reported injuries among those workplaces that had HSCs. This complements earlier findings (e.g., O'Toole, 1999; Reilly et al., 1995) that suggested that the involvement of nonmanagerial employees improves health and safety at work.

Summary of Findings and Potential Mechanisms

Given the different ways in which the studies described in the preceding paragraphs were designed and their contextual differences, a reasonably complex picture emerges. Two studies suggested that having an HSC is better than not having one (O'Toole, 1999; Reilly et al., 1995), and although one study presented counterevidence (Eaton & Nocerino, 2000) it offered explanations grounded in drawbacks in the research design. One could speculate that, apart from the actual changes introduced by the HSC to improve safety, an added value of an HSC could be the positive messages it conveys to the workforce. In essence, an HSC might contribute to a positive safety climate, which research suggests can enhance employees' adherence to safety policies and procedures (see chapter 2 of this book). The finding that HSCs had a larger number of positive effects when they were voluntarily introduced than when they were imposed by law could be due to the fact that the voluntary HSCs were more successful than the others in bringing about change or that their establishment conveyed stronger messages to the workforce about management commitment to safety.

A further theme that emerges from two of the three studies (i.e., Eaton & Nocerino, 2000; Reilly et al., 1995) is that greater employee involvement in HSCs, as well as in general safety-related initiatives, is related to better safety outcomes. This is consistent with the general participation literature, which shows that employee involvement facilitates the successful implementation of new initiatives (Lawler, 1992). Several processes might explain this finding. For example, high employee involvement in HSCs could mean that good solutions are generated because employees possess local expertise that managers do not possess. Another possibility is that the involvement of employees could enhance the likelihood that changes are accepted by the workforce (e.g., employees on the HSC can use their influence to persuade the workforce to change their approach).

AUTONOMOUS WORK TEAMS

A second type of team we wish to discuss in relation to safety, and one that has attracted more research attention than HSCs, is the autonomous work team. Autonomous work teams, as their name suggests, have members that autonomously carry out operational tasks. Team members are collectively responsible for making decisions traditionally assigned to a supervisor. Analogous to the importance of employee involvement in HSC effectiveness, a question of particular interest is whether a team's degree of autonomy affects safety outcomes. Proponents argue that work teams with a high degree of autonomy should have better safety performance than those with a low degree of autonomy because members of the former type of team have a heightened sense of responsibility and ownership for safety, greater knowledge of workplace risks, and increased opportunity to anticipate and act on hazards (Parker & Turner, 2002). In contrast, rival hypotheses suggest that a lack of direct supervision, combined with increased autonomy, either could encourage team members to go beyond their abilities without having an integrated understanding of the work system (e.g., knowing the knock-on effects of one team's actions on another's) or could diffuse responsibility for safety.

Each of the two arguments is plausible, but no clear evidence exists that would support one over the other. Various definitions of autonomous teams, different research designs, variations across contexts, multiple levels of analysis, and a range of safety outcomes all make it hard to draw definitive conclusions. However, in the following paragraphs we explore some of the existing studies in detail.

Key Studies on Autonomous Work Groups and Safety

Some of the seminal work on sociotechnical systems conducted by Eric Trist and colleagues at the Tavistock Institute in the United Kingdom provided the first comprehensive examination of the role of autonomous teams in influencing safety. Trist, Higgin, Murray, and Pollock (1963) presented a thorough case study of the structure and functioning of a UK coal mine that underwent significant changes in work organization and mechanization in the years following World War II. The study examined large work groups (in some cases up to 50 people) operating under different mining conditions and with varying types of pit technology. The research team collected qualitative and quantitative data over several years on aspects of these work groups and their performance. One focus of the investigation was differences between two "panels" of teams involved in the composite longwall method of coal mining.

Over several 5-month phases, in response to varying levels of workload, one panel of teams (termed the *conventionally organized*, or Number 1, panel) restricted its team members to single tasks, whereas the other panel of teams (what the researchers called the *fully composite*, or Number 2, panel) rotated its members between functions and tasks. The panels were adjacent to each other in the pit, physical conditions were identical, and both panels used exactly the same technology. Additionally, the teams consisted of employees with comparable qualifications and experience, membership was self-selected, and neither panel looked to management for guidance on how day-to-day work should be organized.

After 10 months, the researchers found a remarkable rise in days-off caused by injuries at the coal face in the conventionally organized panel. However, the fully composite panel of teams, which rotated its miners systematically through multiple tasks, had no change in days-off due to injuries over the same period. Trist et al. (1963) attributed this to that panel's higher levels of stimulation and responsibility, a reduction in boredom, and a reduction in monotony-caused fatigue. Another potential explanation could be that members of the fully composite panel of teams acquired greater knowledge and understanding of all tasks than the other panel did and were therefore able to coordinate their actions with other members more safely.

For almost a decade after the work of Trist and colleagues (1963), there were no published studies investigating sociotechnical work redesign and employee safety. Although Walton's (1972) often cited case study of self-managing teams is the one exception, neither a specific description of what constitutes safety nor an explanation of how the research problem was analyzed appeared in the article, making it impossible to tell whether the seemingly positive result was derived from working in teams per se or from the host of other factors that co-occurred.

In the early 1970s, Trist and colleagues embarked on another large-scale longitudinal investigation of work reorganization, this time in the Rushton coal mine (Trist, Susman, & Brown, 1977). Trist and colleagues were heavily involved in the planning and implementation of the autonomous teams, and Paul Goodman of Carnegie-Mellon University led the independent evaluation team. Separate publications by the two research teams resulted from the same project (Goodman, 1979; Trist et al., 1977) and are discussed together in the following paragraphs.

One of the main motivations for management and the union in sanctioning the Rushton research was a desire to improve mine safety (Goodman, 1979), and as such safety became the focus of much training and feedback that the experimental groups received. Special efforts were made to increase employees' safety knowledge, potentially increasing employees' felt responsibility for achieving high levels of safety performance. A benefit of this special interest in safety meant that Goodman's (1979) evaluation team had access to many safety-related indicators with which to assess the effects of autonomous work groups. Employees involved in the work redesign felt that the experiment with autonomous groups had a strong, positive effect on safety. More impartial ratings of safety proved to be similarly positive. Both on-site observations of safety behaviors and independent ratings from inspectors showed that there was a marked improvement in safety behaviors such as compliance with procedures in the experimental teams over the study period. However, the number and severity of violations varied across the study period and did not drop significantly for the experimental teams.

Overall, Goodman (1979) stated that although some of the safety indicators were positive, the "public claims . . . about the safety improvements at Rushton have been overstated" (p. 229). First, some of the ratings (e.g., observations of safety behavior) might have been confounded with the socially desirable goal of the intervention to improve safety. Second, archival data tended to highlight infrequent and specific incidents. Third, it was not clear to what extent any improvements in safety behaviors were attributable to the existence of autonomous teams rather than safety-specific initiatives that occurred concurrently. Although positive, these results are far from definitive.

The next important study was by Pearson (1992). Pearson investigated the introduction of autonomous work teams in a heavy engineering shop over 1 year. A particular strength of this study was its Solomon four-group design: the work groups were assigned randomly to either an autonomous (experiment) or a nonautonomous (control) condition and then again randomly split into pre- and posttest conditions. Pearson assigned the experimental groups to Levels 1 and 2, and the control groups to Levels 3 and 4. This design was used over six measurement waves, with each wave lasting 8 weeks. The pretest period lasted for two waves, the implementation of autonomous work practices took place between the second and third waves, and the posttest period covered the waves thereafter. More specifically, Levels 1 and 3 had completed two questionnaires before the main intervention occurred and 8 weeks before (i.e., wave 3) Levels 2 and 4 completed their first questionnaire. Group-level injuries were computed at each wave by aggregating nonfatal events that caused an employee to be absent for at least one work shift.

Pearson (1992) found that injuries for the autonomous work groups stayed the same over the experimental period; however, the group injury rate for the nonautonomous team increased over the same period. The implication is that autonomous work groups in some way prevented or reduced injuries that would have otherwise increased. Pearson observed that the autonomous work groups often discussed safety in their meetings and sought advice from members of the team who were responsible for safety. In contrast, the nonautonomous teams relied on safety representatives or a safety officer, both of whom were external to the team, to detect unsafe practices. These findings suggest that factors such as greater communication and increased felt responsibility among team members helped keep injuries down.

Unlike the Rushton study, however, Pearson's (1992) descriptions did not clearly indicate to what degree the autonomous team initiative also involved coaching team leaders about safety. Was the concertive mechanism that Pearson described caused by the work redesign itself, or was it attributable to active coaching of the team leaders? Zohar (2002), for example, has found strong links between how supervisors behave in relation to safety and employee safety outcomes. It is plausible that the motivation for the autonomous teams to take greater responsibility for their own safety is either an improvement in work design, a change in team leaders' behaviors, or a combination of both. Pearson's study, like that of Trist et al. (1977), highlights the difficulty of isolating the effect of autonomous work teams from other co-occurring initiatives.

In spite of these questions, a benefit of Pearson's (1992) study was the explicit recognition of the effect of context on employee responses to work redesign. Pearson acknowledged that the changes in technology might have limited the success of the teams. The plant had installed a production and maintenance system that limited the autonomy that operators had and made tasks more specialized. These potential yet often unoperationalized influences on work organization need to be considered in future research in this area. Additionally, this point about the limited extent to which autonomous work groups actually achieved an increase in employee job control highlights the need to assess the degree to which this initiative is indeed successful in changing the nature of the work. Without establishing that the work redesign had its intended effects on jobs (e.g., increasing autonomy), one cannot really make any firm conclusions about its effect on outcomes.

Unlike the studies reviewed so far, which have focused on manufacturing environments, Cohen and Ledford's (1994) study examined the relationship between autonomous teams and workplace safety in the service sector. Using a cross-sectional quasi-experimental design, Cohen and Ledford examined differences between traditionally managed teams and self-managing teams in a telecommunications firm. With the assistance of a steering group in the collaborating company, the researchers matched the teams across functions such that pairs of teams consisted of a traditionally managed team (which formed the control group) and a self-managed team (which formed the experimental group).

There were significant negative correlations between some quality-of-work-life variables (i.e., job satisfaction, social satisfaction, and organizational commitment) and number of days missed due to injury or illnesses. However, there were no associations between these same predictor variables and number of injuries sustained. This suggests that the self-managed team initiative might have enhanced individuals' motivation to come back to work faster after injuries, but it does not appear to have led to a reduced occurrence of injuries. A further analysis comparing the matched pairs of teams found no significant differences between the traditionally managed and self-managing teams on either injury frequency or number of days missed.

Cohen and Ledford (1994) attributed the null findings to the fact that the context of this experiment placed no particular emphasis on workplace safety, in contrast to previous studies (e.g., Pearson, 1992; Trist et al., 1977). In this regard, it could be argued that autonomous work groups involve employees' taking on more responsibility for key team goals (e.g., cost, quality, safety) and that, if something is not a salient team goal, then there is no reason to expect that work redesign should change it. For example, one reason autonomy is beneficial is that it allows employees to deal effectively with, and prevent, uncertainties such as breakdowns (Parker & Wall, 1998). If there are few uncertainties, or if the uncertainties do not involve much safety-related risk, then enhanced autonomy is not likely to positively affect safety, at least via this mechanism.

There were, however, weaknesses in Cohen and Ledford's (1994) design that might make such conclusions premature. The study was cross-sectional, so it is not possible to say what the teams' safety levels were like prior to any work redesign. Also, there might have been differences between those teams that chose to become self-managing and those that remained traditionally organized (e.g., because of types of technology or unsupportive supervision) that confound any straightforward comparison. Finally, the researchers gave no clear evidence of how "self-managing" the self-managing teams really were.

Hechanova-Alampay and Beehr's (2002) study is the most recent contribution to this particular area of research. These researchers used third-party ratings of teams' self-management to investigate the relationship between empowerment and safety performance. More specifically, the researchers asked a 10-member representative group of trained assessors to reach consensus on 22 attributes (e.g., decision making on team rules, problem solving, improving work processes, and internal training processes) for assessing the level of empowerment of each of 24 manufacturing teams. The correlational findings suggest that empowerment was negatively related to both unsafe behaviors ($r = -.48$) and team injuries ($r = -.51$). This relationship suggests that greater team self-management is related to safer working. The primary strength of this study is that, unlike much of the existing literature, it assessed the nature of team empowerment, in this case through panel assessment of team characteristics. The limitations, however, as with much of the literature in this area, are the study's inability to draw causal inferences due to the cross-sectional design and the susceptibility of archival injury data to reporting biases and other threats to validity.

The quasi-experimental and correlational studies reviewed so far suggest positive links between autonomous group work and safety. However, these studies do not isolate which features of autonomous work groups are important in promoting safety (or if indeed it is the concurrent introduction of other safety initiatives). For example, is it the expanded autonomy that is key? Or improved group processes? The set of studies we discuss in the following paragraphs allows a closer examination of what particular work characteristics associated with autonomous teams might be important.

Cross-Level Studies of Operational Teams and Safety

Recent studies regarding work teams and safety have made headway in this area by using multiple levels of analysis. In two separate studies, Simard and Marchand (1995, 1997) drew on the same sample of 1,061 work groups (consisting of more than 23,000 employees) across 97 Canadian manufacturing plants. In the first study (Simard & Marchand, 1995), multilevel models were developed to predict the level of work groups' safety initiative, including their propensity to take personal initiative for improving safe work conduct, suggest ideas to supervisors about improving safety, and persist in applying pressures on supervisors to improve safety. In the second study (Simard & Marchand, 1997), the focus was on predicting the level of work groups' safety compliance, including their propensity to comply with rules, follow supervisors' advice, and wear personal protective equipment. The authors distinguished between "microantecedents," which reside at the team level (e.g., work processes and risk, characteristics of the work group, and supervision characteristics), and "macroantecedents," which reside at the organizational level of analysis and beyond (e.g., top management's commitment to safety and socioeconomic context). All of the outcomes and predictors were assessed using supervisory ratings.

In both of these studies, the micro organizational factors emerged as the best predictors. In the case of safety initiative, Simard and Marchand (1995) found that a participatory approach to safety management was the most important predictor. That is, when supervisors and employees had joint involvement in safety-related activities on the shop floor, work groups were more likely to anticipate hazards, make suggestions, and put pressure on the organization to make safety improvements. The study also found that work-group cohesion and cooperation between work-group members and supervisors, as well as elements of task autonomy and uncertainty, were positively associated with work groups' taking safety initiatives. With the safety compliance outcome in Simard and Marchand (1997), a cooperative relationship between work group and supervisor was the strongest predictor of work-group compliance; to a lesser degree, participatory styles of supervisory management of safety and work-group cohesion were significant in the same direction. In contrast to the safety initiative study, no elements of work organization (e.g., task autonomy) were significantly related to safety compliance.

Some limitations of these studies are worth noting. One is that all of the work-group variables as well as the outcome variables were derived from supervisors' responses. Although Simard and Marchand (1995) suggested that there were no systematic response biases in supervisor ratings, it is possible that having supervisors rate their work groups on positive characteristics (e.g., cohesion) might encourage self-enhancing biases, thus inflating the true relationship between group processes and a socially desirable outcome like safety behaviors. For example, supervisors who get along well with their work groups might have a tendency to rate their work groups high on both group processes and safety outcomes (thus creating a halo effect). The second limitation is that, because the sample included only firms with more than 70 employees, the generalizability of the models to small enterprises is restricted.

The final study we report in this section explored group processes in relation to safety outcomes. Like Simard and Marchand (1995, 1997), Hofmann and Stetzer (1996) used a cross-level research strategy. The context was a single organization and consisted of 21 teams in a chemical processing plant. The researchers investigated how several group-level factors (i.e., group process, safety climate, and intentions to approach other team members engaging in unsafe behaviors) and one individual-level factor (i.e., perceptions of role overload) affected work-group safety (i.e., unsafe behavior, injuries). It was hypothesized that the relationship between task-oriented group processes (e.g., planning and coordinating efforts, sharing information about work-related events) and unsafe behaviors would be mediated by the intention to approach team members engaged in unsafe acts. Although this hypothesis was not significant using hierarchical linear modeling, the researchers found support for their model when they reanalyzed the relationship using ordinary least squares in an attempt to disentangle any substantial effects from the influence of low statistical power. That is, group processes such as planning and coordinating were associated with a higher intention to approach team members engaged in unsafe acts, which in turn was associated with fewer injuries and less unsafe behavior. Mechanisms like this, by which team working seems to affect safety, are severely underresearched.

Summary of Findings and Potential Mechanisms

Studies on autonomous teams and safety use a variety of methods, have inconsistent definitions of what autonomous work groups are, and vary in the degree to which work design is combined with interventions directed specifically at improving safety. This diversity makes it difficult to attribute safety performance to autonomous teams. Nevertheless, there are some coherent themes. With the exception of the Cohen and Ledford (1994) study, the key investigations of autonomous work groups (e.g., Hechanova-Alampay & Beehr, 2002; Pearson, 1992; Trist et al., 1963; Trist et al., 1977) have suggested positive safety consequences of this work redesign. Certainly, no study showed negative safety consequences. The research collectively suggests that autonomous work design might promote safer working through mechanisms such as greater felt responsibility and ownership for safety, increasing group member communication, and reducing boredom. Another mechanism, rarely suggested but plausible, is that autonomous work groups increase employees' understanding of one another's tasks and the wider unit, thereby enhancing collective safe working.

None of the above studies, however, throws light on which particular aspects of this work design are most important for safety. This information is more readily gained from the multilevel studies that examined various work characteristics separately. Employee involvement in safety activities; task autonomy (or nonroutine work organization); group processes (e.g., group cooperation, planning and coordination, willingness to approach other members); and leadership behaviors (cooperative relationships between work group and supervisor) were shown to positively influence safety outcomes. This research therefore converges with the research on autonomous work groups insofar as one would expect this form of work redesign to enhance autonomy and participation, as well as to influence group processes, and perhaps even to result in more coaching-oriented leadership. Nevertheless, it also suggests that safety might be improved by other initiatives, such as leadership development, or might even be enhanced by working on a team without its necessarily being autonomous. It also suggests that autonomous work groups will have a greater impact on safety if attention is given to establishing positive group processes such as participative planning and coordination.

COCKPIT CREWS

We now focus on cockpit crews, the third category of work group that has been considered extensively in the safety domain. Like some types of organizational teams, cockpit crews demand tight coordination to process large amounts of information in short spaces of time. The highly interdependent nature of a cockpit crew requires members to transmit information, issue and acknowledge commands, and ask questions about flight operations and conditions. Not surprisingly, given this context, the key focus of research in relation to cockpit crews has been on group processes, such as the crew's ability to communicate effectively. In contrast to the research on HSCs and autonomous teams, which in part assesses whether team structures are more effective than nonteam structures, this literature typically has assumed that there needs to be a team. The focus in cockpit crew safety research has therefore been more squarely on team functioning rather than on team type.

A broad conclusion from the vast cockpit crew research is that group processes influence safety-related outcomes such as errors. Given that comprehensive reviews of this and related literatures (e.g., crew resource management) already exist (e.g., Foushee, 1984; Helmreich & Merritt, 1998; Wiener, Kanki, & Helmreich, 1996; Wiener & Nagel, 1988), our goal here is not to attempt another review. Rather, we showcase some theoretical and methodological innovations used in studying safety in cockpit crews that have implications for teams in other organizational contexts. The cockpit literature is seldom cited in the mainstream applied psychology literature on work teams, despite rich potential for cross-fertilization. Specifically, we describe some safety outcomes that are used in this research, a novel analytic technique for understanding group processes and their consequences, and the potential role of knowledge-based processes in safety.

Safety-Related Outcomes: A Unique Approach

Aviation researchers have established a range of standardized indicators to track cockpit operations and situational responses that help or hinder flight safety. Data from the Line Operations Safety Audit (LOSA), for example, are a part of a large human factors project based at the University of Texas at Austin. So far, researchers there have used expert observers on more than 3,500 domestic and international airline flights to record systematically various threats to safety and how they are managed by the crew (e.g., error detection, exacerbating the error, failures to respond to error), as well as detailed information on the types of the errors (Helmreich, Klinect, & Wilhelm, 1999). *Crew error* has been broadly defined as "action or inaction leading to deviation from team or organizational intentions" (Helmreich, 2000, p. 781) and has been classified into various categories (e.g., intentional noncompliance errors, procedural errors, communication errors, proficiency errors, and operational decision errors; Helmreich & Foushee, 1993).

One of the advantages of this observational approach is that it overcomes some of the problems of using accidents (e.g., airplane crashes) as the only safety outcome. As we discuss earlier in this chapter, outcomes like this are relatively infrequent, which means that such data typically have limited variance. Additionally, these outcomes are distal from potential crew-level determinants and are affected by multiple factors (e.g., weather conditions) in the flight system. Observational data are more proximal, direct, and statistically reliable, assuming the observed events occur reasonably often.

There are therefore potential advantages of using observational techniques in studies of organizational work teams. A researcher who has successfully applied this method to the study of subunit safety in the workplace is Zohar (2002). Zohar used independent observers to monitor the extent to which employees wore earplugs. We advocate consideration of this and related techniques as a way of understanding how group processes affect safety.

Linguistic Analysis of Crew Communication

Crew communication is recognized as vital for safety and performance outcomes (e.g., Foushee, 1984; Foushee & Helmreich, 1988). Sexton and Helmreich (2000) examined how communication is related to performance and error rates. However, unlike many other studies that have evaluated perceptions of communication quality or communication content, this study investigated via linguistic analysis how crew members used language. The researchers were interested in the use of first-person plural (i.e., *we*, *our*, *us*) and first-person singular (i.e., *I*, *my*, *me*) words; they argued that use of first-person plural words provided a gauge of how oriented the speaker was to seeing the crew as a team (in essence, assessing a form of team identity). A further dimension of interest was achievement-oriented language, such as words like *try*, *effort*, and *goal* that indicated the crew was working toward a successful outcome. This latter dimension is analogous to concepts like goal orientation (Locke & Latham, 1990), and the researchers argued that greater achievement-oriented language would be related to better safety outcomes.

Data to test these hypotheses were derived from the transcripts of crew communications of 12 three-person crews (each consisting of a captain, first officer, and flight engineer) flying simulated flights on Boeing 727s. Each simulation contained five flight segments over 2 days, and transcripts were made from four of these five segments. Two of these flight segments were high workload, whereas the other two were low workload. Data on individual performance, errors, and communications skills were collected by an expert observer across each segment. Sexton and Helmreich (2000) argue that, as communication is socially constructed, collapsing data across all crew positions when exploring language use attenuates the effects of individuals' agency in language use across time. They therefore analyzed the language and flight outcome variables at the individual level of analysis.

The study found that language use differs as a function of both position and workload as well as between members of the same cockpit crew. More specifically, captains used the first-person plural (e.g., *we*) more frequently than either first officers or flight engineers did, particularly during high-workload segments. Overall, there was a linear increase in the use of first-person plural in cockpits across time, perhaps indicative of increasing familiarity. Use of first-person plural words and words indicative of achievement were positively related to performance and communication, and negatively related to error rates. Interestingly, the researchers found that language use by an individual in one position at an earlier time was related to the performance or error rate of an individual in a different position at a later time. For example, the captain's use of achievement words in an earlier segment was strongly related to fewer errors by the flight engineer in a later segment.

In sum, this study suggests that language use indicative of team identity and goal orientation is related to important safety-related outcomes. Language use both reflects and shapes important team processes such as team identity and familiarity. This linguistic approach could be highly informative in studies of other types of organizational teams. For example, Pearson's (1992) study suggested better safety performance among autonomous work groups, which was attributed to these groups' assuming greater personal responsibility for safety rather than seeing it solely as the responsibility of safety personnel. One could speculate that members of the autonomous work groups might have made more first-person plural references (e.g., *we*) in relation to safety than members of the conventional groups, who might talk more about safety in terms of *they*. Investigation of team mental models, which we discuss in the next section, could also benefit from this approach.

Knowledge-Based Processes

In addition to communicating, having a mutual understanding of the nature of events relevant to safety and flight efficiency is also critical in cockpit crews. A shared mental model enables individuals to make sense of and engage with their environment (e.g., Cannon-Bowers & Salas, 2001; Mohammed & Dumville, 2001). Researchers (e.g., Mathieu, Heffner, Goodwin, Salas, & Cannon-Bowers, 2000) have used empirical methods to examine how team members' mental models influence team process and performance. One way this has happened in relation to safety performance has been with intervention studies in experimental settings (Blickensderfer, Cannon-Bowers, & Salas, 1998). More specifically, researchers have used cross-training to provide crew members with opportunities to perform one another's roles, thus providing a basis for a greater common understanding.

Much of this research on cross-training so far has used undergraduate student samples on flight simulators to test the effect of cross-training under various conditions of workload. For example, in a study by Volpe, Cannon-Bowers, Salas, and Spector (1996) using a computerized flight simulator, 80 undergraduate male students were assigned randomly to dyads in one of four conditions in a 2 × 2 between-subjects design: high workload with cross-training; low workload with cross-training; high workload without cross-training; and low workload without cross-training. Cross-training took the form of verbally presenting team members with information about the other's roles. Results showed that this form of cross-training was a determinant of greater team processes, better communication, and simulator performance. A follow-up study (Cannon-Bowers, Salas, Blickensderfer, & Bowers, 1998) had a sample of 40 three-person teams drawn from U.S. Navy recruits and used a more cognitively demanding simulator that required greater task interdependence. In this study, cross-training constituted training the team members in one another's duties rather than verbally clarifying their studies. Again, teams that had cross-training demonstrated a greater degree of interpositional knowledge than those that had no cross-training and demonstrated the other positive aspects of team functioning found in the earlier study.

These two studies highlight how gaining greater interpositional knowledge may facilitate greater team functioning and, in this context, safe performance. Interestingly, we speculate earlier in this chapter that one of the ways in which autonomous work groups and related interventions might enhance safety is through expanding team members' knowledge of other team members' jobs and the wider unit—in essence, interpositional knowledge. Again, the literature on flight crews has a potential application to the study of teams in other contexts, in this case, putting greater emphasis on knowledge-based mechanisms than has traditionally been the case.

CONCLUSIONS

Key Findings

The research described in this chapter suggests that team structures within organizations can have positive consequences for safety. However, a shortage of rigorous field studies—especially those focusing on health and safety committees (HSCs) and autonomous work teams—combined with highly diverse methods, measures, and settings, makes it important to qualify such a conclusion.

One reason that teams appear to be important is that they provide a structure for employee involvement in decision making. Earlier in this chapter, we discuss evidence from literature on HSCs that suggests that the presence of an HSC, particularly if it is voluntarily introduced and contains representatives of the workforce rather than just managers, is related to the lowering of organizational injury rates. This finding converges with studies of autonomous work groups; these studies suggest that this work structure, which has as a core feature greater collective employee autonomy to make decisions, is associated with improved safety in day-to-day operations. Consistent with this were the studies that point to the importance of employee participation in safety-related activities and task autonomy as determinants of teams' safety behaviors, particularly safety initiatives. A practical implication of these findings, taken together, is that one way to enhance work safety is to increase employee involvement, either in the form of greater employee participation in safety-related decision making (e.g., HSCs) or in the form of greater task or group autonomy. Later in this section we discuss the potential mechanisms by which these effects might occur.

In the literature on autonomous teams, although the direct evidence is sparse, effective group processes (e.g., group cohesion, cooperation, group planning) have also come to the fore as potential determinants of workplace safety. This finding would be no surprise to researchers investigating cockpit crews who have demonstrated links between group processes, particularly crew communication, and safety-related outcomes such as errors. Indeed, given that the main rationale of teams is to coordinate action around interdependent tasks, it would be surprising if effective team processes like coordination did not impinge on safety outcomes. From a practical perspective, this suggests the importance for safety of training those who need to work together not only in the task-related aspects of the job but also in the interpersonal and process skills that can make or break a team. Teamwork is sometimes introduced with the expectation that individuals will intuitively know how to work collectively as a team. However, the findings we review in this chapter suggest the good sense of providing teamwork training, not just for team performance but for safety as well.

Finally, the quality of interaction between supervisors and employees (i.e., cooperative leadership) is an important thread. This was evident in the studies showing the consequence of joint health and safety committees, and on a more micro level in the type of interaction between supervisors and their work groups in safety activities. This finding complements findings about the importance of safety-focused leadership shown in both group (e.g., Zohar, 2002) and nongroup studies of safe working. Leadership in teams will remain an important topic of research for understanding the antecedents of group-level safety. At a practical level, these findings suggest the importance for work safety of leadership training and development.

Future Directions

Although promising, the conclusions given in the preceding paragraphs must be considered provisional due to the relatively small number of robust studies. A straightforward yet recurrent appeal, therefore, is for more rigorous studies, such as longitudinal designs combined with reliable indicators of safety. In the course of discussing the existing literature, we also identified several methodological issues that would need to be considered in future research. Next we suggest some other important ways forward.

Attention to Context

The only study (i.e., Cohen & Ledford, 1994) of autonomous work groups that did not report positive consequences for safety was one that was conducted in a context (i.e., an organization in which safety was perhaps not particularly salient) very different from that of the studies that were conducted in manufacturing settings. This discrepancy alone alerts us to the need to consider context. Social, environmental, political, cultural, and technological differences are highly likely to affect whether teams are appropriate in the first place, as well as which team characteristics or processes are important for safety and which safety outcomes are most valid. The need to integrate context into all stages of organizational research has been highlighted in the organizational behavior literature more generally (e.g., Johns, 2001; Rousseau & Fried, 2001). Some of the research reviewed in this chapter has made a special effort to do this (e.g., Cohen & Ledford, 1994; Pearson, 1992; Trist et al., 1977) and provides examples for future contributions in this area. In addition, growing interest in multilevel studies in teams and safety research (e.g., Hofmann & Stetzer, 1996; Simard & Marchand, 1995, 1997) implicitly moves studies in this area toward the importance of contextual influences.

The Value of Alternative Modes of Inquiry

We acknowledge that all of the research discussed in this chapter has come from a positivist tradition. For us, this accentuates the importance of research that has looked at teams and safety through other lenses. For example, Weick's (1993) account of the "interactive disintegration" (p. 628) of a fire-fighting team in the Mann Gulch disaster provides a rich account of how role identities and interpersonal dynamics can structure or hamper resilience in ambiguous, safety-critical situations. Although not focusing on safety as a goal, Barker's (1993) seminal study of the development of concertive control in self-managing teams offers a way forward for understanding, for example, the social construction of the safety-production trade-off or how group-level safety climates (e.g., Zohar, 2000) develop. Although some of the essence of these concepts is captured in the group process literature, the notion of relational agency and the social construction of safety captured by the spirit of Weick's and Barker's research is untapped by positivistic approaches.

Attention to Mechanisms

Several researchers have speculated on mechanisms by which teams can affect safety, but these mechanisms have rarely been investigated. One advantage of understanding significant mechanisms is that it will help researchers identify when particular interventions are likely to be appropriate. For example, consider the possible ways in which autonomous work groups might impinge on safety. First, autonomous work groups might enhance safe working via motivational mechanisms, such as employees feeling increased responsibility for, and ownership of, safety. Autonomous work groups might also improve group interaction. For example, by being self-managing, team members might communicate directly with each other rather than communicating via a supervisor. This form of work redesign, as we have already speculated, might enhance knowledge and understanding (e.g., interpositional knowledge), which in turn means that employees make fewer errors and work more safely. These are just some of the ways in which autonomous work groups might result in a safer workplace. If we had a better understanding of such processes, we could more easily ensure that work structures and supports are put in place to maximize the likelihood of the work redesign's success.

Attention to Negative Outcomes

Our review has mostly focused on the potential positive outcomes of teamwork for safety. This follows from the research that has tended to suggest either no consequences or positive consequences of teamwork on safety. However, the possibility exists that working in teams could reduce safety under some circumstances. We identify three possibilities here.

One possibility is that work characteristics might be negatively affected, which could then impinge on safety. For example, in the case of autonomous work groups, there could be negative consequences on safety as a result of approaches to working together that have not yet been considered (e.g., perhaps role clarity would be reduced as a result of diffused responsibility); because this form of working in teams has been poorly implemented or is implemented at the same time as other changes with negative consequences on safety (e.g., downsizing; see chapter 4 of this volume); or because forms of teamwork other than autonomous work groups are introduced.

Regarding the last point, we focus in this chapter on autonomous work teams and show them to have potential positive benefits for safety. However, teamwork is often introduced without expanding autonomy, such as in the case of lean production teams. Lean production teams involve multiskilled team members who carry out standard operating procedures. Typically, there is a frontline supervisor who manages the team, and often there are offline continuous improvement activities in which the team members work on simplifying and standardizing procedures. Views about whether employee involvement is enhanced with lean teams are mixed, although a recent review of lean production by Landsbergis, Cahill, and Schnall (1999) found little evidence that workers are "empowered" under these systems but rather that lean production tends to intensify work demands and work pace. Other researchers have similarly argued that teamwork can be a form of work intensification, serving to increase workload (Delbridge, Turnbull, & Wilkinson, 1992). If teamwork does lead to enhanced workload, then this could conceivably reduce employee compliance with safety procedures. Previous evidence suggests that high workload can reduce compliance (Hofmann & Stetzer, 1996). Different forms of working in teams, or poorly implemented autonomous work groups, might therefore be detrimental to safety at work because of their negative impact on work characteristics.

A second way in which teamwork could negatively affect safety is through its potential negative effects on team processes. For example, group members might engage in social loafing and reduce their efforts (Karau & Williams, 1993), and groupthink might result in poor-quality decision making (Janis, 1982). More generally, the benefits of working in teams typically arise because of the coordination of individual efforts, but these coordination processes can fail. For example, Tjosvold (1990) identified situations in which flight crew teams failed to coordinate effectively, such as when crewmembers were not able to discuss opposing views openly.

A third way in which teamwork might negatively affect safety relates to the composition of the team. Most of the autonomous work groups that have been studied have tended to be rather homogeneous in gender and ethnicity. However, workplaces have become increasingly diverse, and in some circumstances diversity can have negative consequences. In an extensive review of 40 years of research, Williams and O'Reilly (1998) reported that many studies show the negative effects of diversity on group processes and functioning, such as by reducing social integration and increasing affective conflict. It is therefore possible that introducing teams within diverse workplaces might negatively affect safety by having negative effects on groups processes.

In summary, it is possible that teamwork could have detrimental effects on safety under some circumstances. More research is needed to understand how likely these negative effects are, whether they outweigh the potential positive benefits, and whether it is possible to intervene to limit the negative effects (e.g., via training).

More Creative Methods

This point applies to the types of safety outcomes used as well as different ways of assessing team processes. For example, although social network theory is a well-trodden approach to understanding relationships between organizations, there is considerably less research on exploring intraorganizational networks and performance (Flap, Bulder, & Ölker, 1998). In particular, no study to our knowledge exploits this method for exploring social influence on safety-related outcomes over time. For instance, the opportunity exists for researchers to use a social network approach to track communication patterns and boundary management in teams in safety-critical work systems. Additionally, we encourage researchers to draw inspiration from the aviation literature in terms of using simulations to explore team safety in experimental settings.

Integrating Team Safety Research With Wider Literature

There is a vast literature linking team work characteristics, group processes, and other group aspects (e.g., team size) to team effectiveness outcomes (Guzzo & Dickson, 1996). However, to date, safety has not been seen widely as an indicator of team effectiveness. Yet it is highly likely that many of the same determinants of good performance (e.g., team design, team processes) will also positively affect safety. Indeed, several of the aspects identified here as important for safety (e.g., group autonomy) have been shown to affect team performance (Parker & Turner, 2002). It makes sense, therefore, to draw on the team effectiveness literature for inspiration. Likewise, there is a growing literature on error management in groups (e.g., Edmondson, 1996). The general point is that safety has been neglected in organizational literature, but there is no need to re-create the wheel entirely. Theories and methods from other traditions can be usefully employed in exploring teams and safety.

A Final Word

Teams are becoming increasingly prevalent in our workplaces. On the one hand, it seems that teams can provide a structure for increases in employee involvement in decision making, social support, information sharing, and participative leadership—all aspects likely to promote safety at work. On the other hand, introducing teams per se is no automatic guarantee of success, and there is the potential for negative outcomes such as diffused responsibility, miscommunication, and role ambiguity. Although research has made some progress, we still need to know much more about how best to design and structure teams, what sorts of team processes are most important, and how teams should be led if we are to promote workplace safety. To do this, we urge researchers to seek safety by going further afield than has been done before.

REFERENCES

BA jumbo plunges as man storms cockpit. (2000, December 29). *The Times* (London). Retrieved June 22, 2001, from http://www.thetimes.co.uk/article/0,,3-59795,00.html.

Barker, J. R. (1993). Tightening the iron cage: Concertive control in self managing teams. *Administrative Science Quarterly, 38*, 408–437.

Blickensderfer, E., Cannon-Bowers, J. A., & Salas, E. (1998). Cross-training and team performance. In J. A. Cannon-Bowers & E. Salas (Eds.), *Making decisions under stress: Implications for individual and team training* (pp. 299–311). Washington, DC: American Psychological Association.

Cannon-Bowers, J. A., & Salas, E. (2001). Reflections on shared cognition. *Journal of Organizational Behavior, 22*, 195–202.

Cannon-Bowers, J. A., Salas, E., Blickensderfer, E., & Bowers, C. A. (1998). The impact of cross-training and workload on team functioning: A replication and extension of initial findings. *Human Factors, 40*, 92–101.

Cohen, S. G., & Ledford, G. E., Jr. (1994). The effectiveness of self-managing teams: A quasi-experiment. *Human Relations, 47*, 13–43.

Commission on the Future of Worker-Management Relations. (n.d.). *Safety and health programs and employee involvement*. Retrieved June 22, 2001, from http://www.dol.gov/dol/_sec/public/media/reports/dunlop/section7.htm

Delbridge, R., Turnbull, P., & Wilkinson, B. (1992). Pushing back the frontiers: Management control and work intensification under JIT/TQM factory regimes. *New Technology, Work and Employment, 7*, 97–106.

Eaton, A. E., & Nocerino, T. (2000). The effectiveness of health and safety committees: Results of a survey of public-sector workplaces. *Industrial Relations, 39*, 265–290.

Edmondson, A. (1996). Learning from mistakes is easier said than done: Group and organization influences on the detection and correction of human error. *Journal of Applied Behavioral Science, 32*, 5–28.

Flap, H., Bulder, B., & Ölker, B. V. (1998). Intra-organizational networks and performance: A review. *Computational and Mathematical Organization Theory, 4*, 109–147.

Foushee, H. C. (1984). Dyads and triads at 35,000 feet: Factors affecting group process and aircrew performance. *American Psychologist, 39*, 885–893.

Foushee, H. C., & Helmreich, R. L. (1988). Group interaction and flight crew performance. In E. L. Weiner & D. C. Nagel (Eds.), *Human factors in aviation* (pp. 189–225). San Diego, CA: Academic Press.

Frone, M. R. (1998). Predictors of work injuries among employed adolescents. *Journal of Applied Psychology, 83*, 565–576.

Goodman, P. S. (1979). *Assessing organizational change: The Rushton quality of work experiment*. New York: Wiley.

Guzzo, R. A., & Dickson, M. W. (1996). Teams in organizations: Recent research on performance and effectiveness. *Annual Review of Psychology, 47*, 307–338.

Hechanova-Alampay, R. H., & Beehr, T. A. (2002). Empowerment, span of control and safety performance in work teams after workforce reduction. *Journal of Occupational Health Psychology, 6*, 275–282.

Helmreich, R. L. (2000). On error management: Lessons from aviation. *British Medical Journal, 320*, 781–784.

Helmreich, R. L., & Foushee, H. C. (1993). Why crew resource management? Empirical and theoretical bases of human factors training in aviation. In E. Wiener, B. Kanki, & R. Helmreich (Eds.), *Cockpit resource management* (pp. 3–45). San Diego, CA: Academic Press.

Helmreich, R. L., Klinect, J. R., & Wilhelm, J. A. (1999). Models of threat, error, and CRM in flight operations. In *Proceedings of the Tenth International Symposium on Aviation Psychology* (pp. 677–682). Columbus: Ohio State University.

Helmreich, R. L., & Merritt, A. C. (1998). *Culture at work in aviation and medicine.* Aldershot, UK: Ashgate.

Hemingway, M. A., & Smith, C. S. (1999). Organizational climate and occupational stressors as predictors of withdrawal behaviours and injuries in nurses. *Journal of Occupational and Organizational Psychology, 72*, 285–299.

Hofmann, D. A., & Stetzer, A. (1996). A cross-level investigation of factors influencing unsafe behaviors and accidents. *Personnel Psychology, 49*, 307–339.

Janis, I. L. (1982). *Groupthink.* Boston: Houghton Mifflin.

Johns, G. (2001). In praise of context. *Journal of Organizational Behavior, 22*, 31–42.

Karau, S. J., & Williams, K. D. (1993). Social loafing: A meta-analytic review and theoretical integration. *Journal of Personality and Social Psychology, 65*, 681–786.

Landsbergis, P. A., Cahill, J., & Schnall, P. (1999). The impact of lean production and related new systems of work organization on worker health. *Journal of Occupational Health Psychology, 4*, 108–130.

Lawler, E. E., III. (1992). *The ultimate advantage: Creating the high involvement organization.* San Francisco: Jossey-Bass.

Locke, E. A., & Latham, G. P. (1990). *A theory of goal setting and task performance.* Englewood Cliffs, NJ: Prentice Hall.

Mathiason, T. (2001, August 12). Safety crackdown as work deaths rise. *The Observer*, p. 4.

Mathieu, J. E., Heffner, T. S., Goodwin, G. F., Salas, E., & Cannon-Bowers, J. A. (2000). The influence of shared mental models on team process and performance. *Journal of Applied Psychology, 85*, 273–283.

McVeigh, T. (2000, November 12). Repair bungles add to chaos on railways. *The Observer*, p. 6.

Mohammed, S., & Dumville, B. C. (2001). Team mental models in a team knowledge framework: Expanding theory and measurement across disciplinary boundaries. *Journal of Organizational Behavior, 22*, 89–106.

Occupational Safety and Health Administration. (1991). *Hearing report on OSHA reform.* Retrieved June 24, 2003, from http://www.osha.gov/pls/oshaweb/

owadisp.show_document?p_table=INTERPRETATIONS&p_id=20443&p_text_version=FALSE

O'Toole, M. F. (1999). Successful safety committees: Participation not legislation. *Journal of Safety Research*, *30*, 39–65.

Parker, S. K., & Turner, N. (2002). Work design and individual work performance: Research findings and an agenda for future inquiry. In S. Sonnentag (Ed.), *The psychological management of individual performance: A handbook in the psychology of the management of organizations* (pp. 69–93). Chichester, UK: Erlbaum.

Parker, S. K., Turner, N., & Griffin, M. A. (2003). Designing healthy work. In D. A. Hofmann & L. E. Tetrick (Eds.), *Individual and organizational health* (pp. 91–130). San Francisco: Jossey-Bass.

Parker, S. K., & Wall, T. D. (1998). *Job and work design: Organizing work to promote well-being and effectiveness*. London: Sage.

Pearson, C. A. L. (1992). Autonomous workgroups: An evaluation at an industrial site. *Human Relations*, *45*, 905–936.

Reilly, B., Paci, P., & Holl, P. (1995). Unions, safety committees and workplace injuries. *British Journal of Industrial Relations*, *33*, 275–288.

Rousseau, D. M., & Fried, Y. (2001). Location, location, location: Contextualizing organizational research. *Journal of Organizational Behavior, 22*, 1–13.

Sexton, J. B., & Helmreich, R. L. (2000). Analyzing cockpit communications: The links between language, performance, error, and workload. *Human Performance in Extreme Environments*, *5*, 63–68.

Simard, M., & Marchand, A. (1995). A multilevel analysis of organizational factors related to the taking of safety initiatives by work groups. *Safety Science*, *21*, 113–129.

Simard, M., & Marchand, A. (1997). Workgroups' propensity to comply with safety rules: The influence of micro–macro organisational factors. *Ergonomics*, *40*, 172–188.

Sonnentag, S. (1996). Work group factors and individual well-being. In M. A. West (Ed.), *Handbook of work group psychology*. Chichester, UK: Wiley.

Tjosvold, D. (1990). Flight crew collaboration to manage safety risks. *Group and Organization Studies*, *15*, 11–19.

Trist, E. L., Higgin, G. W., Murray, H., & Pollock, A. B. (1963). *Organizational choice*. London, England: Tavistock.

Trist, E. L., Susman, G. I., & Brown, G. R. (1977). An experiment in autonomous working in an American underground coal mine. *Human Relations*, *30*, 201–236.

Volpe, C. E., Cannon-Bowers, J. A., Salas, E., & Spector, P. E. (1996). The impact of cross-training on team functioning: An empirical investigation. *Human Factors*, *38*, 87–100.

Walton, R. E. (1972, November/December). How to counter alienation in the plant. *Harvard Business Review*, 70–81.

Weick, K. E. (1993). The collapse of sensemaking in organizations: The Mann Gulch disaster. *Administrative Science Quarterly, 38*, 628–652.

Wiener, E. L., Kanki, B. G., & Helmreich, R. L. (1996). *Cockpit resource management.* San Diego: Academic Press.

Wiener, E. L., & Nagel, D. C. (Eds.). (1988). *Human factors in aviation.* San Diego, CA: Academic Press.

Williams, K. Y., & O'Reilly, C. (1998). The complexity of diversity: A review of forty years of research. In B. Staw & R. Sutton (Eds.), *Research in organizational behavior* (Vol. 21, pp. 77–140). Greenwich, CT: JAI Press.

Zohar, D. (2000). A group-level model of safety climate: Testing the effect of group climate on microaccidents in manufacturing jobs. *Journal of Applied Psychology, 85*, 587–596.

Zohar, D. (2002). Modifying supervisory practices to improve sub-unit safety: A leadership-based intervention model. *Journal of Applied Psychology, 87*, 156–163.

4

JOB INSECURITY: EXPLORING A NEW THREAT TO EMPLOYEE SAFETY

TAHIRA M. PROBST

Virtually every sector of the American economy is experiencing extreme financial pressures due to commercial rivalries around the globe, government deregulation of industry, and the increasing pace of organizational technology change. As a result, massive layoffs, conversion of full-time jobs to part-time positions, and the increase of temporary workers are but a few of the realities facing organizations and workers today. Organizational restructuring in the form of corporate downsizing, mergers and acquisitions, plant closings, and workforce reorganizations affects millions of workers each year. Annually, more than 500,000 U.S. employees can expect to lose their jobs as a result of these transitions (Simons, 1998). Recent estimates suggest that Fortune 500 companies alone eliminated 2.5 million jobs from their corporate rolls in the decade between 1983 and 1993 (Simons, 1998).

Although these are impressive statistics, they do not capture the number of employees who might be concerned that their jobs may be the next ones slated for "right-sizing." As evidence that the effects of these organizational trends are felt beyond the corporate borders of the transitioning organizations themselves, a recent nationwide survey in the United States found that despite a record low 4.3% jobless rate in the United States and a relatively strong economy, 37% of workers were fearful of involuntary job loss—a significant increase from 11% during the recession years of the early 1990s (Belton, 1999).

Today's pervasive climate of job insecurity has been shown to have multiple negative effects on affected employees. Employees with low job security report lower job satisfaction (Ashford, Lee, & Bobko, 1989; Davy, Kinicki, & Scheck, 1991); a greater incidence of physical health conditions (Cottington, Mathews, Talbot, & Kuller, 1986; Dooley, Rook, & Catalano, 1987; Kuhnert, Sims, & Lahey, 1989; Roskies & Louis-Guerin, 1990); and higher levels of psychological distress (Dekker & Schaufeli, 1995) when compared to employees with secure jobs. There are important job-related outcomes as well. The more dissatisfied employees are with their perceived job security, the less committed they are to the organization (Ashford et al., 1989; Davy et al., 1991); the more frequently they engage in work withdrawal behaviors such as absenteeism, tardiness, and work task avoidance (Probst, 2002c); and the more likely they are to quit their job (Ashford et al., 1989; Davy et al., 1991).

Although there is much research documenting the negative effects of job insecurity on employee attitudes, turnover, and mental and physical health, until recently there was a dearth of research assessing the potential effects of job insecurity on employee safety outcomes—this despite a call from the National Institute for Occupational Safety and Health for more research to be conducted assessing the effects of today's changing organization of work on employee accident rates and the incidence of workplace injury (Sauter, 1993). Therefore, the purpose of this chapter is to examine the theoretical and empirical evidence for the proposition that employee job insecurity can have detrimental effects on worker safety and health. The first part of the chapter focuses on the development of a model illustrating the proposed relationships between job insecurity and employee safety outcomes. In addition, the results from several recent empirical tests of this model are presented. These data—collected in the field and in the laboratory—consistently confirm the negative effects of job insecurity on worker safety outcomes. Following the test results is a section devoted to an exploration of potential organizational implications for prevention. Two questions are addressed: Are there organizational changes that can be made to circumvent the negative relationship between job insecurity and worker safety? In particular, what role does an organization's safety climate play in attenuating this relationship? Finally, the chapter offers methodological challenges and suggested directions for future research.

INTEGRATING JOB INSECURITY AND SAFETY THEORY AND RESEARCH

Theory and research in the area of job insecurity and those in the area of worker safety have largely progressed independently of each other. Researchers in the area of job insecurity have focused on delineating the attitudinal, behavioral, and job-related consequences of such insecurity but have largely ignored safety as a potential outcome. One notable exception was a study conducted by Landisberger, Cahill, and Schnall (1999), in which the researchers found detrimental effects on employee health and injury rates in a variety of industries that were implementing lean production cultures. Although this study did not address job insecurity per se, it is accepted that one of the hallmarks of lean production is the implementation of organizational downsizing (American Management Association, 1997; Landisberger et al., 1999).

Similarly, researchers in the area of workplace safety have identified numerous antecedents of worker safety, addressing the impact of such factors as ergonomic conditions (Melamed, Luz, Najenson, Jucha, & Green, 1989); employee personal characteristics such as gender, education, and personality (e.g., Fergusen, McNally, & Both, 1984; Hansen, 1989; Leigh, 1986; Leveson, Hirschfeld, & Hirschfeld, 1980); and organizational characteristics such as safety climate (e.g., Brown & Holmes, 1986; Dedobbeleer & Beland, 1991; Zohar, 1980). Yet there has been little research to date considering employee job insecurity as a predictor of worker safety.

Despite the relative independence of these two areas, a number of recent research findings looking at individual- and organizational-level variables have suggested not only the existence of a link between job insecurity and employee safety but also potential mechanisms explaining this relationship.

Individual Factors Predicting Safety Outcomes

Research has shown that individuals who have high levels of negative affect, anxiety, and depression are less likely to properly use their personal protective equipment (Dunbar, 1993). Therefore, one explanation for a proposed link between job insecurity and negative safety outcomes would be that job insecurity causes negative job attitudes in the form of anxiety regarding job security and lowered job satisfaction. These attitudes, in turn, may result in a reduction in adherence to safety policies.

Research also suggests that safety knowledge and safety motivation are additional individual factors that are important in predicting safety compliance (Hofmann, Jacobs, & Landy, 1995; Neal, Griffin, & Hart, 2000). Using a contingency approach, Hofmann et al. predicted that employees would be less motivated to comply with safety policies to the extent that they were not rewarded for performing in a safe manner. They expected that perceived contingencies suggesting that greater rewards will follow from completing the work as quickly as possible would lead employees to take unnecessary shortcuts. In addition, they expected that a perception that management is not committed to safety would translate to a lack of motivation on the part of employees to comply with safety policies (Hofmann et al., 1995).

Theory and research on job stress suggests that, during times of employee job insecurity, the acquisition or maintenance of safety knowledge and motivation may be compromised. Stress has been shown to lead individuals to focus narrowly on a few specific aspects of their environment (Barthol & Ku, 1959; Mandler, 1982). According to Hofmann and Stetzer (1996), one consequence of this may be that employees tend to focus their attention on performance rather than on safety during times of stress—as might be precipitated by perceived job insecurity.

Organizational Variables Influencing Safety

Socioeconomic characteristics of organizations such as the tendency to provide "worse working conditions" have also been shown to be negatively associated with worker propensity to safety compliance (Simard & Marchand, 1997). Simard and Marchand found that secondary labor market organizations (i.e., those that traditionally offer low wages, low job security, and low status) tended also to have lower managerial participation in safety issues, which in turn was related to a reduction in accident prevention and compliance. Therefore, macro-level organizational variables were shown to impact micro-level outcomes such as employee safety compliance. Thus, another explanation for the linkage between job insecurity and reduced safety outcomes is that organizations that typically have job insecurity (i.e., secondary labor market organizations) also tend not to foster organizational cultures supportive of safety due to increased pressures for production in order to survive economically (Simard & Marchand, 1997). (See chapter 5 of this volume for a related discussion of this topic.)

It has also been suggested that workers face a conflict between safety and production (Faverge, 1980; Janssens, Brett, & Smith, 1995; Kjellen, 1984; Leplat & Rasmussen, 1984). The more an organization places an emphasis on production, the more employees perceive that safety is subordinated to the demands of production (Janssens et al., 1995). Hofmann and Stetzer (1996) summarize this trade-off and its consequences succinctly:

> Perceptions of performance pressure can lead workers to perceive that engaging in short cut behavior is an expected, or required, part of the job (Wright, 1986). Specifically, workers who perceive a high degree of

performance pressure will focus their attention on completing the work and less on the safety of their work procedures. As a result, they may begin to perceive short cut methods as required to meet the performance demands (e.g., Wright, 1986). (p. 309)

The degree to which employees choose production over safety in the safety–production trade-off will be in part determined by their perceptions regarding organizational reward contingencies. Hofmann and Stetzer (1996) predicted that employees who are experiencing job stress in the form of increased job demands will focus on performance rather than safety because performance is more likely to result in salient rewards for the employee. In addition, unsafe behavior may actually be perceived to be rewarding if it allows the employee to perform work tasks more quickly (Slappendal, Laird, Kawachi, Marshall, & Cryer, 1993). Thus, during times of job insecurity, employees desiring to retain their jobs may choose to focus more on performance and less on safety.

DEVELOPMENT OF A MODEL LINKING JOB INSECURITY TO SAFETY OUTCOMES

Given the theoretical and empirical evidence suggesting a possible link between job insecurity and negative employee safety outcomes, a model was developed to delineate the process by which the purported effects may occur (Probst & Brubaker, 2001).[1] The following sections describe the development of this model (see Figure 4.1), which illustrates the possible relationships between job insecurity, safety motivation, knowledge, and compliance and job-related injuries and accidents. From this model, several testable hypotheses can be developed.

Job Insecurity and Job Satisfaction

Decreased perceptions of job security have consistently been found to be related to decreased job satisfaction in many studies (e.g., Ashford et al., 1989; Davy et al., 1991; Probst, 2000). Job satisfaction has been found to have five facets: satisfaction with work, pay, promotion opportunities, supervision, and coworkers (Smith, Kendall, & Hulin, 1969). In addition, more recent work (Probst, 1998) has found that a sixth facet of job satisfaction, namely, job security satisfaction, is also important to consider in today's work environment. Job insecurity would likely have a negative impact on many of these dimensions. Based on this finding, the first part of the model

[1]The following section describing the development of the job-insecurity/safety model is largely based on Probst and Brubaker (2001). For further details regarding the development of the model, readers are directed to that article.

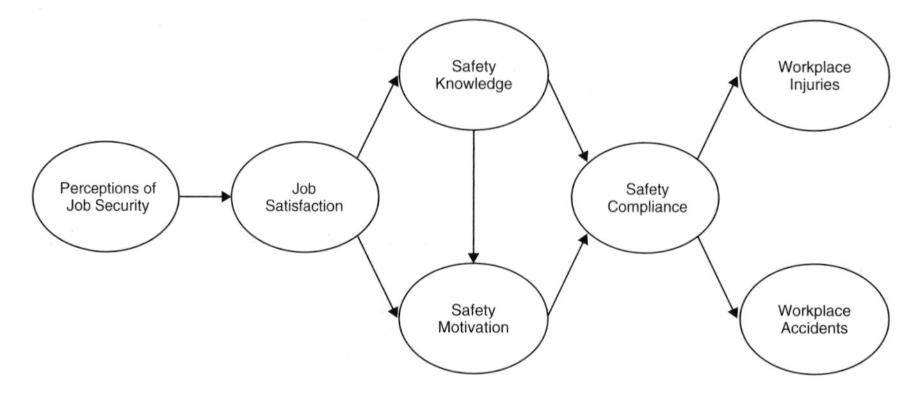

Figure 4.1. Theoretical model of the effects of job insecurity on employee safety outcomes. From "The Effects of Job Insecurity on Employee Safety Outcomes: Cross-sectional and Longitudinal Explorations," by T. M. Probst and T. L. Brubaker, 2001, *Journal of Occupational Health Psychology, 6*, pp. 139–159. Copyright 2001 by the American Psychological Association. Adapted with permission of the publisher and author.

postulates that employee perceptions of job security are positively related to employee job satisfaction such that greater job security leads to more positive job-related attitudes.

Recent research also suggests that many of the effects of job insecurity on individual and organizational outcomes are mediated via employee levels of job satisfaction. For example, in a path analysis modeling the effects of being a layoff survivor, Davy et al. (1991) found that while job security positively affected job satisfaction, it did not directly influence employees' behavioral intent to withdraw or organizational commitment. Rather, the relationship between these variables and job security existed only through the intervening variable of job satisfaction. In other words, when job attitudes were held constant, there was no relationship between job insecurity and turnover intentions or organizational commitment. On the basis of this and similar findings (e.g., Probst, 2002c), job insecurity is not expected to directly influence safety motivation or knowledge levels. Rather, it is predicted that an individual must experience job dissatisfaction as a result of the insecurity for negative outcomes to occur. Therefore, it is expected that job satisfaction mediates the relationship between job insecurity and safety motivation and safety knowledge.

Job Satisfaction and Safety Knowledge and Motivation

Two key variables that predict employee safety compliance are safety knowledge and safety motivation. *Safety motivation* has been defined as an employee's "motivation to perform a job in a safe manner" (Hofmann et al.,

1995, p. 133); "safety initiative" (Andriessen, 1978, p. 367); and the motivation to perform safety behaviors (Griffin & Neal, 2000; Neal et al., 2000). Safety motivation can also be operationalized using an expectancy-valence motivational approach. According to valence-instrumentality-expectancy (VIE) theory (Vroom, 1964), individuals will expend effort on activities that lead to desired rewards. Therefore, if an individual is rewarded for adhering to safety policies, one would expect that person's motivational force to be high for those behaviors. If the reward structure is such that individuals are "rewarded" for noncompliance, then one would expect their motivational force to comply with safety policies to be low.

It is important to note that safety motivation is not necessarily inclusive of safety knowledge or compliance. Safety knowledge is characterized by an employee's understanding of safe operating procedures and adequate safety training and instruction (Hofmann et al., 1995). An employee who is knowledgeable of safety rules may not always be motivated to comply. Likewise, an employee who does not understand all of the safety rules may be motivated, but this motivation may or may not translate into actual compliance due to the lack of proper knowledge or skill. Griffin and Neal (2000) similarly have asserted that safety knowledge and motivation are both unique and necessary predictors of safety compliance. Therefore, it is important to measure these two constructs separately.

When an individual perceives that his or her job security is threatened and is dissatisfied by that perception, it is expected that both safety motivation and safety knowledge will be adversely affected. In particular, job dissatisfaction is predicted to be related to low levels of safety knowledge and safety motivation. This prediction is generated from a cognitive resources framework (Kanfer & Ackerman, 1989) and from Eysenck and Calvo's (1992) model of anxiety and performance.

According to Kanfer and Ackerman (1989), each individual has a set amount of available cognitive resources when engaged in the completion of any given task. These finite cognitive resources can be allocated to on-task, off-task, or self-regulatory activities. In a work setting, on-task activities include those behaviors related to production, quality assurance, and safety compliance. Off-task activities include behaviors such as chatting with coworkers, thinking about family, or planning for the weekend. Self-regulatory activities include the monitoring of one's environment. During times of organizational change, the monitoring of job security would be aptly classified as a self-regulatory activity. This involves estimating the chances that one might be affected by the impending organizational transition, how one's job might change as a result, keeping up-to-date with organizational rumors, and the like.

When individuals have high job security, these self-regulatory activities can be disengaged, leaving more resources available for the on-task activities of production, quality, and safety. However, when job security is perceived to be low, some of those cognitive resources may be funneled into self-regulatory activities aimed at monitoring progress toward the goal of job retainment. Thus, during times of organizational transition, valuable cognitive resources may be consumed in the monitoring of job security and in the maintenance of production schedules that would otherwise be utilized to maintain (or increase) safety knowledge, motivation, and compliance. During times of organizational stability and job security, these cognitive resources can be solely devoted to the demands of safety and production.

Other theories also suggest that the stress and dissatisfaction resulting from job insecurity will lead to lowered safety knowledge and motivation. Eysenck and Calvo (1992) suggested that anxiety (such as is expected to result from job insecurity) can either (a) drain working memory resources, thereby leading to a decrease in performance, or (b) increase cognitive arousal, thereby serving as a motivational source that results in performance improvement. Where one will see performance decrements versus performance improvements depends on the employee's perception of the organization's prioritization of safety, quality, and production. Wickens (1992) suggests that safety, in particular, represents an additional task that can compete with performance-related tasks when attention or performance capacities are exceeded. In addition, Sanders and Baron (1975) in their distraction-conflict theory suggest that arousal and anxiety can lead employees to relegate workplace hazards to the background. On the basis of these theories, it can be expected that employees dissatisfied with their job security would report a lowered emphasis on safety knowledge and report lower safety motivation than employees relatively unconcerned about their job security.

Safety Compliance

A crucial component of the job-insecurity/safety model concerns actual compliance with or, alternatively, violation of organizational safety policies. *Safety compliance* has been defined as the extent to which employees adhere to safety procedures and carry out work in a safe manner (Neal et al., 2000). According to Campbell (1992), performance on any task is a function of three components: skill, knowledge, and motivation. Griffin and Neal (2000) conceptualized skill and knowledge as somewhat overlapping constructs in this setting. Nonetheless, they contended that safety compliance (or "safety performance" as they termed it) is a function of both knowledge and skills, as well as individual motivation to comply with safety policies and procedures. Therefore, compliance with safety policies is expected to be predicted both by safety motivation and knowledge.

On the basis of the theory of reasoned action (Fishbein & Azjen, 1975) and other theories of motivation (e.g., expectancy theory; Vroom, 1964), it is assumed that employees who report low motivation to comply with safety rules and regulations would also report a greater incidence rate of violating those safety rules. Thus, the model postulates that lower levels of safety motivation are predictive of more violations of the organizational safety rules.

All things (e.g., motivation) being equal, individuals who have received safety-related training and understand organizational safety incentive systems (and therefore presumably have greater knowledge and skill regarding appropriate safety behavior than other employees) are expected to adhere to proper safety protocol more frequently than individuals who have less normative knowledge. Employees may or may not be motivated to comply with safety policies, but if they do not possess the requisite knowledge to do so, they are not expected to be able to consistently comply with safety rules. Thus, the model also predicts that employees who report less safety knowledge will report more safety violations.

Workplace Injuries and Accidents

Finally, it is expected that lowered safety knowledge and reduced motivation to comply with safety policies would result in a higher incidence of workplace injuries and accidents due to increased noncompliance with organizational safety policies. *Accidents* have been defined as including actual reported accidents, unreported accidents, and near misses (incidents that could cause an injury but fortunately do not; Smecko & Hayes, 1999). It can be predicted that noncompliance with organizational safety policies would be related to higher levels of workplace accidents and injuries.

Theorists studying the relationship between attitudes, behaviors, and outcomes (e.g., Fishbein & Azjen, 1975) have suggested that one must make a careful distinction between outcomes (e.g., accidents and injuries) and behaviors (e.g., safety compliance). Attitudes (e.g., safety motivation) can only predict behavior (e.g., compliance), whereas outcomes such as injuries are best predicted by both behavior and extraneous factors not under the control of the individual. Therefore, attitudes should not be expected to have a direct influence on outcomes such as injuries and accidents; rather, the effects of attitudes on outcomes are mediated through their effect on behavior (e.g., compliance). As such, the model predicts that compliance mediates the relationships between safety knowledge and safety motivation and injuries and accidents.

EMPIRICAL TESTS OF THE JOB-INSECURITY/SAFETY MODEL

Although there have been relatively few empirical tests of the relationships between job insecurity and employee safety outcomes, the available research to date does suggest that the propositions of the model developed above are largely supported.

Initial Test of the Job-Insecurity/Safety Model

In one of the first studies to test the propositions set forth in the job-insecurity/safety model, Probst and Brubaker (2001) conducted two cross-sectional structural equation modeling tests of the model and one longitudinal test of the model using a hierarchical multiple regression analysis. Participants in the study were sampled from two geographically distinct plants of a large food processing company located in the Pacific Northwest of the United States. Both plants were similar with respect to technologies employed, production, job categories, and organizational structure. Both plants had undergone major organizational changes affecting the job security of the organization's employees. In the first plant, an entire shift of workers had been laid off; during focus group interviews, employees reported a general feeling that the plant was being slowly phased out of existence in favor of a larger plant located elsewhere. In the second location, the swing shift was being eliminated in favor of a night shift. Employees who could not accommodate the new shift schedule were expected to lose their jobs. Interviews with employees, the general plant manager, and the human resources manager indicated that production was expected to remain at former levels during these organizational transitions. Thus, even though there would be fewer employees, overall plant production levels were expected to remain constant.

Employees were asked to participate in the study at two different times: (1) immediately after the shift changes and elimination were announced and (2) six months following the organizational restructuring. The first time was selected to gather data while job insecurity was presumably at its highest levels. The second time was selected to allow employees to adjust to the organizational restructuring while also potentially showing the long-term effects of such insecurity on safety outcomes.

The results of the two cross-sectional structural equation modeling analyses and the longitudinal multiple regression analyses were quite consistent and largely supported the model. In both cross-sectional structural equation modeling analyses and in the longitudinal multiple regression analysis, job insecurity was consistently shown to have a strong negative relation to job satisfaction. In addition, safety motivation and safety knowledge were consistently positively predicted by job satisfaction. This was true in both of the cross-sectional analyses and in the longitudinal results. This finding suggested that job insecurity can have an important effect on safety motivation and safety knowledge through its impact on such job attitudes as job security satisfaction; coworker and supervisor satisfaction; and satisfaction with pay, promotions, and the work itself.

The latter portion of the model assessed behavioral and physical outcomes of job insecurity and decreased safety motivation and knowledge. Participants consistently reported less compliance with safety policies when their motivation to comply with safety rules was reduced (i.e., when they perceived few external incentives to comply). Surprisingly, safety knowledge did not appear to be an important predictor of safety compliance in any of the three analyses—perhaps suggesting that safety training programs are bounded in their effectiveness by the employee's motivation to comply with safety policies. With respect to physical health outcomes, greater numbers of workplace accidents were consistently found when safety compliance was reduced. Finally, partial support was found for the hypothesis that safety compliance is related to experienced injuries on the job. Of the three analyses, only one of the cross-sectional analyses showed this predicted relationship.

Follow-Up Behavioral Laboratory Experiment

The results of the Probst and Brubaker (2001) study indicate that, compared with employees whose jobs are secure, employees who perceive their jobs to be insecure report lower levels of safety knowledge and reduced motivation to comply with safety policies. In turn, these variables are related to a decrease in safety compliance and an increase in job-related accidents and injuries. Probst and Brubaker theorized that safety is compromised when employees feel insecure about their jobs because employees may devote more of their time and energy to meeting production demands in the effort to retain their employment. Although there is theoretical support for this notion (e.g., Eysenck & Calvo, 1992; Kanfer & Ackerman, 1989), this proposition could not be empirically tested in Probst and Brubaker's data set due to the lack of experimental manipulation. Therefore, it is not clear whether job insecurity causes poor safety outcomes or whether a poor safety record causes job insecurity. Either alternative could be plausibly argued.

To provide complementary evidence in support of the job-insecurity/safety model developed by Probst and Brubaker (2001), a follow-up laboratory experiment was designed (Probst, 2001, 2002a). In this experiment, the purpose was to examine the effects of threatened job loss on employee adherence to safety policies, work quality, and productivity levels in a controlled laboratory setting. Thirty-seven upper-level undergraduates from a nontraditional university setting participated in the experiment. Participants were told that they were to be employed by an organization that produced paintings for children's rooms. Their task was to produce as many high-quality paintings within a number of work periods as possible while adhering to a number of safety rules designed to minimize their exposure to the potentially irritating paint products. Halfway through the work sessions, students in the experimental condition were told that 50% of them would be laid off due to declining sales of the paintings. Layoff decisions would be based on their overall performance (i.e., productivity, work quality, and safety adherence) in the following work period. Employees who were laid off would lose any compensation they had accrued to that point as well as future possibilities for compensation. In the control sessions, no mention of layoffs was made and the sessions continued as earlier.

There were several pertinent findings from this experiment. First, the results appear to indicate that product quality, safety, and production do, as theorized by Probst and Brubaker (2001), compete for employee resources. As employee production numbers (i.e., number of completed paintings) increased, employee adherence to safety policies decreased and the rated quality of work products declined. More important, while individuals threatened with layoffs displayed higher productivity than individuals who were not threatened with layoffs, the threatened individuals also violated more safety policies and produced lower-quality paintings than their secure counterparts. In other words, under conditions of job insecurity, productivity was higher but safety compliance and product quality were significantly lower. Consistent with the results of the Probst and Brubaker (2001) study, these results suggest that job insecurity can have a negative impact on employee safety behaviors. The underlying explanation for the effect may be that employees perceive production demands to be more important than safety adherence when it comes to retaining their jobs during times of job insecurity.

PRACTICAL IMPLICATIONS FOR PREVENTION: CIRCUMVENTING THE NEGATIVE JOB-INSECURITY/SAFETY CYCLE

Although the results of the two studies described in the preceding section may appear discouraging in that there is mounting evidence for the adverse effects of job insecurity on safety outcomes, research also seems to suggest that there may be steps organizations can take to circumvent this cycle. There has been a lot of attention paid recently to the impact that an organization's safety climate (Zohar, 1980) can have on employee safety behaviors (see chapter 1 of this book for more on this issue). Research (e.g., Griffin & Neal, 2000; Neal et al., 2000) consistently has shown that an organization's safety climate is a strong predictor of the level of safety knowledge and motivation exhibited by employees, as well as of their level of safety compliance and participation. Thus, it seems clear that organizations with strong safety climates can exert a positive influence on employee behaviors.

Taken in conjunction with each other, research on job insecurity and research on organizational safety climate may suggest that there are ways for an organization to attenuate the negative relationship observed between job insecurity and safety outcomes at work. Although organizations may not be able to control employee levels of job insecurity, organizations can control the message they send to employees during times of organizational transition. They can say, "Look, safety does count in this organization. We care about your safety." Better yet, organizations can reward employees for carrying out safety procedures. Ideally, this would be something an organization does as a matter of policy and not just during times of change. However, organizations need to provide incentives and rewards for employee safety so that employees can see that important job-related outcomes (e.g., salary, promotions) are contingent on their adherence to safety procedures and not just on good production numbers. In this fashion, one might see that a strong organizational climate for safety attenuates the relationship between job insecurity and negative safety outcomes because employees receive the message that they need not and should not subordinate safety to the demands of production in order to retain their jobs.

Empirical research appears to support this notion (Probst, 2002b). In a field study conducted within a small company engaged in light manufacturing in the Pacific Northwest, Probst (2002b) found that multiple aspects of the organization's safety climate (as defined by Griffin & Neal, 2000, and Neal et al., 2000) moderated the impact that job insecurity had on employee safety outcomes. In particular, among employees who perceived that management places a strong value on safety, the relationships between job insecurity and safety knowledge, safety motivation, reported accidents, and near accidents were significantly attenuated when compared to individuals who did not perceive that management valued safety. An even stronger effect was found for the perceived effectiveness of the organizational safety systems. When the organization's safety systems were perceived to be positive, job insecurity had almost no relation to safety knowledge, safety compliance, reported accidents, near accidents, health conditions, or number of worker injuries. However, when the safety systems were not viewed in a favorable light, strong relationships between these safety outcomes and job insecurity were observed. Similar effects were found for the level of safety communication within the organization and the overall perception of the organizational safety environment.

Although the organizational safety climate attenuated the negative relationship between job insecurity and safety outcomes, the organization's emphasis on production and perceived pressure for production exacerbated the negative effects of job insecurity on safety outcomes. When production pressure was perceived to be high, job insecurity had an adverse impact on employee safety knowledge, the number of reported accidents, the number of near accidents, and employee health conditions. When production pressure was perceived to be low, these relationships were not observed.

Although these data were based on individual differences in perceptions regarding the organization's safety climate, they nonetheless suggest the important role that the organizational safety climate can play in attenuating the negative effects of job insecurity on safety outcomes.

CRITICAL MEASUREMENT AND RESEARCH DESIGN ISSUES AND DIRECTIONS FOR FUTURE RESEARCH

Although the implications of the research reviewed above are numerous, there are nonetheless several critical measurement and research design issues that weaken the inferences that can be made from this research. The major issues that need to be confronted are (a) the heavy reliance on correlational data collected in field studies; (b) the ethical and practical impossibility of manipulating job insecurity in the field, compounded by (c) the lack of realism that can be achieved in the laboratory; and (d) the use of employee self-reports of safety outcomes.

Although the results of the first empirical test of the job-insecurity/safety model were supportive, correlational field data cannot confirm the hypothesized directionality of the model. Simply establishing a significant relationship between job insecurity and safety outcomes does not warrant the conclusion that job insecurity caused the poor safety outcomes. It could be plausibly argued that poor safety outcomes cause job insecurity because employees recognize that organizations want to retain employees who comply with safety policies. Stronger statements regarding causality can be made only through the collection of experimental data or data from carefully designed longitudinal research. However, experimenting with employee job insecurity and observing its effects on employee safety outcomes are ethically objectionable and practically difficult to implement in a field setting. Although such data can be collected in the laboratory (as was the case in the Probst, 2002a, study), there are limitations to the level of realism that can be achieved in a laboratory setting. One must question the extent to which participants believe that safety hazards are present, the extent to which participants are motivated by the chance to earn money, and the extent to which they are affected by the threat of "job" loss. Nonetheless, the use of such a multimethod approach to the examination of the effects of job insecurity on safety outcomes may be the only feasible way to truly advance the field by providing complementary evidence from multiple methods demonstrating the robustness of such effects.

One possible method of increasing the realism of laboratory experiments would be to make use of recent technological advances in the area of virtual reality simulators. Highly realistic simulators such as those used by hazardous materials training programs could also be effectively made use of in a laboratory setting, where the threat of layoffs can be manipulated and the effects of such a manipulation on one's performance in the virtual reality trainer could be observed.

The last measurement issue that must be considered in future research is the use of self-report (versus objective and independently gathered) data on employee safety outcomes. Self-report data raise concerns about the possible operation of single-source bias as an alternative explanation to the substantive findings. Although there are statistical methods for determining the extent to which such a bias may be operating, using archival data on accidents, injuries, and safety violations in conjunction with self-report measures may be preferable because research suggests that the accurate recall of workplace accidents may extend back only four weeks (Landen & Hendricks, 1995). However, although archival data regarding accidents and injuries may be available, this data cannot be connected back to the employee survey responses if surveys are anonymously collected. Anonymous surveys are arguably preferable in order for participants to be open and honest regarding their workplace experiences. Therefore, one must carefully weigh the benefits and disadvantages associated with gathering archival data at the expense of respondent anonymity. One possible compromise might be the use of experience sampling methods for the collection of data. Collecting real-time data regarding employee safety behaviors, accidents, and injuries alleviates the accuracy-of-recall problem (Landen & Hendricks, 1995) without compromising respondent anonymity. With the advent of inexpensive handheld devices, this may be a promising method for future data collections.

CONCLUSIONS

Recent research has clearly suggested that employee job insecurity can have an adverse impact on employee levels of safety knowledge and motivation, safety compliance, and workplace injuries and accidents. However, the research has also suggested that organizations can play a key role in circumventing this negative relationship by placing a strong emphasis on the organizational safety climate. Future research in this area faces many measurement and research design challenges. However, recent advances in research methodology and technology may equip researchers with the needed tools to meet the difficulties associated with these challenges.

REFERENCES

American Management Association. (1997). *Corporate job creation, job elimination, and downsizing*. New York: Author.

Andriessen, J. H. (1978). Safe behaviour and safety motivation. *Journal of Occupational Accidents, 1*, 363–376.

Ashford, S., Lee, C., & Bobko, P. (1989). Content, causes, and consequences of job insecurity: A theory-based measure and substantive test. *Academy of Management Journal, 32*, 803–829.

Barthol, R. P., & Ku, N. D. (1959). Regression under stress to first learned behavior. *Journal of Abnormal and Social Psychology, 59*, 134–136.

Belton, B. (1999, February 17). Fed chief: Tech advances raise job insecurity. *USA Today*, p. B2.

Brown, R. L., & Holmes, H. (1986). The use of a factor-analytic procedure for assessing the validity of an employee safety climate model. *Accident Analysis and Prevention, 18*, 455–470.

Campbell, J. P. (1990). Modeling the performance prediction problem in industrial and organizational psychology. In M. Dunnette & L. M. Hough (Eds.), *Handbook of industrial and organizational psychology* (2nd ed., Vol. 1, pp. 687–731). Palo Alto, CA: Consulting Psychologists Press.

Cottington, E. M., Mathews, K. A., Talbot, E., & Kuller, L. H. (1986). Occupational stress, suppressed anger, and hypertension. *Psychosomatic Medicine, 48*, 249–260.

Davy, J., Kinicki, A., & Scheck, C. (1991). Developing and testing a model of survivor responses to layoffs. *Journal of Vocational Behavior, 38*, 302–317.

Dedobbeleer, N., & Beland, F. (1991). A safety climate measure for construction sites. *Journal of Safety Research, 22*, 97–103.

Dekker, S. W., & Shaufeli, W. B. (1995). The effects of job insecurity on psychological health and withdrawal: A longitudinal study. *Australian Psychologist, 30*, 57–63.

Dooley, D., Rook, K., & Catalano, R. (1987). Job and non-job stressors and their moderators. *Journal of Occupational Psychology, 60*, 115–132.

Dunbar, E. (1993). The role of psychological stress and prior experience in the use of personal protective equipment. *Journal of Safety Research, 24*, 181–187.

Eysenck, M. W., & Calvo, M. G. (1992). Anxiety and performance: The processing efficiency theory. *Cognition and Emotion, 6*, 409–434.

Faverge, J. M. (1980). Le travail en tant qu' activité de recuperation. *Bulletin de Psychologie, 33*, 203–206.

Ferguson, J. L., McNally, M. S., & Both, R. F. (1984). Individual characteristics as predictors of accidental injuries in naval personnel. *Accident Analysis and Prevention, 16*, 47–54.

Fishbein, M., & Azjen, I. (1975). *Belief, attitudes, intentions, and behavior: An introduction to theory and research.* Boston: Addison-Wesley.

Griffin, M. A., & Neal, A. (2000). Perceptions of safety at work: A framework for linking safety climate to safety performance, knowledge, and motivation. *Journal of Occupational Health Psychology, 5*, 347–358.

Hansen, C. P. (1989). A causal model of the relationship among accidents, biodata, personality, and cognitive factors. *Journal of Applied Psychology, 74*, 81–90.

Hofmann, D. A., Jacobs, R., & Landy, F. (1995). High reliability process industries: Individual, micro, and macro organizational influences on safety performance. *Journal of Safety Research, 26*, 131–149.

Hofmann, D. A., & Stetzer, A. (1996). A cross-level investigation of factors influencing unsafe behaviors and accidents. *Personnel Psychology, 49*, 307–339.

Janssens, M. J., Brett, J. M., & Smith, F. J. (1995). Confirmatory cross-cultural research: Testing the viability of a corporation-wide safety policy. *Academy of Management Journal, 38*, 364–382.

Kanfer, R., & Ackerman, P. L. (1989). Motivation and cognitive abilities: An integrative/aptitude-treatment interaction approach to skill acquisition. *Journal of Applied Psychology, 74*, 657–690.

Kjellen, U. (1984). The deviation concept in occupational accident control. *Accident Analysis and Prevention, 16*, 289–323.

Kuhnert, K., Sims, R., & Lahey, M. (1989). The relationship between job security and employees' health. *Group and Organization Studies, 14*, 399–410.

Landen, D. D., & Hendricks, S. (1995). Effect of recall on reporting of at-work injuries. *Public Health Reports, 110*, 350–354.

Landisberger, P. A., Cahill, J., & Schnall, P. (1999). The impact of lean production and related new systems of work organization on worker health. *Journal of Occupational Health Psychology, 4*, 108–130.

Leplat, J., & Rasmussen, J. (1984). Analysis of human errors in industrial incidents and accidents for improvement of work safety. *Accident Analysis and Prevention, 16*, 77–88.

Leigh, J. P. (1986). Individual and job characteristics as predictors of industrial accidents. *Accident Analysis and Prevention, 18*, 209–216.

Leveson, H., Hirschfeld, M. L., & Hirschfeld, A. H. (1980). Industrial accidents and recent life events. *Journal of Occupational Medicine, 22*, 53–57.

Mandler, G. (1982). Stress and thought process. In L. Goldberger & S. Breznitz (Eds.), *Handbook of stress* (pp. 88–104). New York: Free Press.

Melamed, S., Luz, J., Najenson, T., Jucha, E., & Green, M. (1989). Ergonomic stress levels, personal characteristics, accident occurrence, and sickness absence among factory workers. *Ergonomics, 32*, 1101–1110.

Neal, A., Griffin, M. A., & Hart, P. M. (2000). The impact of organizational climate on safety climate and individual behavior. *Safety Science, 34*, 99–109.

Probst, T. M. (1998). Antecedents and consequences of job insecurity: An integrated model (Doctoral dissertation, University of Illinois at Urbana-Champaign, 1998). *Dissertation Abstracts International*, 59(11-B), 6102.

Probst, T. M. (2000). Wedded to the job: Moderating effects of job involvement on the consequences of job insecurity. *Journal of Occupational Health Psychology, 5*, 63–73.

Probst, T. M. (2001, June). *Assessing the effects of job insecurity on worker safety.* Paper presented at the 3rd biannual NIOSH/CDC National Occupational Research

Agenda Symposium: Leading Research in Occupational Safety and Health, Washington, DC.

Probst, T. M. (2002a). Layoffs and tradeoffs: Production, quality, and safety demands under the threat of job loss. *Journal of Occupational Health Psychology, 7*, 211–220.

Probst, T. M. (2002b, June). *Organizational safety climate and production pressure: Attenuating and exacerbating job insecurity's toll on employee safety.* SHRM Research Award invited address at the annual conference of the Society for Human Resource Management, Philadelphia, PA.

Probst, T. M. (2002c). The impact of job insecurity on employee work attitudes, job adaptation, and organizational withdrawal behaviors. In J. M. Brett & F. Drasgow (Eds.), *The psychology of work: Theoretically based empirical research* (pp. 141–168). Mahwah: Erlbaum.

Probst, T. M. & Brubaker, T. L. (2001). The effects of job insecurity on employee safety outcomes: Cross-sectional and longitudinal explorations. *Journal of Occupational Health Psychology*, 6, 139–159.

Roskies, E., & Louis-Guerin, C. (1990). Job insecurity in managers: Antecedents and consequences. *Journal of Organizational Behavior, 11*, 345–359.

Sanders, G. S., & Baron, R. S. (1975). The motivating effects of distraction on task performance. *Journal of Personality and Social Psychology, 32*, 956–963.

Sauter, S. L. (1993). Introduction to the NIOSH proposed national strategy. In G. P. Keita & S. L. Sauter (Eds.), *Work and well-being: An agenda for the 1990s* (3rd ed., pp. 11–16). Washington, DC: American Psychological Association.

Simard, M., & Marchand, A. (1997). Workgroups' propensity to comply with safety rules: The influence of micro–macro organisational factors. *Ergonomics, 40*, 172–188.

Simons, J. (1998, November 18). Despite low unemployment, layoffs soar—corporate mergers and overseas turmoil are cited as causes. *The Wall Street Journal*, p. A2.

Slappendal, C., Laird, I., Kawachi, I., Marshall, S., & Cryer, C. (1993). Factors affecting work-related injury among forestry workers: A review. *Journal of Safety Research, 24*, 19–32.

Smecko, T., & Hayes, B. (1999, April). *Measuring compliance with safety behaviors at work.* Paper presented at the 14th annual conference of the Society for Industrial and Organizational Psychology, Atlanta, GA.

Smith, P. C., Kendall, L. M., & Hulin, C. L. (1969). *The measurement of satisfaction in work and retirement*. Chicago: Rand-McNally.

Vroom, V. H. (1964). *Work and motivation*. New York: Wiley.

Wickens, C. D. (1992). *Engineering psychology and human performance*. New York: HarperCollins.

Wright, C. (1986). Routine deaths: Fatal accidents in the oil industry. *Sociological Review, 4*, 265–289.

Zohar, D. (1980). Safety climate in industrial organizations: Theoretical and applied implications. *Journal of Applied Psychology, 65*, 96–102.

5

CONTINGENT WORK AND OCCUPATIONAL SAFETY

MICHAEL QUINLAN AND PHILIP BOHLE

THE NATURE AND EXTENT OF CONTINGENT WORK

Contingent work (coined by Audrey Freeman in 1985 and widely used in North America; see Hipple, 2001) and *precarious employment* (French in origin but gaining international acceptance) are omnibus terms used to describe similar sets of employment arrangements. Central to both concepts is uncertainty about continuity of employment. The term *contingent work* implies that labor is performed only when required, whereas the term *precarious employment* connotes the job and income insecurity inherent in many such arrangements. Typically included under both labels are casual, short-term contract or temporary (and leased) workers; subcontractors and franchisees; teleworkers; home-based workers; sole traders/owner-managers of micro small businesses; and the self-employed.

We would like to thank Peter Tergeist, Employment Analysis and Policy Division, Directorate for Education, Employment, Labour and Social Affairs, Organization for Economic Cooperation and Development, for supplying unpublished data.

Contingent work and *precarious employment* have displaced two earlier terms: *nonstandard employment* and *atypical employment*. This shift was arguably a response to limitations of the earlier terms. Permanent full-time work was never standard or typical for many workers, especially women. All four terms tried to encapsulate significant changes in work arrangements, but ambiguities remain and even the newer terms have proved problematic.

Shiftwork was included as an atypical form of employment but has not been seen as falling under the rubric of contingent work or precarious employment. Many shiftworkers hold relatively secure jobs, although the extensive literature on the occupational health and safety (OHS) effects of shiftwork also largely ignores significant numbers of shiftworkers who are self-employed or hold temporary/short-term posts in industries like retailing and personal services.

Part-time work was also included within the category of atypical work and is also sometimes found included under the label of contingent or precarious employment. Like shiftwork, part-time work may be contingent in terms of its timing, but not all holders of part-time jobs have short or insecure tenure (although about two thirds of part-time employees in Australia are also temporary workers; Burgess & de Ruyter, 2000, p. 252). One solution to the problem of how to classify part-time workers has been to differentiate temporary and permanent part-time workers. While this solution is logical, from an OHS perspective there are risks associated with part-time work that are not confined to casual or temporary part-timers, including limited access to training, limited participation in OHS decision making, inexperience, and split shifts. Whether these risks should be viewed as a consequence of contingent employment or something else remains unclear. Similar ambiguities can be raised in connection with home-based work and telework/telecall center work.

The connection between working hours and contingent work, too, requires further investigation. Contingent workers occupy the extremes in terms of the number of hours worked, ranging from temporary workers employed only a few hours a week to groups of self-employed subcontractors whose unregulated hours far exceed those of employees undertaking similar tasks. The expansion of self-employment in some countries has almost certainly contributed to an increase in working hours. However, there is also evidence that the proportion of noncontingent workers working long hours (i.e., more than 48 hours per week) has grown due to the introduction of systems that involve flexible scheduling and longer shifts as well as increasing levels of paid and unpaid overtime (see Hetrick, 2000).

There is also the question of including micro small business as a contingent work arrangement, something usually not done by statistical agencies. Some researchers have argued there is little to differentiate sole independent subcontractors from small firms with a working owner-manager and five or fewer employees in a dependent subcontracting arrangement. Even rejecting this argument, evidence in the United States and other countries indicates that small businesses employ a more than proportionate share of part-time and temporary workers (Headd, 2000). At the very least, small businesses need to be recognized as a potentially relevant variable in discussing contingent work.

Despite ongoing attempts at refinement, the terms *contingent work* and *precarious employment* are not entirely adequate labels for the forms of work they are used to describe (Bourhis & Wils, 2001). We suspect that resolution of some definitional ambiguities will hinge as much on empirical evidence as on conceptual clarification. Accordingly, this chapter will use *contingent work* as a catchall term encompassing phenomena such as short-tenure jobs, contract-specific engagement, and job insecurity.

Debates over conceptual clarity should not detract attention from the substantial changes in work arrangements revealed by researchers. Over the past two decades, industrialized countries have experienced a significant growth in short-term, flexible, and insecure work arrangements. The precise mix of temporary and short-contract employment, leased labor, part-time work, home-based work, independent subcontracting, and micro small business employment has varied across countries (although different statistical recording conventions also play a part here). For example, a greater proportion of the workforce is self-employed in Canada than in the United States, and this gap widened during the 1990s (Manser & Picot, 1999). Notwithstanding these variations and gaps in the data, the clear trend in Western Europe, North America, and Australasia is that contingent work arrangements have grown substantially (Bureau of Labor Statistics, 1995a; Burgess & de Ruyter, 2000; De Grip, Hoevenberg, & Williams, 1997).

Between 1973 and 1999 in 21 countries in the Organization for Economic Cooperation and Development (OECD), the proportion of the workforce engaged in part-time jobs almost doubled, rising from an average of 8.15% to 15.8% (OECD, 2000, Table E). The average proportion of workforce holding temporary jobs in these countries showed a more modest increase, rising from 9.48% to 11.15% between 1983 and 1994 (Quinlan & Mayhew, 1999, p. 492). Statistical recording conventions in some countries (notably the United States) mean that the latter figure almost certainly underestimates the change. It is also likely that the level of security bestowed on "permanent" jobs by regulatory regimes in different countries affects the level of temporary employment. Even so, a substantial increase (from 31.08% in 1983 to 42.25% in 1994) in the number of workers ages 16 to 19 holding temporary jobs was recorded.

Combining a study by Campbell and Burgess (2001) with unpublished OECD statistics, we have been able to update temporary-employment data for Australia and 14 European Union (EU) countries (excluding Austria).

The average proportion of the workforce in temporary employment across 15 countries grew from 9.57% in 1983 to 13.75% in 1999, an overall increase of 43.68% in 16 years. In some EU countries, notably Belgium and Finland, there has been a rapid expansion of temporary employment in the last 5 years. Unpublished OECD (2001) data for several other countries indicated levels of temporary employment roughly comparable to the EU average; these countries included Iceland (11.1% in 1999), Norway (10.12%), Switzerland (11.84%), Canada (12.09%), and Japan (11.91%). The level of temporary employment was lower in Hungary (5.2%) and the Czech Republic (8.67%) but higher in two developing countries, Mexico (21.1%) and Turkey (20.73%).

As far as we are aware, the OECD, the International Labor Organization (ILO), and similar agencies do not produce comparable data on the prevalence of other contingent work arrangements like telework, home-based work, or labor leasing. Many countries either do not collect these data or collect them only on an irregular basis, and achieving definitional consistency is likely to prove more challenging than in relation to temporary work. Nonetheless, available evidence indicates significant growth in most categories. For example, a survey by the U.S. Bureau of Labor Statistics (1995a, 1995b) found that the number of workers employed by firms supplying temporary help (with 20 or more employees) grew by 43% between 1989 and 1994. Similarly, a recent EU workforce survey (Paoli & Merllie, 2001) found that teleworkers accounted for 10% of self-employed workers and 4% of all employees.

Figures combining available data on various categories of contingent work in particular countries illustrate the magnitude of change. In Australia, those holding a casual or temporary job and those who are considered nonemployees (self-employed, subcontractors, etc.) constituted less than 30% of the workforce in 1982 but approximately 40% in 1999 (Burgess & de Ruyter, 2000, p. 252); if permanent part-time workers are added, the figure rises to 48%. Similar significant shifts have been identified in some EU countries; Canada (see Lowe, 2001); and the United States, where around 30% of the workforce holds part-time, temporary, on-call, day-hire, or short-term contract positions or is self-employed (Hipple, 2001).

Evidence of Effects on Safety

There is growing evidence indicating that some types of contingent work are associated with a significant deterioration in worker safety, health, and well-being. Before focusing on safety, we wish to summarize this evidence.

As part of an ongoing process of collecting published research on the relationship between contingent work and OHS, we undertook several reviews. The first (Quinlan, Mayhew, & Bohle, 2001) examined 93 studies in 12 countries plus the European Union published between 1984 and 2000. The studies used a variety of research methods and OHS indices (e.g., cardiovascular disease, sickness-related absence, injury rates, occupational violence and psychological distress), and covered a broad range of industries. Eleven studies were deemed indeterminate, mainly because they lacked a benchmark or control group. Analysis of the remaining 82 studies revealed a pattern of results that was remarkably consistent irrespective of location, research methods, or OHS indices used. Of these 82 studies, 76 (more than 90%) found that, compared with secure full-time work, contingent work was associated with worse OHS outcomes, such as higher injury rates or psychological strain.

Allocating the studies into five categories of contingent work—organizational restructuring/job insecurity, outsourcing/home-based work, temporary work, small-business work, and part-time work—revealed similar outcomes, except for part-time work. Of the admittedly small group of 7 published studies on part-time work (the next smallest group was small-business work, with 16 studies), only 1 found OHS outcomes to be worse than in full-time employment (consistent with our earlier caution about lumping part-time work in with temporary workers and the like). Subsequent, more exhaustive reviews of the literature (totaling 150 studies)—including analysis of 68 organizational restructuring/job insecurity studies published between 1966 and early 2001 (Bohle, Quinlan, & Mayhew, 2001)—merely reaffirmed the results of the first review.

Turning to the effects of contingent work on safety, we identified 62 studies[1] from our database of 150 studies that considered safety-related indices, including injury rates, occupational violence, and worker knowledge of safety-related risks or regulation (including workers' compensation). Research on this topic is recent, with most of the studies being published after 1996. The 62 studies were undertaken in nine countries: the United States (27 studies), Australia (12), France (9), Canada (4), Sweden (4), the United Kingdom (3), Brazil (2), Finland (2), and Germany (2). Of these studies, 24 examined outsourcing or home-based workers, 21 examined temporary workers (including leased workers), 11 examined downsizing/job insecurity, 12 examined small business, 3 examined call center/teleworkers, and 2 examined part-time workers. In terms of methods, 26 studies analyzed existing data sets, 24 used cross-sectional surveys, 7 were qualitative case studies, and 5 used longitudinal design. With regard to OHS indices, 40 studies examined fatality or injury data, 9 investigated worker legal knowledge or access to entitlements, 9 assessed OHS risk-control measures, and 7 examined occupational violence (from verbal abuse or harassment to physical assault). These totals exceed 62 because some studies investigated two or

[1]A list of these publications is available from the authors.

more countries, included two or more categories of contingent work, or employed multiple OHS indices or research methods.

The findings for this subset of studies were consistent with those from earlier reviews. Of the 62 studies, 14 were deemed indeterminate because they lacked a control group or baseline. Of the remainder, 44 studies indicated that contingent work was associated with poorer safety outcomes than noncontingent work. Only four studies found either improved safety outcomes or no effect.

As an aside it should be noted that the separation of safety outcomes and health outcomes is a somewhat arbitrary process because contingent work may affect both simultaneously, with complex interactions that have implications for treatment and preventive interventions. For example, the overload and psychological distress associated with job insecurity may be conducive to occupational violence. Equally, fatigue from the long hours worked by self-employed subcontractors can entail simultaneous risks of illness and injury (with compounding effects where chronic illness or injury affects their capacity to work safely). Fatigue-induced injury risks may also extend to other workers or members of the public. The use of drug stimulants to combat fatigue by long-haul truck drivers provides a further illustration of these interactions, with both short- and long-term effects on health and safety as well as indirect effects via the impact of drug use on family relationships (Quinlan, 2001). Analyzing the second European workforce survey, Benavides, Benach, and Diez-Roux (2000) found that fatigue, backache, and muscular pain were positively associated with precarious employment—suggesting that such associations are far from atypical.

Our reviews revealed gaps in existing research. Downsizing/job-insecurity studies rarely consider safety effects (for an exception see Snyder, 1994), and there have been few studies of teleworkers, permanent part-time workers, call-center workers, leased workers, or temporary workers in major areas of employment such as hospitality and retailing. In the United States, Canada, Australia, and other countries, immigrants and racial minorities are often concentrated in contingent jobs. It has been argued that immigrants' outsider status (most acute for recent arrivals and illegal immigrants) and low levels of union membership make it difficult for them to secure legal entitlements to safe working conditions (Taylor, 1999), but these issues have been seldom investigated. Adverse safety effects may extend to fellow workers (e.g., through the growth of occupational violence; Neuman & Baron, 1998; Snyder, 1994) or spread outside the workplace. Other issues deserving attention include gender effects and the impact of contingent work on the provision of occupational health/return-to-work services, worker and manager knowledge of legal responsibilities and entitlements, and OHS management more generally.

A number of methodological problems help explain the number of indeterminate studies and some gaps in the literature. Locating a suitable control group may be difficult where few noncontingent workers perform similar tasks. Thus, several studies of telecall centers linked musculoskeletal injury to higher work intensity (see Ferreira, de Souza Conceicao, & Saldiva, 1997; Sznelwar, Mascia, Zilbovicius, & Arbix, 1999), but all lacked a baseline/benchmark or control group. Even when jobs appear similar—as with, say, home care workers and those performing the same role in an institutional setting—there may be significant differences between the work methods and conditions. Similarly, high participant attrition can invalidate longitudinal cohort studies of temporary, leased, or subcontract workers. Furthermore, in industries like long-haul trucking competition between self-employed and employee drivers for tasks can lead to a deterioration in safety for both groups, thereby masking some health effects of subcontracting (Mayhew & Quinlan, 1999).

Another criticism of existing research is that it ignores variations in work arrangements within particular categories (e.g., between self-employed and home-based employee workers); worker characteristics such as age, education, family commitments, and prior experience of job insecurity; and whether temporary work was a voluntarily chosen or not. Writers on psychological contracts (e.g., McLean Parks, 2000) have argued for even finer gradation of categories of contingent work such as temporary/fixed-contract workers. An additional complexity is the potential for one group of workers to combine several different categories of contingency such as part-time home care workers whose employer has restructured or downsized (see Brulin, Winkvist, & Langendoen, 2000). At the same time, several of these factors may be far more important to understanding health and well-being than safety outcomes. Although the degree to which workers have chosen contingent work may affect job satisfaction (Feldman, Doerpinghaus, & Turnley, 1995), and job satisfaction may in turn affect psychological well-being, the available evidence does not indicate a strong causal link between job satisfaction and occupational injuries (Frone, 1998). Likewise, ongoing debates about how to measure job insecurity; the need to take account of repeated waves of downsizing (Hellgren, Sverke, & Isaksson, 1999; Kinnunen, Mauno, Naetti, & Happonen, 1999); differences due to personality or age effects; and the role of moderating variables (like providing information and social support) seem more relevant to understanding effects on health and well-being than to safety.

Acknowledging all these problems, the rapidly expanding body of research is providing increasingly compelling evidence regarding the deleterious effects of contingent work on safety. The findings of psychological studies on contingent work and safety are consistent with the wider literature.

Why Do Contingent Work Arrangements Undermine Safety?

Why contingent work is associated with inferior OHS outcomes is poorly understood. More than half of the studies we reviewed overlooked causality or accorded it only cursory consideration. Nevertheless, existing theory and empirical evidence provide clues about the characteristics and correlates of contingent work that lead to OHS disadvantages.

In general, psychological research has focused on the impact of contingent work on health and well-being rather than safety. An example of this is research on the influence of precarious employment on psychological contracts (Smithson & Lewis, 2000). The notion of psychological contracts might be used to examine some safety effects of contingent work. For example, a breach of the psychological contract may explain a connection several writers have drawn between downsizing/contingent work and increased workplace aggression (Neuman & Baron, 1998). However, this approach may not be sufficiently encompassing. In a study of the impact of hospital downsizing on assaults on staff, Snyder (1994) argued that the increasing rate of admissions in conjunction with ward closures meant that only the most psychiatrically needful patients were accommodated. This, in combination with the discontinuation of special programs for aggressive or chronically impaired subgroups, resulted in a patient population more likely to engage in assaults than prior to downsizing. It is possible to suggest that changes to workload, intensity, and work processes like those just described can be analyzed in terms of their effect on the psychological contract, but whether this is the best way of understanding how such changes affect occupational safety is a moot point.

Psychologists have proposed other conceptual frameworks to analyze contingent work and safety. Drawing on Atkinson's (1987) flexible firm model and dual labor market theory, Aronsson (2000) adopted a center–periphery perspective by which companies are differentiated hierarchically in networks shaped in accordance with conditions of power and dependence. Employers at the center reinforce their benefits or security at the expense of a periphery composed of subcontractors and their employees. This division is mirrored by differences in the job security, conditions, and bargaining power of workers in the core and periphery. This approach can readily consider both health and safety effects, including effects on worker knowledge, compliance, and access to entitlements under OHS legislation (these are discussed later in this chapter).

Psychological theory on risk perception and safety can also provide insights into behavioral factors contributing to higher levels of injury among contingent workers. The model of rule-related behavior proposed by Reason, Parker, and Lawton (1998) is a good example. Contingent workers are likely to be disadvantaged in relation to several of the 10 categories of behavior described in the model.

Two of the categories, correct improvisations and correct violations of inappropriate rules or procedures, refer to positive adaptations to hazardous environments. Reason et al. (1998) stated that hazardous situations routinely require violation of formal safety rules, noting that "in many hazardous technologies, the important issue is not whether to violate, but when to violate" (p. 290). Organizational failings may make violations an essential part of getting work done, whereas in other circumstances workers must improvise because there are no relevant formal rules. In both cases, selection of "correct" or "incorrect" actions depends on the worker's appraisal of local hazards, including evaluation of not just the hazards themselves but also the appropriateness of formal safety rules. Relevant occupational experience, particularly in the specific organizational context within which the hazard is encountered, is critical because experienced workers "know where the 'edge' between safety and disaster lies and do not exceed it, except in extreme circumstances" (p. 290).

Because of their limited relevant experience, insufficient training, and limited communication networks, contingent workers may not be in a position to improvise or to violate unsafe rules correctly. In addition, the job insecurity inherent in much contingent work may discourage these workers taking the initiative for fear it will be seen as rocking the boat. For their part, organizations and managers may be reluctant to train or empower contingent workers to think or respond independently to emerging conditions.

Contingent workers are also likely to be at greater risk of displaying three negative behaviors: mispliances (psychologically unrewarding compliance with inappropriate or inaccurate safety rules); incorrect but psychologically rewarding compliance with bad rules, and psychologically rewarding violation of good rules due to faulty hazard assessment. (Here, "psychologically rewarding" is defined as the extent to which the behavior achieves the personal goals of the worker[s] involved; Reason et al., 1998.) Once again, job insecurity is likely to be a significant factor. It may encourage compliance with bad rules because of perceptions that it would be "more than the job is worth" not to comply. In any case, workers must know that rules and procedures exist and must understand them thoroughly before they can comply. When rules proliferate in an uncoordinated fashion (a common situation), systematic processes of initial training, refresher training, testing, and retesting may be required for workers to develop this understanding (Reason et al., 1998). Short periods of employment in multiple workplaces and employer reluctance to train "low-cost" temporary workers often limit contingent workers' access to ongoing training or exclude them from training altogether. Ultimately, safe behavior on the part of individual workers requires accurate hazard perception, appropriate task training, and an ability to learn from experience—all areas in which contingent workers are likely to be disadvantaged.

Reason et al. (1998) also recognize that many organizational factors affect safety, such as workloads; role structures; person–environment fit; supervisory guidance; safety briefings; and other forms of training, communication, and coordinated teamwork. Once again, the often brief and tenuous connection between contingent workers and their workplaces can be expected to disadvantage them. For example, little organizational effort is often expended on selecting and placing temporary employees to ensure they have the skills and experience necessary to perform their allocated tasks safely. This problem is compounded if these employees also have little access to formal and informal communication regarding safety because they are excluded from formal training or because limited working hours or organizational tenure constrain them from establishing effective communication networks with supervisors and other workers.

Reason et al.'s (1998) model provides a valuable framework for more systematic empirical investigation of the behavioral factors contributing to the higher injury rates among contingent workers. However, the empirical literature on contingent work also implicates a range of factors beyond the control of individual workers. They can be grouped into three broad categories: economic pressures and reward systems, disorganization, and regulatory failure (Quinlan et al., 2001).

Economic Pressures and Reward Systems

Evidence indicates contingent workers often work under considerable economic pressure due to job and income insecurity, intense competition for work (leading to underbidding on contracts), and task-based payment. Consistent with the wider evidence on incentive payment systems (see Brisson, Vinet, & Vezina, 1989; Sunderstrom-Frisk, 1984), these pressures can encourage hazardous work practices, including working too long or too fast; cutting corners (e.g., construction workers not using safety harnesses); and accepting hazardous tasks (e.g., last-minute deliveries in road transport).

One study (Mayhew & Quinlan, 1999) found that home-based garment workers reported almost three times as many acute and chronic injuries as factory-based workers undertaking similar tasks. This difference appeared to arise from long working hours, itself a response to low earnings and task-based payment. Low payment or insecure income flows can induce self-employed subcontractors, such as building tradespeople, to take on too much work, to work long hours, and to try to continue working when injured, leading to more chronic injuries (Mayhew & Quinlan, 1997b). Evidence from the United States indicates that contingent employees earn significantly less on average than their noncontingent counterparts, and those with part-time jobs in particular (contingent workers are twice as likely to be part-timers) are likely to enter into multiple jobholding to supplement their income (Hipple, 2001). Multiple jobholding can pose safety risks (due to travel time, task reorientation, and added stress) over and above those associated with longer hours in the same job. Another U.S. study found that over the past 20 years the burden of working at night and the risk of occupational injury have been increasingly borne by low-wage workers (Hamermesh, 1999). These connections are worthy of further investigation.

Economic pressures may be most acute for self-employed subcontractors, but outsourcing and labor-leasing arrangements can translate pressure to employees because their employers are operating on small profit margins or face uncertain work flows. There are grounds for investigating links between contingent work and the growth of what Heery (2000) has termed "contingent pay systems," which afford little security and stability and transfer risks to the worker. At the very least, both appear to weaken the psychological contract between workers and their employers.

A number of studies have indicated that the economic pressures and work intensity associated with downsizing, outsourcing, and temporary work are conducive to occupational violence such as that inflicted by brokers known as "middlemen" on home-based clothing workers or by delivery-system-dependent clients on truck drivers. In the clothing industry, middlemen resort to abuse, threats, and even assaults on clothing outworkers in their homes as a means of exerting pressure to meet tight-production demands or in response to disputes about payment or the quality of work (Mayhew and Quinlan, 1999). The particular vulnerability of the workers (the vast majority are female and recently arrived immigrants) and the isolation of the setting (from government monitoring) is clearly conducive to the practice. In trucking, the mixture of tight schedules/low inventories and unpaid delays/queuing at warehouses with trip-based payment and uncertainty about future work (amongst owner drivers) provides a potent mix for frustration, aggression, and violence in freight yards (Mayhew & Quinlan, 2000, 2001).

Contingent work may also reduce the additional resources, or slack (see Cyert & March, 1963), in organizations that enable them to adapt to internal and external pressures, including unexpected interruptions to work flows. In a study of outsourcing in refuse collection, Gustafsson and Saksvik (2001) found that reduction in slack had adverse effects on psychosocial work conditions, resulting in much higher levels of absence. In an analogous vein, Stoop and Thissen (1997) argued that highly articulated transport systems with narrow windows for service or delivery, such as just-in-time (JIT) systems, are not conducive to safety. Given the close association between JIT and outsourcing (or contingent work in road and maritime transport; Quinlan, 2001) it is reasonable to suggest that this relationship may apply more generally.

Disorganization

Using contingent workers can contribute to hazardous forms of disorganization at work, including underqualification (lack of training or inexperience), attenuated interworker communication and management control, and diminished capacity of workers to participate in safety.

Lack of familiarity with work arrangements has been identified as a significant risk factor in connection with outsourcing and temporary or leased workers. These suggestions are consistent with earlier French (see Dwyer, 1994) and U.S. research on additional risks when workers are new to jobs, change tasks, or are young and inexperienced (for a related discussion of the risks to young workers, see chapter 6 of this volume). For example, a study of sawmill workers by Cooke and Blumenstock (1979, pp. 118–119) found that younger workers and those assigned to temporary jobs were significantly more likely than other workers to experience serious injuries. These findings are also consistent with research identifying inexperience, lack of training, and lack of supervision as major safety risks for young workers (see, e.g., Castillo, Davis, & Wegman, 1999; Knight, Castillo, & Layne, 1995).

Unfortunately, although several studies highlight the concentration of young workers in industries like retailing and hospitality, few take the added step of analyzing the interaction of age, employment status, and tenure. The association of these three factors is by no means simple. In a study of work injuries among adolescents, Frone (1998) noted there are two conflicting hypotheses; one suggests that the experience associated with increased job tenure should reduce the occurrence of injuries, whereas the other argues that longer tenure can entail assignment to more skilled, demanding, and potentially riskier tasks. Studies suggest that both hypotheses will find support in different work situations. Frone found that job tenure was positively related to work injuries, although he suggested that close supervision may increase worker compliance, especially for younger workers; this view has been supported by a study of young workers in a fast-food chain (Mayhew & Quinlan, 2002).

Rebitzer's (1995) large study of contract and direct-hire workers in the U.S. petrochemical industry suggests several complications in the relationship between contingent work, task allocation, training, and experience. Rebitzer found that the higher "accident" probability among contract workers compared to direct hires was partially explained by the greater likelihood of contract workers to engage in high-risk tasks, be less experienced, and receive less safety training. Rebitzer also found that closer supervision by the host plant led to lower injury rates among contract workers. Much of this is in agreement with Frone (1998), but Rebitzer also found that the training and experience of contract workers was "less effective in reducing the risk of accidents than is that of direct-hire workers" (cited in Kochan, Smith, Wells, & Rebitzer, 1994, p. 70). This finding may indicate the need to more closely examine how training and experience shape the behavior of different categories of workers, or what factors moderate the effects of training and experience (including issues like economic and reward pressures to which we referred earlier in this chapter).

Relocating tasks to isolated, small, or less structured working environments where management control is reduced and regulatory standards are less likely to have an effect may also be seen as a form of disorganization. Outsourcing to small businesses could be viewed as inviting this form of disorganization, although a less ambiguous example would be shifting work to home-based settings. Aside from the risk of injury, there is growing literature on the risks of occupational violence or sexual harassment that workers such as home care providers may face (Barling, Rogers, & Kelloway, 2001). Home care providers appear to face a greater risk than their counterparts in hospital settings, although more research is needed to confirm this. Despite presumptions to the contrary, some women working at home may be at greater risk of occupational violence than those doing the same work in a factory or office because of the isolation of the home and the lower chance that the incident will be witnessed or reported (see our earlier discussion of clothing outworkers; Mayhew & Quinlan, 1999).

Disorganization should not be seen simply as an outcome of oversight but rather as a characteristic feature of the relationship between contingent workers and their employers. Use of temporary workers affects employer attitudes to induction, training, participation in workplace committees, and other activities with implications for safety. In a U.S. study of leased workers, Feldman et al. (1995, p. 139) concluded that workers surveyed frequently reported not receiving "clear instructions about their job duties or even cursory training in relevant procedures and equipment." The insecurity of temporary work also makes it more difficult for workers to raise or negotiate safety issues. A study by Aronsson (1999) based on a stratified sample (n = 1,564) of the Swedish Labour Market Survey found that temporary workers were more likely than permanent employees to report deficiencies in training and OHS knowledge. Temporary workers also believed their employment status inhibited their ability to raise issues and have concerns treated seriously. Contingent work may have critical reporting effects in a number of areas, some affecting worker safety. For example, Jones and Arana (1996) argued the pessimism and negativity associated with downsizing led not only to more mistakes by health care workers but also to a tendency to cover them up. Reporting effects are potentially critical and require careful investigation.

Aronsson's (1999) study supported the hypothesis that the weak labor market position of women, due to their concentration in contingent jobs, also rendered them liable to exclusion from discussion and negotiation over OHS. In the United States, Australia, and other countries, women and vulnerable minority groups (e.g., recent immigrants) as well as contingent workers (e.g., part-time workers) are overrepresented in small business (Manser & Picot, 1999) where union membership is exceptional and contact with government OHS agencies less frequent than is the case with larger employers. Available evidence in countries such as the United States and Australia indicate that contingent workers are less likely than others to be union members (see Campbell, 1996; Hipple, 2001). This denies them an additional source of information and support in relation to safety.

Use of temporary or leased workers and subcontractors may promote higher levels of labor turnover and associated problems of worker commitment, perhaps reinforced by management expectations (Lowry, 1998; Saks, Mudrack, & Ashforth, 1996; Wilcox & Lowry, 2000). These factors could presumably contribute to safety problems, but we are of unaware of research exploring this.

Regulatory Failure

There is growing recognition that contingent work presents a challenge to existing regulatory regimes designed to safeguard workers and provide compensation and rehabilitation for the injured (Johnstone, Mayhew, & Quinlan, 2002; Quinlan & Mayhew, 2000). Safety laws and compliance programs in most countries were designed primarily for relatively secure and directly employed workers in large single-employer worksites. More attenuated and complex chains of command created by the use of subcontractors, leased labor, home-based workers, and more ambiguous employment relationships present major challenges to both the form and implementation of laws. Regulators are now slowly responding to this challenge.

Although inadequate payment or excessive hours can be associated with unsafe work practices, in most countries there is no effective legal recourse to address the issue in relation to the groups such as self-employed subcontractors and home-based workers. Furthermore, regulatory coverage of some groups of contingent workers, such as teleworkers, is ambiguous or complicated by differences in legal status (some are employees, others self-employed) and the location of work (some are home based, but others work from centers or even from multiple locations). The growing practice of subcontracting telework services overseas, to India or Ireland for example, raises significant transnational regulatory issues (see European Foundation for the Improvement of Living and Working Conditions, 1997).

There is growing evidence that contingent workers have less knowledge than permanent full-time employees of their legal entitlements and responsibilities in terms of prevention, workers' compensation, and rehabilitation (Quinlan & Mayhew, 2000). Complex outsourcing arrangements are also associated with confusion about regulatory requirements among employers. Contingent work arrangements can exacerbate reporting problems that, in turn, affect prevention, provision of treatment and occupational health services, and rehabilitation by employers and government agencies. Although psychologists are aware of serious reporting problems in relation to occupational injuries, these problems may be especially acute in elaborate subcontracting chains and other contingent work arrangements (Quinlan & Mayhew, 1999). A study of the employment and compensation claim records for 769 contractors and 32,000 workers who constructed the Denver International Airport (Glazner et al., 1998; Lowery et al., 1998) revealed injury rates more than double those published by the Bureau of Labor Statistics. The discrepancy was most pronounced in relation to small firms (it was also found that injury rates were highest at the beginning of contracts—see our discussion of disorganization earlier in this chapter). An earlier study of the U.S. petrochemical industry (Kochan et al., 1994) found that Occupational Safety and Health Administration (OSHA) inspectors and most plant managers lacked data on injuries to contract workers and that there were reliability problems with the data collected by some managers.

Underreporting is but one symptom of the problematic relationship of contingent workers to workers' compensation regimes. Even allowing for important regulatory and institutional differences between countries, available evidence suggests that in comparison with noncontingent workers, contingent workers are less knowledgeable about workers' compensation, less likely to make a claim when injured, more likely to been denied access or have their claim disputed, and less likely to receive rehabilitation (Quinlan & Mayhew, 1999). One result of this is that contingent workers in some countries are more likely to have their medical costs and income support provided by public health insurance and social security rather than workers' compensation providers. In the United States the absence of a universal health system (and lower membership of employer-provided health care schemes among contingent workers) makes such transfers less likely. Several U.S. studies (e.g., Foley, 1998) found that compensation claim rates or costs are significantly higher for contingent workers than for noncontingent workers. A study by Park and Butler (2001) of 5,125 workers' compensation claims in the state of Minnesota found that compensation for leased/agency staff was around three times that of permanent full-time workers; the difference was even greater when the researchers controlled for worker characteristics (e.g., age). Notwithstanding these findings, there is still evidence that, as in other countries, the administration of workers' compensation for contingent workers has proved problematic (Quinlan & Mayhew, 1999).

Provision of occupational health services and rehabilitation by government agencies or private operators is more problematic for contingent workers because rehabilitating temporary workers presents logistical difficulties. Employers, labor leasing firms, and other interested parties are also less likely to see a need to provide rehabilitation or health services to temporary workers (Rondeau Du Noyer & Lasfargues, 1990). These issues are clearly critical to psychologists involved in workers' compensation counseling, vocational reassignment, and the like.

Remedies

In the past decade government agencies, employers, industry associations, unions, and OHS professionals have increasingly accepted the need to address risks posed by contingent work. In the following paragraphs we briefly describe and evaluate a number of these initiatives, although generalization is difficult since particular contingent work arrangements may require different remedies.

Governments and their OHS agencies have responded to the risks posed by contingent work in a number of ways, including the issuing of new standards and directives on particular contingent work arrangements (see the European Union's initiatives on self-employed workers; European Commission, 2002), the production of guidance material and targeted compliance programs or prosecutions (as occurred in relation to subcontracting in the United Kingdom and Australia), joint industry initiatives on small business (requiring them to produce safe work plans and the like), and setting of minimum OHS and labor standards for government tenders (as in the United States). Although of value, these responses are largely reactive and apply to only some types of contingent work. For example, in most of the countries with which we are familiar government initiatives have targeted small businesses, subcontractors, labor-leasing organizations, telework/telecall centers, and home-based workers. Little attention has been given to downsizing, reorganization, or temporary or short-term contract work (and initiatives on behalf of young workers generally ignore the fact that many, if not most, are employed on a temporary or short-term basis). An exception to the neglect of temporary work arrangements is found in France, where the government adopted a broad but arguably effective approach of minimizing the gap in entitlements between temporary and permanent workers.

Moreover, in most countries the coverage of those initiatives that have been undertaken is restricted or patchy. For example, in the United States the OSHA Process Safety Management of Highly Hazardous Chemicals standard contains detailed provisions on subcontractor safety, but these provisions apply to only a small number of employers (Johnstone, Mayhew, & Quinlan, 2001). The same point applies to the safety case regime (that could address issues of staffing-level changes and downsizing) in European Union and Australian hazardous industry regulation.

Even where government agencies have developed programs or compliance regimes targeting contingent work arrangements, they often fail to address the underlying causes of unsafe work practices identified earlier in this chapter. That is, they fail to address the pressures or incentives for noncompliance. Some industry-specific interventions do avoid the latter problem. One example is the "Behind the Label" strategy for regulating OHS among home-based clothing workers recently initiated in some Australian jurisdictions. This entails an integrated web of controls (setting minimum standards in relation to wages, OHS, and workers' compensation); targeting of retailers as the key influence in the production chain; and contractual tracking mechanisms to prevent regulatory evasion (Johnstone et al., 2001).

In the United States, Australia, and elsewhere, regulatory agencies have also sought to build alliances or collaborative ventures with industry associations (of employers) to establish induction, training, or safety planning requirements for subcontractors or small operators (Quinlan, 2003). Industry-specific remedies have the advantage of being tailored to the specific needs of that industry (including taking account of the specific configuration of contingent work arrangements). At present the development of such programs or modification of existing industry OHS interventions appears to fall well short of those industries in which contingent work has become significant or pervasive.

Notwithstanding the last problem, and other difficulties such as the resource demands of monitoring complex work organization associated with contingent work arrangements (multiple employers or worksites, greater workforce volatility) and the weakening of statutory participatory mechanisms (where these are central to regulatory regime), there appears to be a clear trend of increasing regulatory activity in most countries with which we are familiar. Furthermore, despite some potentially serious reporting effects, workers' compensation insurers and government agencies have become aware of the impact of contingent work arrangements on claims behavior (see Butler, Park, & Zaidman, 1998; Foley, 1998; Park & Butler, 2001; Quinlan, 2003). Changes to premium settings, coverage provisions, and other policies (e.g., those relating to the job security of claimants and pooling rehabilitation in the case of small business) in response to this could provide an additional incentive for employers to improve the management of OHS in relation to contingent workers. There is evidence that some private insurers and government agencies are beginning to implement these options in the United States and Australia if not elsewhere (Quinlan, 2001, 2003; Quinlan & Mayhew, 1999).

For their part, employer responses have ranged from complete apathy or active attempts to minimize legal responsibility through to the development of specific policies (sometimes minimalist and in other cases encompassing) in relation to some groups of contingent workers. Although we are aware of no detailed survey evidence, it appears that subcontractor safety policies are now relatively common among medium to large employers (public and private) in North America, Australasia, and the European Union (Quinlan, 2003). These subcontractor safety programs often entail mandatory induction and training, tender or contract requirements, special rules, and discipline procedures. However, these controls are unlikely to work unless informed by an understanding of how and why risks arise, effective monitoring or auditing, and a strategic assessment of all outsourcing decisions in terms of their OHS implications (Mayhew & Quinlan, 1997a).

It is possible to cite cases in which even small companies have adopted the latter measures, such as a small-niche home-building company that developed a set of subcontractor management rules (in relation to using "preferred" subcontractors, task sequencing, and the maximum number of subcontractors permitted on any given site) that addressed critical risks and were enforced by daily site inspection. Just how typical such programs are is a moot point. There has been little systematic research into the nature and effectiveness of subcontractor safety management programs, so it is difficult to generalize.

Many of the points just made in relation to management responses to subcontracting apply to leased and temporary workers. In their review of the petrochemical industry, Kochan et al. (1994) found that one or more positive outcomes were achieved by a company that, ignoring the advice of its lawyers, managed its (leased) contract workers in the same way as those directly hired.

Pervasive controls and extensive training may be especially valuable in connection to young casual workers in settings with high labor turnover such as fast food and retailing. A study of a large multinational fast-food company found that the risks involving young temporary workers were minimized by an OHS management program vigorously implemented in both company-owned and franchise outlets that entailed a detailed induction program and ongoing training of all workers, comprehensive risk assessment, careful task specification, and the integration of OHS into the design of all tasks. The company also provided mechanisms for worker feedback, although this aspect was relatively weak. Despite deficiencies in relation to managing occupational violence and their workers' compensation entitlements, these young temporary workers had a good understanding of the risks to which they were exposed and ways of addressing them (Mayhew & Quinlan, 2002). Yet such practices appear exceptional. The fragmentary evidence available (for Australia, see Quinlan, 2003) indicates that, like government agencies, relatively few employers have directed attention to the safety risks posed by their use of temporary workers. The same appears to apply to downsizing. A recent Australian study (Quinlan, 2003) found that the view prevailing among regulators, OHS managers, and union officials interviewed was that, despite regulatory requirements to the contrary, most organizations failed to investigate the potential for adverse safety effects, with senior management seeing this aspect essentially as so much noise that could be addressed after the change had taken place.

In Australia and other countries a number of industry associations (e.g., those representing the construction or mining industry) have played a role in promoting measures to assist their members manage the safety risks associated with contingent workers. For example, in Australia the Minerals Council in a number of states has produced guidance material on subcontractors (including explicit reference to risk factors such reward pressures) and has also developed a mandatory induction scheme for contract and leased workers. In a few cases (such as hospitality and labor leasing) industry association material has even been used as an initial model for government guidance material. As noted above, industry associations have also collaborated with regulators in the development of more effective remedies.

In Europe, North America, and Australia unions have expressed increasing concern at the safety risks posed by contingent work arrangements. Responses have included efforts to publicize these safety risks, to push for additional government intervention and prosecutions (and in some countries like Australia to join with industry and regulators in collaborative initiatives), and to negotiate directly with employers to adopt additional safeguards (including limits on the use of contingent workers). Again, these initiatives have not been the subject of systematic investigation. As with governments and employers, union activity appears to have been of some value but too fragmented, partial, and ad hoc to have a substantial effect (Quinlan, 2003).

Finally, OHS professionals, such as those involved in providing occupational health services or rehabilitation, appear to be increasing concerned at the problems posed to their activities by the growth of contingent work (Quinlan et al., 2001), but we are unaware of any specific remedies they have developed to address this. Overall, although there is evidence in most countries with which we are familiar that government, employers, unions, and others are giving increased attention to the OHS risks associated with contingent work, our impression (admittedly based on fragmentary evidence) is that the development of remedies is at an early stage and falls far short of rectifying the problems identified at the beginning of this chapter.

CONCLUSIONS

There is now a large and fairly persuasive body of international evidence that contingent work is associated with inferior OHS outcomes. Although more research is needed into particular categories and subcategories of contingent work, the overall trend in the pattern of results, using a range of methods and indices, has been remarkably consistent. Psychologists have made an important contribution to this research. Attempts to understand why contingent work arrangements should lead to deterioration in OHS are in their early stages. There is still scope for adapting more general theories, such as that proposed by Reason et al. (1998), to help explain the processes leading to this deterioration. A growing body of literature highlights the contributions of rewards and work intensity, disorganization, and regulatory failure.

Most research by psychologists into the OHS effects of contingent work has focused on indices of health and well-being rather than safety. Some conceptual tools, such as the psychological contract, are more attuned to this focus. Notwithstanding considerable overlap, future research may need to recognize some critical distinctions in analyzing the features of contingent work that give rise to safety problems on the one hand and health and psychological well-being on the other. It is also vital to recognize a number of critical reporting effects.

Although it is difficult to generalize about ways of remedying the risks associated with contingent work arrangements, this chapter points to a number of critical requirements. First is the need for regulatory agencies to develop a more comprehensive and strategic set of compliance measures, such as both generic and industry-specific guidance material, mechanisms for tracking the reassignment of work, and targeted enforcement. Second, employers using contingent workers need to give more explicit recognition to this arrangement in their OHS programs with regard to induction and training; worksite rules (governing task design and scheduling, interworker communication, discipline, and consultation); and hazard monitoring. To be effective these measures would need to factor in the cost of additional control measures, the impact of cumulative decisions (to outsource, for example), the possibility of subtle changes to work practices, and other risks associated with particular categories of contingent work. Third, industry associations, unions, and OHS professionals also have an important role to play in promoting remedial measures. Although this chapter has pointed to a number of positive developments, the overall conclusion is that remedial responses thus far fall a long way short of meeting the problems posed by contingent work.

REFERENCES

Aronsson, G. (1999). Contingent workers and health and safety. *Work, Employment and Society, 15*(3), 439–460.

Aronsson, G. (2000, May). *Forms of employment, work environment and health in a center–periphery perspective*. Paper presented to Just in Time Employment: Psychological and Medical Perspectives Research Workshop, European Foundation for Improvement in Living and Working Conditions, Dublin, Ireland.

Atkinson, J. (1987). Flexibility or fragmentation? The United Kingdom labour market in the eighties. *Labour and Society, 12*(1), 87–105.

Barling, J., Rogers, A., & Kelloway, E. K. (2001). Behind closed doors: In-home workers' experience of sexual harassment and workplace violence. *Journal of Occupational Health Psychology, 6*(3), 255–269.

Benavides, F., Benach, J., & Diez-Roux, A. (2000). How do types of employment relate to health indicators? Findings from the second European Survey of working conditions. *Journal of Epidemiology and Community Health, 54*(7), 494–501.

Bohle, P., Quinlan, M., & Mayhew, C. (2001). The health and safety effects of job insecurity: An evaluation of the evidence. *Economic and Labour Relations Review, 12*(1), 32–60.

Bourhis, A., & Wils, T. (2001). The fragmentation of traditional employment: Challenges raised by the diversity of typical and atypical jobs. *Relations Industrielles, 56*(1), 66–91.

Brisson, C., Vinet, A., & Vezina, M. (1989). Disability among female garment workers. *Scandinavian Journal of Work Environment and Health, 15*, 323–328.

Brulin, C., Winkvist, A., & Langendoen, S. (2000). Stress from working conditions among home care personnel with musculoskeletal symptoms. *Journal of Advanced Nursing, 31*(1), 181–189.

Bureau of Labor Statistics. (1995a). *New data on contingent and alternate employment* (Report 900). Washington, DC: U.S. Department of Labor.

Bureau of Labor Statistics. (1995b). *New survey reports on wages and benefits for temporary help service workers*. Washington, DC: U.S. Department of Labor.

Burgess, J., & de Ruyter, A. (2000). Declining job quality in Australia: Another hidden cost of unemployment. *Economic and Labour Relations Review, 11*(2), 246–269.

Butler, R., Park, Y., & Zaidman, B. (1998). Analyzing the impact of contingent work on workers' compensation. *Employee Benefits Practices Quarterly, 4*, 1–20.

Campbell, I. (1996). Casual employment, labour regulation and Australian trade unions. *Journal of Industrial Relations, 38*(4), 571–590.

Campbell, I., & Burgess, J. (2001). Casual employment in Australia and temporary employment in Europe: Developing a cross national comparison. *Work, Employment and Society, 15*(1), 171–184.

Castillo, D., Davis, L., & Wegman, D. (1999). Young workers. *Occupational Medicine: State of the Art Reviews, 14*(3), 519–536.

Cooke, W., & Blumenstock, M. (1979). The determinants of occupational injury severity: The case of Maine sawmills. *Journal of Safety Research, 11*(3), 115–120.

Cyert, R., & March, J. (1963). *A behavioural theory of the firm*. Englewood Cliffs, NJ: Prentice Hall.

De Grip, A., Hoevenberg, J., & Williams, E. (1997). Atypical employment in the European Union. *International Labour Review, 136*(1), 49–71.

Dwyer, T. (1994, July). *Precarious work, powerless employees and industrial accidents: A sociological analysis*. Paper presented to International Sociological Conference, Bielefeld, Germany.

European Commission. (2002). *Proposal for a Council Recommendation concerning the application of legislation governing health and safety work to self-employed workers.* Brussels: Author.

European Foundation for the Improvement of Living and Working Conditions. (1997). *The social implications of telework.* Dublin, Ireland: Author.

Feldman, D., Doerpinghaus, H., & Turnley, W. (1995). Employee reactions to temporary jobs. *Journal of Managerial Issues, 7*(2), 127–141.

Ferreira, M., de Souza Conceicao, G., & Saldiva, P. (1997). Work organization is significantly associated with upper extremeties musculoskeletal disorders among employees engaged in interactive computer-telephone tasks of an international bank subsidiary in Sao Paulo, Brazil. *American Journal of Industrial Medicine, 31*, 468–473.

Foley, M. (1998). Flexible work, hazardous work: The impact of hazardous work arrangements on occupational health and safety in Washington State, 1991–1996. In I. Sirageldin (Ed.), *Research in human capital and development* (Vol. 12, pp. 123–147), Greenwich, CT: JAI Press.

Frone, M. (1998). Predictors of work injuries among employed adolescents. *Journal of Applied Psychology, 83*(4), 565–576.

Glazner, J., Borgerding, J., Lowery, J., Bondy, J., Mueller, K., & Kreiss, K. (1998). Construction injury rates may exceed national estimates: Evidence from the construction of Denver International Airport. *American Journal of Industrial Medicine, 34*, 105–112.

Gustafsson, O., & Saksvik, P. (2001, June). *Productivity, commitment and psychosocial work environment effects of outsourcing in the refuse collection sector.* Paper presented at the 6th European Industrial Relations Congress, Oslo, Norway.

Hamermesh, D. (1999). Changing inequality in work injuries and work timing. *Monthly Labor Review, 122*(10), 22–30.

Headd, B. (2000). The characteristics of small-business employees, *Monthly Labor Review, 123*(14), 13–18.

Heery, E. (2000). The new pay: Risk and representation at work. In D. Winstanley & J. Woodall (Eds.), *Ethical issues in contemporary human resource management* (pp. 172–188). Houndmills, England: Macmillan.

Hellgren, J., Sverke, M., & Isaksson, K. (1999). A two-dimensional approach to job insecurity: Consequences for employee attitudes and well-being. *European Journal of Work and Organizational Psychology, 8*(2), 179–195.

Hetrick, R. (2000). Analyzing the recent upward surge in overtime. *Monthly Labor Review, 123*(2), 30–33.

Hipple, S. (2001). Contingent work in the late 1990s. *Monthly Labor Review, 124*(3), 3–27.

Johnstone, R., Mayhew, C., & Quinlan, M. (2001). Outsourcing risk? The regulation of OHS where contractors are employed. *Comparative Labor Law and Policy Journal, 22*(2&3), 101–143.

Jones, L., & Arana, G. (1996). Is downsizing affecting incident reports? *The Joint Commission Journal on Quality and Improvement, 22*(8), 592–594.

Kinnunen, U., Mauno, S., Naetti, J., & Happonen, M. (1999). Perceived job insecurity: A longitudinal study among Finnish employees. *European Journal of Work and Organizational Psychology,* 8(2), 243–260.

Knight, E. Castillo, D., & Layne, L. (1995) A detailed analysis of work-related injury among youth treated in emergency departments. *American Journal of Industrial Medicine, 27,* 793–805.

Kochan, T., Smith, M. Wells, J., & Rebitzer, J. (1994). Human resource strategies and contingent workers: The case of safety and health in the petrochemical industry. *Human Resource Management, 33*(1), 55–77.

Lowe, G. (2001). *The quality of work: A people-centred agenda.* Don Mills, Ontario, Canada: Oxford University Press.

Lowery, J., Borgerding, J., Zhen, B., Glazner, J., Bondy, J., & Kreiss, K. (1998). Risk factors for injury among construction workers at Denver International Airport. *American Journal of Industrial Medicine, 34,* 113–120.

Lowry, D. (1998). *Towards improved employment relations practices of casual workers in the highly casualised firm.* Unpublished doctoral dissertation, Monash University.

Manser, M., & Picot, G. (1999). The role of self-employment in U.S. and Canadian job growth. *Monthly Labor Review, 122*(4), 10–25.

Mayhew, C., & Quinlan, M. (1997a). The management of occupational health and safety where subcontractors are employed. *Journal of Occupational Health and Safety—Australia and New Zealand, 13*(2), 161–169.

Mayhew, C., & Quinlan, M. (1997b). Subcontracting and OHS in the residential building sector. *Industrial Relations Journal, 28*(3), 192–205.

Mayhew, C., & Quinlan, M. (1999). The effects of outsourcing on occupational health and safety: A comparative study of factory-based and outworkers in the Australian clothing industry. *International Journal of Health Services, 29*(1), 83–107.

Mayhew C., & Quinlan, M. (2000). The relationship between precarious employment and patterns of occupational violence: Survey evidence from seven occupations. In K. Isaksson, C. Hogstedt, C. Eriksson, & T. Theorell (Eds.), *Health effects of the new labour market* (pp. 183–205). New York: Kluwer/Plenum.

Mayhew, C., & Quinlan, M. (2001). Occupational violence in the long distance transport industry: A case study of 300 truck drivers. *Current Issues in Criminal Justice, 13*(1), 36–46.

Mayhew, C., & Quinlan, M. (2002). Fordism in fast food: Pervasive management control and occupational health and safety risks for young temporary workers. *Sociology of Health and Illness, 24*(3), 261–284.

McLean Parks, J. (2000). *An overview of current research on contingent work and psychological contracts.* Paper presented to Just in Time Employment: Psychological and Medical Perspectives Research Workshop, European Foundation for Improvement in Living and Working Conditions, Dublin, Ireland.

Neuman, J. H., & Baron, R.A. (1998). Workplace violence and workplace aggression: Evidence concerning specific forms, potential causes, and preferred targets. *Journal of Management, 24* (3), 391–419.

Organization for Economic Cooperation and Development. (2000). *Employment outlook.* Paris: Author.

Organization for Economic Cooperation and Development. (2001). *Unpublished database on permanent/temporary employment.* Paris: Author.

Paoli, P., & Merllie, D. (2001). *Third European survey of working conditions.* Dublin, Ireland: European Foundation for the Improvement of Living and Working Conditions.

Park, Y.-S., & Butler, R. J. (2001). The safety risks of contingent work: Evidence from Minnesota. *Journal of Labor Research, 22*(4), 831–849.

Quinlan, M. (2001). *Report of inquiry into safety in the long haul trucking industry.* Sydney, Australia: Authority of Motor Accidents. Retrieved May 19, 2002, from http://www.maa.nsw.gov.au/roadsafety36reports.htm

Quinlan, M. (2003). *Developing strategies to address OHS and workers' compensation responsibilities arising from changing employment relationships.* Research project commissioned by the WorkCover Authority of New South Wales, Sydney, Australia.

Quinlan, M., & Mayhew, C. (1999). Precarious employment and workers' compensation. *International Journal of Law and Psychiatry, 22*(5&6), 491–520.

Quinlan, M., & Mayhew, C. (2000). Precarious employment, work re-organisation and the fracturing of OHS management. In K. Frick, P. Jensen, M. Quinlan, & T. Wilthagen (Eds.), *Systematic occupational health and safety management: Perspectives on an international development* (pp. 175–198). Oxford: Pergamon.

Quinlan, M., Mayhew, C., & Bohle, P. (2001). The global expansion of precarious employment, work disorganisation, and consequences for occupational health: A review of recent research. *International Journal of Health Services, 31*(2), 335–414.

Reason, J., Parker, D., & Lawton, R. (1998). Organizational controls and safety: The varieties of rule-related behaviour. *Journal of Occupational and Organizational Psychology, 71*(4), 289–298.

Rebitzer, J. (1995). Job safety and contract workers in the petrochemical industry. *Industrial Relations, 34*(1), 40–57.

Rondeau Du Noyer, C., & Lasfargues, G. (1990) Aptitude du travail des salaries en situation precaire. *Archives des maladies professionelles: de medicine du travail et de securite sociale, 52*, 105–106.

Saks, A., Mudrack, P., & Ashforth, B. (1996). The relationship between the work ethic, job attitudes, intentions to quit, and turnover for temporary service employees. *Canadian Journal of Administrative Sciences, 13*(3), 226–236.

Smithson, J., & Lewis, S. (2000). Is job insecurity changing the psychological contract? *Personnel Review, 29*(6), 680–702.

Snyder, W. (1994). Hospital downsizing and increased frequency of assaults on staff. *Hospital and Community Psychiatry, 45*(4), 378–380.

Stoop, J., & Thissen, W. (1997). Transport safety: Trends and challenges from a systems perspective. *Safety Science, 26*(1/2), 107–120.

Sunderstrom-Frisk, C. (1984). Behavioural control through piece-rate wages. *Journal of Occupational Accidents*, 6, 9.

Sznelwar, L., Mascia, F., Zilbovicius, M., & Arbix, G. (1999). Ergonomics and work organization: The relationship between Taylorisitc design and workers' health in banks and credit card companies. *International Journal of Occupational Safety and Ergonomics*, 5(2), 291–301.

Taylor, A. (1999). Organizing marginalized workers. *Occupational Medicine: State of the Art Reviews, 14*(3), 687–695.

Wilcox, T., & Lowry, D. (2000). Beyond resourcefulness: Casual workers and the human-centered organisation. *Business and Professional Ethics Journal, 19*(3&4), 29–53.

6

YOUNG WORKERS' OCCUPATIONAL SAFETY

CATHERINE LOUGHLIN AND MICHAEL R. FRONE

> It is important . . . that work experiences be structured in ways to protect the health of youth and optimize their chance to become healthy and successful adults. (Wegman & Davis, 1999, p. 580)

Young people are participating in the paid labor force at unprecedented rates (Loughlin & Barling, 2001). This increased participation can be explained in part by expanding opportunities in the service sector of the economy and the rise in nonstandard employment (both of which tend to favor young workers). However, given that hallmarks of adolescence include identity development and the striving for autonomy and achievement (Adams, Montemayor, & Gullota, 1996; Feldman & Elliott, 1990; Vondracek, 1994), it is not surprising that many young people seek entry into the paid labor force at this time. Occupational researchers are now recognizing this and interest is growing in the work characteristics and developmental outcomes of young people's paid employment (see Frone, 1999; Institute of Medicine, 1998; Loughlin & Barling, 1998, 1999). One developmentally significant health outcome from paid employment is work-related injury.

Despite the importance of workplace safety among young workers, our understanding of it is underdeveloped (Frone, 1998; Institute of Medicine, 1998). This lack of knowledge is puzzling because when nonfatal injury rates at work are examined across the life span, young workers (ages 15–24) typically represent the age group with the highest rate of risk (see the following section for more detail). Moreover, as Layne, Castillo, Stout, and Cutlip (1994) have pointed out, "adolescent occupational injuries can be prevented only once hazards have been identified and age-specific intervention strategies have been developed and incorporated into . . . safety and training programs" (p. 660). We therefore have three goals in writing this chapter. First, we briefly summarize what is known about the prevalence of workplace injuries among young workers, as well as the risk factors associated with these injuries. Second, we summarize several areas of research on general risk taking among adolescents and on safety behaviors among employed adults in an effort to highlight unexplored ways to expand research on workplace injuries among young workers. Third, we highlight implications for safety management and training to improve workplace safety for young workers based on work injury research conducted to date and on general research looking at adolescent risk taking.

WORKPLACE INJURIES AMONG YOUNG WORKERS

Prevalence of Workplace Injuries

Occupational health researchers in the United States, Canada, and Europe have highlighted a consistent trend showing that the prevalence of nonfatal occupational injuries decreases with increasing age (e.g., Castillo, 1999; Centers for Disease Control and Prevention [CDC], 2001; Dupre, 2000; Human Resources Development Canada [HRDC], 2000; Kraus, 1985; National Institute for Occupational Safety and Health [NIOSH], 1995, 1997). In other words, adolescent workers are at a higher risk of experiencing an injury at work than adult workers are.[1] As for absolute levels of workplace injuries, approximately 60,000 young workers are involved in lost-time injuries on the job each year in Canada (HRDC, 2000). In the United States, it was estimated that 64,100 workers ages 14 to 17 were seen in hospital emergency departments for work-related injuries in 1992 (Layne et al., 1994). However, because only about one third of work injuries are treated in hospital emergency departments, the National Institute for Occupational Safety and Health (NIOSH) estimates that about 200,000 adolescents are injured on the job each year (Institute of Medicine, 1998). Miller and Waehrer (1998) estimated that in 1993 there were 371,000 teenagers

[1]Although workers at the other end of the age distribution are also vulnerable to injury at work, older workers tend to compensate for their reduced physical capacity by being more safety conscious on the job (Ringenbach & Jacobs, 1995). This, in conjunction with their greater on-the-job experience (Tsang, 1992), tends to lead to safety outcomes better than those for young workers.

injured in the United States, with incurred costs of $5 billion. In Britain, depending on the occupations considered, it is estimated that from 20% to 35% of employed adolescents are injured at work (Hobbs & McKechnie, 1997).

The rate and cost of adolescent injuries is particularly surprising when one considers that most adolescents are concentrated in occupations not traditionally considered dangerous. For example, more than 50% of youth work injuries occur in restaurants and grocery stores (Institute of Medicine, 1998; NIOSH, 1997). Typical injuries include lacerations, strains and sprains, contusions, burns, and fractures (Institute of Medicine, 1998; NIOSH, 1997).

Despite the research conducted to date, several issues make it difficult to estimate the absolute prevalence of injuries to employed youth and the specific types of injuries experienced by them. First, definitions of what constitutes a *work injury* (e.g., any injury, any lost-time injury, injuries resulting in three or more lost workdays) can vary widely across studies. Second, definitions of what constitutes *employment* differ across studies, especially as they relate to informal employment (e.g., baby-sitting, lawn cutting). Third, many studies use samples that fail to adequately cover the population of employed youth and fail to cover all potential injuries experienced by adolescents. Fourth, official records may underestimate the number of adolescent work injuries because many injuries either go unreported or are undocumented (Conway & Svenson, 1998; Parker, Carl, French, & Martin, 1994; Veazie, Landen, Bender, & Amandus, 1994). For example, Parker et al. (1994) estimated that during a 12-month period, two thirds of adolescent work injuries were not reported to the Minnesota Department of Labor and Industry. Finally, self-reports of workplace injuries may suffer from underreporting due to recall errors (Landen & Hendricks, 1995). Collectively, these limitations are likely creating published prevalence estimates that are underestimates. Nonetheless, even with a fair amount of potential underestimation, the research reviewed earlier suggests that workplace injuries among adolescent workers are a serious problem that crosses national boundaries. Occupational health researchers therefore need to understand more fully the factors that increase young workers' risk of being injured at work.

Risk Factors for Workplace Injuries

After examining the literature on work injuries among adolescents and adults, Frone (1998) identified five general categories of risk factors that have received some attention: demographics, personality, employment characteristics, emotional and physical health, and substance use. In the following subsections we discuss research exploring the link between each category of risk factors to work injuries among young workers.

Demographics

Researchers have consistently documented that adolescent males are more likely to be injured at work than adolescent females (e.g., Belville, Pollack, Godbold, & Landrigan, 1993; Brooks, Davis, & Gallagher, 1993; Frone, 1998; Layne et al., 1994; NIOSH, 1997; Schober, Handke, Halperin, Moll, & Thun, 1988). However, little research has attempted to explain this gender difference. One possible explanation is that adolescent males are more likely to engage in risky behaviors than adolescent females (Brynes, Miller, & Schafer, 1999). However, Dunn, Runyon, Cohen, and Schulman (1998) speculated that the higher rates of work injuries among adolescent males is due to their job experiences, exposure to work-related hazards, and employers' expectations for male workers rather than to their risk-taking behavior. Only one study has attempted to explain gender differences in work injuries among adolescents. Frone (1998) explored 20 possible mediating variables, ranging from personality characteristics, work characteristics, and physical and emotional health to substance use. He found that a significant gender difference in work injuries became nonsignificant after controlling for exposure to hazardous work environments and on-the-job substance use and impairment. In other words, adolescent males may be more likely to be injured at work compared to adolescent females because they are more likely to work in a hazardous work environment and are more likely to engage in on-the-job substance use. Thus, both the work experiences and risky behavior of adolescent males may increase their likelihood of experiencing a work injury compared to adolescent females.

As noted earlier, when looking across a wide span of age groups (e.g., ranging from 14 to 65 years old) for employed persons, age is negatively related to work injuries (e.g., Castillo, 1999; CDC, 2001; Dupre, 2000; HRDC, 2000; Kraus, 1985; NIOSH, 1995, 1997). However, when studies are restricted to the narrower age range defining adolescence, findings show that age is positively related to work injuries (e.g., Banco, Lapidus, & Braddock, 1992; Belville et al., 1993; Brooks et al., 1993; NIOSH, 1997). Although no research has explicitly tried to explain this positive relation, it is consistent with labor laws that generally allow adolescents access to increasingly risky jobs with increasing age.

Higher socioeconomic status, as indicated by income, education, or occupational status, is known to be associated with better health outcomes (e.g., Adler et al., 1994) and fewer occupational injuries among adults (e.g., Cubbin, LeClere, & Smith, 2000). In contrast, other researchers have failed to find a relation between socioeconomic status and injury incidence among young workers (Anderson et al., 1994; Williams, Currie, Wright, Elton, & Beattie, 1996).

For certain types of injuries, physical stature is an important risk factor. For example, Parker et al. (1994) assessed adolescents' physical stature by both body weight and body mass. After controlling for age, both measures of stature revealed that small workers were more likely than large workers to experience back injuries while lifting.

Personality

Personality has been implicated as a potentially important risk factor for work injuries among adolescents (NIOSH, 1997). It seems plausible that adolescents with certain personality traits may have a higher risk of work injuries because they are more careless, reckless, or distractible. A number of specific personality characteristics have been suggested as risk factors for work injuries. *Sensation seeking* represents "the need for varied, novel, and complex sensations and experiences and the willingness to take physical and social risks for the sake of such experiences" (Zuckerman, 1979a, p. 10). Compared with those low on sensation seeking, those high on sensation seeking report less anxiety when faced with risks, and they appraise novel situations as less risky and more pleasurable (Zuckerman, 1979b). This suggests that adolescents high in sensation seeking may be more likely to cut corners and ignore safety rules and regulations, thereby increasing the risk of work injury. *Negative affectivity* refers to the chronic experience of negative emotional states and a lack of emotional stability that may lead to lapses in attention or to higher levels of distractibility, thereby increasing the risk of work injury. *Rebelliousness* represents the extent to which individuals are frustrated and defiant when they are exposed to regulations, cannot freely govern their behavior, or cannot initiate independent decisions (e.g., McDermott, 1988). It seems likely that rebellious adolescents may consciously ignore rules and regulations regarding health and safety. Such behavior would increase the likelihood of experiencing injuries at work. *Impulsivity* represents the propensity to get things done quickly and to act suddenly with little forethought for the consequences of one's behavior (Plutchik & van Pragg, 1995). Therefore, impulsive employees may rush to complete a task without adequate consideration of safe operating procedures, resulting in increased risk of injury.

There is much evidence showing that a number of personality dimensions are related to risk-related behaviors among adolescents, such as dangerous driving practices, drinking and driving, alcohol use, sex without contraception, and illicit drug use (Arnett, 1992; Harré, 2000; Jonah, 1997; Stanford, Greve, Boudreaux, Mathias, & Brumbelow, 1996). Moreover, there is some evidence among adults that personality is predictive of work injuries (e.g., Cooper & Sutherland, 1987; Iverson & Erwin, 1997; Sutherland & Cooper, 1991). However, we are aware of only one study that explored the relation between personality and work injuries among employed adolescents. Frone (1998) reported that higher levels of negative affectivity were related to being injured at work in a sample of employed adolescents, though there was no multivariate relation between rebelliousness and impulsivity with work injuries. Clearly, more research needs to explore the relation between personality characteristics and work injuries among adolescents before any firm conclusions can be reached.

Employment Characteristics

Employment characteristics represent a fundamental set of risk factors for understanding the etiology of work injuries. The most obvious dimension is exposure to physical hazards, such as working with dangerous equipment or chemicals, exposure to extreme temperatures and noise, and unsafe work conditions. Hayes-Lundy et al. (1991) found that grease burns among adolescents were the result of working in proximity to dangerous equipment (e.g., grills and fryers), without protective equipment (e.g., cleaning grill without gloves), and under unsafe conditions (e.g., slipping on wet or greasy floors). Drawing on past research with adult employees, Frone (1998) explored the relation of exposure to physical hazards, supervisor monitoring, workload, job boredom, role ambiguity, interpersonal conflict with supervisors and coworkers, work–school conflict, and job satisfaction to work injuries. When these risk factors were examined simultaneously, and after controlling for gender, age, and several personality characteristics, Frone (1998) found that exposure to physical hazards, heavy workloads, and job boredom were major predictors of work injuries among employed adolescents. In addition, there was a marginally significant relation showing that supervisor monitoring was related to lower levels of injuries. The size of the relation between supervisor monitoring and work injuries may have been underestimated in this study because a general measure was used that assessed overall monitoring of task performance. A more specific measure (i.e., assessing supervisor monitoring of compliance with policies and procedures regarding job safety) may reveal a more robust relation in future research (as some research now suggests; Barling, Loughlin, & Kelloway, 2002). In terms of heavy workloads, Barling et al. (2002) also found that heavy workloads were negatively related to safety outcomes for young workers. In addition, these researchers found that management's leadership on the job concerning safety was related to safety outcomes for young workers in a variety of jobs. When managers were role models in terms of safety, when they inspired their workers to work safely, when they challenged young people to think creatively about safety, and when they took a personal interest in their employees' well-being, including their physical safety, safety-related outcomes were higher. These findings have interesting implications for workplace safety interventions that go beyond ergonomic design or regulatory approaches (as we discuss shortly).

Emotional and Physical Health

Overall, emotional and physical health are two factors that may influence the safety of individuals at work. Regarding emotional health, prior research among adults has found that high levels of depression adversely affect an individual's ability to process information (e.g., Sullivan & Conway, 1989) and interfere with both general role functioning (e.g., Broadhead, Blazer, George, & Tse, 1990; Wells et al., 1989) and job performance (e.g., Martin, Blum, Beach, & Roman, 1996). Consistent with these general findings, prior research supports a positive relation between depression and work injuries among adults (Cooper & Sutherland, 1987; Holcom, Lehman, & Simpson, 1993; Zwerling et al., 1996). Poor physical health may also be associated with fatigue and may interfere with one's ability to concentrate at work. However, the results of two studies that have examined the relation between overall physical health and work injuries among adults are inconsistent. Savery and Wooden (1994) failed to find a relation, whereas Zwerling et al. (1996) did find some support for a positive relation.

In the only study to explore the relation of emotional and physical health to work injuries among adolescents, Frone (1998) found significant positive correlations relating both depression and poor physical health to work injuries among adolescents. However, when examined simultaneously, and after controlling for demographic, personality, and employment characteristics, only poor physical health remained predictive of work injuries.

Because much of the research on links between emotional and physical health and safety outcomes has been correlational, the causal ordering of these relationships must also be questioned. It is plausible to suggest that injuries may also be predictive of both emotional and physical health, and future research would benefit from addressing these relationships from a longitudinal perspective.

Substance Use

Past research has generally failed to support a consistent relation between substance use and work injuries among adults (e.g., Dawson, 1994; Feinauer, 1990; Macdonald, 1995, 1997). This is likely the result of failing to differentiate between overall (i.e., context-free) and on-the-job (i.e., context-specific) substance use. Almost all prior studies have relied on measures of overall substance use. These measures mainly assess the use of and impairment from psychoactive substances outside the workplace and outside an individual's hours of employment. Dawson (1994) noted that, by including extraneous (i.e., non-work-related) occasions of use, measures of overall use are likely to underestimate the size of the relation between substance use and work injuries. In other words, if work injuries are partially the result of psychomotor and cognitive impairment while an individual is at work, measures of on-the-job substance use should be more predictive of work injuries than measures of overall substance use.

Frone (1998) examined the relation of both overall and on-the-job substance use (alcohol and marijuana) to work injuries among adolescents. He found that when examined simultaneously, and after controlling for demographic, personality, employment, and health characteristics, on-the-job substance use was positively related to work injuries among adolescents. In contrast, supporting Dawson's (1994) concern regarding the representation of non-work-related occasions of substance use, the measure of overall substance use was unrelated to work injuries.

NEW DIRECTIONS FOR WORKPLACE SAFETY RESEARCH AMONG YOUNG WORKERS

As demonstrated earlier in this chapter, very little empirical research has attempted to identify the risk factors for work injuries among younger workers. Without this knowledge, the development of sound prevention efforts is greatly impaired. More research attention is needed on the risk factors that have been identified. However, in this section, we would like to highlight new avenues for future research on the risk factors for young workers' work injuries. Toward this end, we draw on two areas of research activity. First, we draw on the growing area of research on safety climate as a risk for work injuries among adults (and preliminary research among young workers). Second, we draw on the large research literature exploring various risk-related behaviors among young people, such as dangerous driving, seatbelt use, motorcycle accidents, engaging in unprotected sex, illegal drug use, and general delinquency. On the basis of this research, we present several psychological processes that may increase the risk of workplace injury among young workers. Although care must be exercised in making broad generalizations, emerging knowledge from research on adolescent risk taking in other areas may provide directions for future research and plausible interventions related to workplace safety among young workers. For example, a great deal of research has focused on dangerous driving among youth. Because motor-vehicle-related injury is one of the leading causes of workplace injury among youth (Castillo, 1999), it seems reasonable to assume that useful lessons may be gleaned from this general literature.

Workplace Safety Climate

Workplace safety climate refers to perceptions of enacted policies regarding workplace safety that inform employees of the kinds of behaviors that are expected, rewarded, and supported at work (Hofmann & Stetzer,

1998; Neal, Griffin, & Hart, 2000; Zohar, 2003). Workplace safety climates that support and reward safe working behavior are positively related to higher levels of employee safety performance (e.g., Griffin & Neal, 2000; Neal, Griffin, & Hart, 2000) and lower levels of injuries (e.g., Zohar, 2000). Young workers in organizations with a safety climate that does not value and reward safe working behavior will likely have lower levels of safety performance and higher levels of injuries. Although little systematic research has been conducted on safety climates in jobs held by young workers, a study by Zakocs, Runyan, Schulman, Dunn, and Evensen (1998) suggests that adolescents often receive inadequate safety training and work for managers who do not value and reward safe working behavior. Further, recent research by Barling et al. (2002) suggests that safety climate as well as management's attitudes toward safety are indeed related to safety outcomes among young workers. Because of the potential for constructive interventions in the workplace, much research is needed in this area to explore links between workplace safety climates and young people's injuries, particularly longitudinal research that can directly address issues related to causality.

Perceptions of Relative Invulnerability

Adolescents, as well as adults, demonstrate a tendency to underestimate their risk of injury relative to their peers (Harré, 2000; Quadrel, Fischoff, & Davis, 1993). This tendency has also been referred to as "unrealistic optimism" (Weinstein, 1980), and it appears to be particularly pronounced among young male drivers (Harré, 2000). For example, a study showed that relative to older males, younger males perceive themselves as having a lower probability of being in an accident than their peers (Finn & Bragg, 1987). Much of this research stems from Elkind's (1967) concept of egocentrism, whereby adolescents feel that they themselves are protected from harm but that their peers are not protected. Research on condom use among college students (Thompson, Anderson, Freedman, & Swan, 1996) also supports adolescents' perceptions of invulnerability. It would be interesting to explore the relation of perceptions regarding relative risk for work injuries to workplace safety behaviors and injuries. One would expect that lower levels of perceived relative risk would be related to lower levels of safety performance and higher levels of injuries at work.

Hazard Perception

Hazard perception represents an individual's ability to correctly gauge the "objective" level of hazard in a particular situation. In terms of driving behavior, evidence suggests that when compared to experienced drivers, inexperienced drivers demonstrate a poorer ability to assess hazards, and that this difference is greater among men than women (Harré, 2000). There is evidence that inexperienced drivers scan a smaller range of the environment and do so less frequently, perceive hazards less holistically, and detect hazards less quickly (i.e., hazard perception latency) when compared to experienced drivers. By extension, differences among adolescents in the ability to assess hazards in the work environment quickly and accurately may be related to being injured at work.

Risk Perception

Risk perception refers to the "subjective" perception of risk in a particular situation (Deery, 1999). Individual differences in risk perception have been linked to safe driving behaviors and driving accidents (Deery, 1999). To the extent that adolescents differ in their perception of subjective risk for a specific workplace hazard, they are likely to differ in their level of caution and in the likelihood of experiencing a workplace accident or injury. Given the potential for interventions related to this characteristic, it would be worthwhile to explore whether these findings could be generalized to the case of young workers' injuries.

Self-Assessment of Skill

Self-assessment of skill represents a person's estimation of his or her ability to navigate the environment (Deery, 1999). The accuracy of this assessment is an important part of the safety equation. Overconfidence is a strong source of bias in evaluating risk and has been related to unsafe driving behavior. Although no research has looked at the role of overconfidence in work injuries, one might expect that adolescents who overestimate their skills and abilities may be more likely to take risks and become injured at work.

Acceptance of Risk as a Cost

Research suggests that, relative to older workers, youth have a higher level of risk acceptance, or risk threshold (Deery, 1999). In studies designed to examine motivational models of behavior, results indicate that young drivers may be prepared to accept the risks of dangerous driving in order achieve other goals (Harré, 2000). Moreover, in a study of female university students ages 18 to 23, Shapiro, Siegel, Scovill, and Hays (1998) found that the post hoc justifications used for a wide range of risk-related behaviors were primarily purposeful, that is, used to achieve a personal goal (e.g., to meet people) or to fulfill a personal need (e.g., to relieve stress). It has been argued that youth risk behaviors are "functional, purposive, instrumental, and goal-directed" and that these goals are consistent with normal psychosocial development (Jessor, 1992, p. 378). It is easy to see how goals such as gaining peer acceptance, establishing autonomy from parents, coping with anxiety, or marking transition into adulthood may lead to risk behaviors—including reckless driving, illegal drug use, or unprotected sexual activity—that enable the achievement of those goals (Jessor, 1992). By extension, one might speculate that unsafe work behaviors may be motivated by the desire to appear more productive, obtain supervisor approval, obtain peer approval, or gain other work-related rewards.

SAFETY MANAGEMENT AND TRAINING AMONG EMPLOYED YOUTH

Young workers are entering the workforce at the same time as they are undergoing significant developmental changes from a psychological, social, and career perspective. Our discussion of both identified and plausible risk factors suggests certain safety interventions with regard to young people in the workplace. We will now discuss these interventions in turn.

With respect to gender differences, the majority of young worker injuries occur among male adolescents. This pattern suggests that interventions targeting male adolescents' safety behavior are critical. Frone (1998) found that the excess risk of injury for male adolescents compared to female adolescents was due to male adolescents' higher exposure to dangerous equipment and their elevated propensity to be impaired by alcohol or drugs at work. Thus, better training in terms of the use of dangerous equipment and understanding the dangers of working under the influence of psychoactive substances are certainly needed. It should also be noted that a few studies on general female risk-taking behavior suggest that this may be less of an exclusively male activity than is commonly believed (e.g., Shapiro et al., 1998). The possibility exists that we are not detecting the kind of injuries likely to occur among female adolescents at work. This could occur because of the nature of the work in which they are more likely to be engaged (e.g., domestic) or the kinds of injuries they are more likely to experience (e.g., strain-type injuries that are less likely to force them into an emergency room for treatment). These possibilities should be investigated in future research.

In terms of the work environment, past research found that exposure to physical hazards, heavy workloads, and boredom were related to an elevated risk of being injured at work among adolescents (Frone, 1998). Young workers need to be trained to work with dangerous equipment and protect themselves from other hazardous work conditions. The workloads that adolescents have to manage need to be tailored to their cognitive and physical capacities if they are to avoid injuries at work. Finally, young workers' jobs should have a reasonable, and developmentally appropriate, amount of task variety and responsibility. Such job enrichment is likely to decrease the risk of injuries by decreasing job boredom and inattention.

Programs to reduce injuries at work should target the promotion of healthy lifestyle behaviors that address physical activity and alcohol and drug consumption. Frone (1998) found that adolescents reporting poor physical health and on-the-job substance use were more likely than other adolescents to report being injured at work. The promotion of healthy lifestyle behaviors must begin early in a person's life. However, because of the general importance of overall health for all areas of achievement, including work, employers should work with schools and other social institutions to promote positive physical and mental health among young people. It is in the self-interest of employers to do so because having a labor pool that is physically and mentally healthy decreases the likelihood of injuries, increases role performance, and minimizes health care and other financial costs.

Past workplace safety climate research among adults (e.g., Zohar, 2003) indicates that employers need to unambiguously promote and reward workplace safety if they want to reduce work injuries among employed adolescents. One component of promoting workplace safety is assuring that all employees get adequate safety training. In one study of employed adolescents, Zakocs et al. (1998) found that most (95%) adolescents wanted safety training before beginning their job. However, only 55% received training to avoid injury and only 53% received training to deal with an angry customer. In addition, the adolescents felt that when safety training was received, it was of low quality and their managers did not value safety. This study suggests that to reduce work injuries among adolescents, more attention needs to be paid to promoting a strong and positive safety climate. Preliminary research among young workers (Barling et al., 2002) would support management interventions in this regard.

The tendency for adolescents to perceive that they are relatively invulnerable compared to their peers suggests that a media campaign could portray peers acting in a safe manner rather than a reckless manner, illustrating that most young workers act safely most of the time. The campaign could also describe the precautions that young workers can take to modify high-risk behaviors (Harré, 2000). Still, given limited support for the efficacy of education campaigns in changing behavior, peer group support for safer behavior may be more effective. Risk reduction programs may need to "directly address peer norms concerning certain behaviors in order to increase effectiveness" (Madray & van Hulst, 2000, p. 209). Changes in this regard may be as straightforward as adopting new peer norms (e.g., similar to "don't let a friend drive drunk" campaigns). Group interactive training sessions may be useful for addressing reduced perception of risk in the environment. This format allows young workers to analyze the hazards of specific conditions and avoid risky behaviors and decisions by choosing alternative courses of action (Harré, 2000). In terms of the workplace, this presents a particular challenge because until one is familiar with an environment it is difficult to judge its inherent risks. With little exception, young workers are constantly moving in and out of the labor market relative to the adult worker. They are involved in frequently changing jobs, negotiating fluid work schedules, and responding to changing employer demands and labor market conditions (Wegman & Davis, 1999). In a North Carolina study, researchers (Dunn et al., 1998) found that 67% of teens (ages 14–17) had already worked in two or more locations during their short occupational careers. The transient nature of the youth labor market is likely to make employers reluctant to make the necessary investments in training. One possible remedy to this problem is to provide this training in "work safe" programs in schools where all young people have access to this form of training. However, recent findings underscore the importance of management's leadership on the job in terms of safety (e.g., Barling et al., 2002). Thus, one might question the ultimate impact of training interventions if they are not supported in the workplace.

Interestingly, in terms of self-assessed overestimation of skill, Deery (1999) cited research suggesting that skill-based training related to safety behavior may actually increase young workers' risk. There is some evidence that skill-based training emphasizing an individual's limitations results in a better match between expected and actual performance of a task. Our discussion of young workers' acceptance of risk as a cost suggested that youth risk behavior is instrumental and goal directed. The challenges for occupational psychologists in this regard are to anticipate both the general goals (e.g., peer acceptance) and vocational goals (e.g., job promotion) of young workers and to provide a means for them to achieve these goals without compromising their health and safety. For example, if risky work habits are undertaken in order to impress supervisors, then the means for earning approval should be targeted. Supervisors might praise young workers, indicating their satisfaction when safety procedures are closely followed, or develop a system that publicly recognizes safe work behaviors. Preliminary research suggests that praise and recognition will have the most positive effect on injury-related outcomes (Barling et al., 2002). Finally, it should be noted that reducing the perceived value of engaging in unsafe practices may influence behavior to a greater degree than increasing fear of punishment (Lehto, James, & Foley, 1994).

SUMMARY

Injury at work is a common experience among adolescent employees that transcends national boundaries. The risk factors reviewed here indicate that both personal factors and workplace factors play roles in youth work injuries. Because young workers typically lack control in the workplace and are less likely to have union representation than adults (Gallagher, 1999), they will be at a disadvantage when it comes to improving their safety at work. Therefore, unless enough parents, employers, and policymakers become concerned about the prevalence of work injuries among young people, it will be difficult to gain the momentum necessary to motivate change in the workplace. To some extent, every society has precisely the accident rate it is willing to accept. When one thinks of societal norms about seat-belt use or smoking not so long ago, we see that collective will can be a powerful force for change. Ultimately, the development of a safe workplace through employee education, job design, and a positive safety climate is in the interests of young workers, their employers, and society as a whole.

REFERENCES

Adams, G. R., Montemayor, R., & Gullota, T. P. (1996). *Psychosocial development during adolescence: Progress in developmental contextualism.* Thousand Oaks, CA: Sage.

Adler, N. E., Boyce, T., Chesney, M. A., Cohen, S., Folkman, S., Kahn, R. L., et al. (1994). Socioeconomic status and health: The challenge of the gradient. *American Psychologist, 49,* 15–24.

Anderson, R., Dearwater, S. R., Olsen, T., Aaron, D. J., Kriska, A. M., & LaPorte, R. E. (1994). The role of socioeconomic status and injury morbidity risk in adolescents. *Archives of Pediatric and Adolescent Medicine, 148,* 245–249.

Arnett, J. (1992). Reckless behavior in adolescence: A developmental perspective. *Developmental Review, 12,* 339–373.

Banco, L., Lapidus, G., & Braddock, M. (1992). Work-related injuries among Connecticut minors. *Pediatrics, 89,* 957–960.

Barling, J., Loughlin, C., & Kelloway, E. K. (2002). Development and test of a model linking transformational leadership and occupational injuries. *Journal of Applied Psychology, 87*(3), 488–496.

Belville, R., Pollack, S. H., Godbold, J. H., & Landrigan, P. J. (1993). Occupational injuries among working adolescents in New York State. *Journal of the American Medical Association, 269,* 2754–2759.

Broadhead, W. E., Blazer, D. G., George, L. K., & Tse, C. H. (1990). Depression, disability days, and days lost from work in a prospective epidemiologic survey. *Journal of the American Medical Association, 264,* 2524–2528.

Brooks, D. R., Davis, L. K., & Gallagher, S. S. (1993). Work-related injuries among Massachusetts children: A study based on emergency department data. *American Journal of Industrial Medicine, 24,* 313–324.

Byrnes, J. P., Miller, D. C., & Schafer, W. D. (1999). Gender differences in risk taking: A meta-analysis. *Psychological Bulletin, 125,* 367–383.

Castillo, D. (1999). Occupational safety and health in young people. In J. Barling & E. K. Kelloway (Eds.), *Young workers: Varieties of experience* (pp. 159–200). Washington, DC: American Psychological Association.

Centers for Disease Control and Prevention. (2001). Nonfatal occupational injuries and illnesses treated in hospital emergency departments, United States, 1998. *Morbidity and Mortality Weekly Report*, 50(16), 313–317.

Conway, H., & Svenson, J. (1998). Occupational injury and illness rates, 1992–1996: Why they fell. *Monthly Labor Review, 121*, 36–58.

Cooper, C. L., & Sutherland, V. J. (1987). Job stress, mental health, and accidents among offshore workers in the oil and gas extraction industries. *Journal of Occupational Medicine, 29*, 119–125.

Cubbin, C., LeClere, F. B., & Smith, G. S. (2000). Socioeconomic status and the occurrence of fatal and nonfatal injury in the United States. *American Journal of Public Health*, 90(1), 70–77.

Dawson, D. A. (1994). Heavy drinking and the risk of occupational injury. *Accident Analysis and Prevention, 26*, 655–665.

Deery, H. A. (1999). Hazard and risk perception among young novice drivers. *Journal of Safety Research*, 30(4), 225–236.

Dunn, K. A., Runyan, C. W., Cohen, L. R., & Schulman, M. D. (1998). Teens at work: A statewide study of jobs, hazards, and injuries. *Journal of Adolescent Health Care, 22*, 19–25.

Dupre, D. (2000). Accidents at work in the EU. *Statistics in focus: Population and social conditions* (Catalogue Number: CA-NK-00-004-EN-I). Luxembourg: Eurostat.

Elkind, D. (1967). Egocentrism in adolescence. *Child Development*, 4(3–4), 233–244.

Feinauer, D. M. (1990). The relationship between workplace accident rates and drug and alcohol abuse: The unproven hypothesis. *Labor Studies Journal, 15*, 3–15.

Feldman, S. S., & Elliott, G. R. (1990). Capturing the adolescent experience. In S. S. Feldman & G. R. Elliott (Eds.), *At the threshold: The developing adolescent* (pp. 1–13). Cambridge, MA: Harvard University Press.

Finn, P., & Bragg, D. W. E. (1987). Perception of risk of an accident by young and older drivers. *Accident Analysis and Prevention, 18*, 289–298.

Frone, M. R. (1998). Predictors of work injuries among employed adolescents. *Journal of Applied Psychology, 83*, 565–576.

Frone, M. R. (1999). Developmental consequences of youth employment. In J. Barling & E. K. Kelloway (Eds.), *Young workers: Varieties of experience* (pp. 89–128). Washington, DC: American Psychological Association.

Gallagher, D. G. (1999). Youth and labor representation. In J. Barling & E. K. Kelloway (Eds.), *Young workers: Varieties of experience* (pp. 235–258). Washington, DC: American Psychological Association.

Griffin, M. A., & Neal, A. (2000). Perceptions of safety at work: A framework for linking safety climate to safety performance, knowledge, and motivation. *Journal of Occupational Health Psychology, 5*, 347–358.

Harré, N. (2000). Risk evaluation, driving, and adolescents: A typology. *Developmental Review, 20*, 206–226.

Hayes-Lundy, C., Ward, R. S., Saffle, J. R., Reddy, R., Warden, G. D., & Schnebly, W. A. (1991). Grease burns at fast-food restaurants: Adolescents at risk. *Journal of Burn Care and Rehabilitation, 12*, 203–208.

Hobbs, S., & McKechnie, J. (1997). *Child employment in Britain*. Edinburgh, Scotland: Stationery Office.

Hofmann, D. A., & Stetzer, A. (1998). The role of safety climate and communication in accident interpretation: Implications for learning from negative events. *Academy of Management Journal, 41*, 644–657.

Holcom, M. L., Lehman, W. E. K., & Simpson, D. D. (1993). Employee accidents: Influences of personal characteristics, job characteristics, and substance use in jobs differing in accident potential. *Journal of Safety Research, 24*, 205–221.

Human Resources Development Canada. (2000). Work safely for a healthy future. *Statistical analysis: Occupational injuries and fatalities Canada, 2000/03/06*. Ottawa: Author.

Institute of Medicine. (1998). *Protecting youth at work*. Washington, DC: National Academy Press.

Iverson, R. D., & Erwin, P. J. (1997). Predicting occupational injury: The role of affectivity. *Journal of Occupational and Organizational Psychology, 70*, 113–128.

Jessor, R. (1992). Risk behavior in adolescence: A psychosocial framework for understanding and action. *Developmental Review, 12*, 374–390.

Jonah, B. A. (1997). Sensation seeking and risky driving: A review and synthesis of the literature. *Accident Analysis and Prevention, 29*(5), 651–665.

Kraus, J. F. (1985). Fatal and nonfatal injuries in occupational settings: A review. *Annual Review of Public Health, 6*, 403–418.

Landen, D. D., & Hendricks, S. (1995). Effect of recall on reporting of at-work injuries. *Public Health Reports, 110*, 350–354.

Layne, L. A., Castillo, D., Stout, N., & Cutlip, P. (1994). Adolescent occupational injuries requiring hospital emergency department treatment: A nationally representative sample. *American Journal of Public Health, 84*, 657–660.

Lehto, M. R., James, D. S., & Foley, J. P. (1994). Exploratory factor analysis of adolescent attitudes toward alcohol and risk. *Journal of Safety Research, 25*(4), 197–213.

Loughlin, C., & Barling, J. (1998). Teenagers' part-time employment and their work-related attitudes and aspirations. *Journal of Organizational Behavior, 19*, 197–207.

Loughlin C., & Barling, J. (1999). The nature of youth employment. In J. Barling & E. K. Kelloway (Eds.), *Young workers: Varieties of experience* (pp. 17–36). Washington, DC: American Psychological Association.

Loughlin, C., & Barling, J. (2001). Young workers' work values, attitudes, and behaviours. *Journal of Occupational and Organizational Psychology, 74*(3), 543–558.

Macdonald, S. (1995). The role of drugs in workplace injuries: Is drug testing appropriate? *Journal of Drug Issues, 25*, 703–722.

Macdonald, S. (1997). Work-place alcohol and other drug testing: A review of the scientific evidence. *Drug and Alcohol Review, 16*, 251–259.

Madray, H., & van Hulst, Y. (2000). Reducing HIV/AIDS high-risk behavior among injection drug users: Peers vs. education. *Journal of Drug Education, 30*(2), 205–211.

Martin, J. K., Blum, T. C., Beach, S. R. H., & Roman, P. M. (1996). Subclinical depression and performance at work. *Social Psychiatry and Psychiatric Epidemiology, 31*, 3–9.

McDermott, M. R. (1988). *Measuring rebelliousness: The development of a Negative Dominance scale*. In M. J. Apter, J. H. Kerr, & M. P. Cowles (Eds.), *Progress in reversal theory* (pp. 297–312). Amsterdam: North-Holland, Elsevier.

Miller, T. R., & Waehrer, G. M. (1998). Costs of occupational injuries to teenagers, United States. *Injury Prevention, 4*, 211–217.

National Institute for Occupational Safety and Health. (1995). *Preventing deaths and injuries of adolescent workers* (DHHS Publication No. NIOSH 95-125). Washington, DC: U.S. Government Printing Office.

National Institute for Occupational Safety and Health. (1997). *Special hazard review: Child labor research needs* (DHHS Publication No. NIOSH 97-143). Washington, DC: U.S. Government Printing Office.

Neal, A., Griffin, M. A., & Hart, P. M. (2000). The impact of organizational climate on safety climate and individual behavior. *Safety Science, 34*, 99–109.

Parker, D., Carl, W. R., French, L. R., & Martin, F. B. (1994). Characteristics of adolescent work injuries reported to the Minnesota department of labor and industry. *American Journal of Public Health, 84*, 606–611.

Plutchik, R., & van Pragg, H. M. (1995). The nature of impulsivity: Definitions, ontology, genetics, and relations to aggression. In E. Hollander & D. J. Stein (Eds.), *Impulsivity and aggression* (pp. 7–24). New York: Wiley.

Quadrel, M. J., Fischoff, B., & Davis, W. (1993). Adolescent (in)vulnerability. *American Psychologist, 48*, 102–116.

Ringenbach, K. L., & Jacobs, R. R. (1995). Injuries and aging workers. *Journal of Safety Research, 26*(3), 169–176.

Savery, L. K., & Wooden, M. (1994). The relative influence of life events and hassles on work-related injuries: Some Australian evidence. *Human Relations, 47*, 283–305.

Schober, S. E., Handke, J. L., Halperin, W. E., Moll, M. B., & Thun, M. J. (1988). Work-related injuries in minors. *American Journal of Industrial Medicine, 14*, 585–595.

Shapiro, R., Siegel, A. W., Scovill, L. C., & Hays, J. (1998). Risk-taking patterns of female adolescents: What they do and why. *Journal of Adolescence, 21*, 143–159.

Stanford, M. S., Greve, K. W., Boudreaux, J. K., Mathias, C. W., & Brumbelow, J. L. (1996). Impulsiveness and risk-taking behavior: Comparison of high school and college students using the Barratt Impulsiveness Scale. *Personality and Individual Differences, 21*(6), 1073–1075.

Sullivan, M. J. L., & Conway, M. (1989). Negative affect leads to low-effort cognition: Attributional processing for observed social behavior. *Social Cognition, 7*, 315–337.

Sutherland, V. J., & Cooper, C. L. (1991). Personality, stress, and accident involvement in the offshore oil and gas industry. *Personality and Individual Differences, 12*, 195–204.

Thompson, S. C., Anderson, K., Freedman, D., & Swan, J. (1996). Illusions of safety in a risky world: A study of college students' condom use. *Journal of Applied Social Psychology, 26*, 189–210.

Tsang, P. S. (1992). A reappraisal of aging and pilot performance. *International Journal of Aviation Psychology, 2*(3), 193–212.

Veazie, M. A., Landen, D. D., Bender, T. R., & Amandus, H. E. (1994). Epidemiologic research on the etiology of injuries at work. *Annual Review of Public Health, 15*, 203–221.

Vondracek, F. W. (1994). Vocational identity development in adolescence. In R. K. Silbereisen & E. Todt (Eds.), *Adolescence in context: The interplay of family school, peers, and work in adjustment* (pp. 285–303). New York: Springer-Verlag.

Wegman, D., & Davis, L. K. (1999). Protecting youth at work. *American Journal of Industrial Medicine, 36*, 579–583.

Weinstein, N. D. (1980). Unrealistic optimism about future life events. *Journal of Personality and Social Psychology, 39*, 806–820.

Wells, K. B., Stewart, A., Hays, R. D., Burnam, M. A., Rogers, W., Daniels, M., et al. (1989). The functioning and well-being of depressed patients: Results from the medical outcomes study. *Journal of the American Medical Association, 262*, 914–919.

Williams, J. M., Currie, C. E., Wright, P., Elton, R. A., & Beattie, T. F. (1996). Socioeconomic status and adolescent injuries. *Social Science and Medicine, 44*(12), 1881–1891.

Zakocs, R. C., Runyan, C. W., Schulman, M. D., Dunn, K. A., & Evensen, C. T. (1998). Improving safety for teens working in the retail trade sector: Opportunities and obstacles. *American Journal of Industrial Medicine, 34*, 342–350.

Zohar, D. (2000). A group-level model of safety climate: Testing the effect of group climate on micro-accidents in manufacturing jobs. *Journal of Applied Psychology, 85*, 587–596.

Zohar, D. (2003). Safety climate: Conceptual and measurement issues. In J. C. Quick & L. E. Tetrick (Eds.), *Handbook of occupational health psychology* (pp. 123–142). Washington, DC: American Psychological Association.

Zuckerman, M. (1979a). *Sensation seeking: Beyond the optimal level of arousal.* Hillsdale, NJ: Erlbaum.

Zuckerman, M. (1979b). Sensation seeking and risk taking. In C. E. Izard (Ed.), *Emotions in personality and psychopathology* (pp. 163–197). New York: Plum.

Zwerling, C., Sprince, N. L., Wallace, R. B., Davis, C. S., Whitten, P. S., & Heeringa, S. G. (1996). Risk factors for occupational injuries among older workers: An analysis of the health and retirement study. *American Journal of Public Health, 86*, 1306–1309.

7

ALCOHOL, DRUGS, AND WORKPLACE SAFETY OUTCOMES: A VIEW FROM A GENERAL MODEL OF EMPLOYEE SUBSTANCE USE AND PRODUCTIVITY

MICHAEL R. FRONE

The use of alcohol and other drugs in the workforce represents an important social policy issue because it may undermine employee health, productivity, and safety (e.g., Roman & Blum, 1995; Sindelar, 1998). Such outcomes may interfere with employers' ability to compete effectively in an increasingly competitive domestic and global economic environment. Research exploring the nexus of employee substance use and workplace safety outcomes (i.e., noninjury accidents, fatal and nonfatal physical injuries) is important because each occurs in a significant proportion of workers; each can be costly to individuals, employers, and society; and a shared belief exists among researchers, policymakers, and employers that employee substance use is a risk factor for injuries and accidents at work.

Preparation of this chapter was funded, in part, through grant R01-AA12412 from the National Institute on Alcohol Abuse and Alcoholism.

In the United States, national surveys reveal that among employed adults (ages 18–49) who work full-time, approximately 78% have used alcohol, 9% have used marijuana, and 5% have used other illicit drugs during the preceding 12 months. Furthermore, within the preceding 30 days, 64% have used alcohol, 35% have engaged in binge drinking (a single occasion where five or more drinks were consumed), 12% reported heavy drinking (at least five occasions of binge drinking), 5% used marijuana, and 2% used other illicit drugs (e.g., Department of Health and Human Services, 1999a). It is important, however, to point out that these are estimates of alcohol and drug use across all contexts. Thus, they largely reflect the prevalence of substance use off the job (i.e., use away from work and outside an individual's normal work hours). Unfortunately, few credible data exist on the prevalence of alcohol and drug use or impairment during the workday. Nonetheless, two reviews have summarized what is known about the prevalence of alcohol and drug use at work from a handful of empirical studies (Ames, 1993; Newcomb, 1994). These reviews show that past research differs widely on several critical dimensions: (a) the nature and quality of the sample used; (b) the time frame (past month, past 6 months, past year) evaluated; (c) the specific substance (e.g., alcohol, marijuana, cocaine) under investigation; and (d) the dimension of on-the-job substance use and impairment assessed (e.g., use just before work, use during lunch, use while working, use during breaks, being at work impaired, or some unknown combination). Therefore, it is not surprising that little consistency exists in reported prevalence rates for on-the-job substance use or impairment, which range from less than 1% to about 39%. Finally, the total employment-related cost of alcohol and drug abuse due to lowered productivity and employability in 1992 was estimated to be $80.9 billion ($66.7 billion due to alcohol abuse and $14.2 billion due to drug abuse; Harwood, Fountain, & Livermore, 1998).

Turning to workplace injuries, Leigh, Markowitz, Fahs, Shin, and Landrigan (1997) conducted a comprehensive study of the number of workplace injuries and their cost in 1992. They estimated that 6,529 workers died of work-related injuries (lower bound = 6,060 employees; upper bound = 10,000 employees) and that 13.25 million (lower bound = 8.75 million; upper bound = 19.80 million) nonfatal work injuries occurred. The total cost resulting from work injuries in 1992 was estimated to be $145 billion (lower bound = $97 billion; upper bound = $217 billion). Finally, the U.S. Department of Labor (2001) estimated that, in 2000, the incidence rate for nonfatal occupational injuries in the private sector was 5.8 cases per 100 equivalent full-time workers.

Although past research provides some support for the claim that employee alcohol and other drug use is related to poor employee productivity, including workplace injuries and accidents, several conceptual and methodological issues limit our ability to draw specific conclusions that are useful to managers, treatment providers, and policymakers. Therefore, I address four primary goals in this chapter. First, I provide a brief overview of the literature on employee substance use and productivity. Although the focus of this chapter is on the relation of employee substance use to workplace injuries and accidents, an overall understanding of this relation can be achieved only by exploring these outcomes within the broader context of employee productivity. Second, I present a general model of employee substance use and productivity that illuminates and eliminates past conceptual and measurement problems that have undermined past research in this area. The model, therefore, will help explain prior inconsistent findings and provide a framework to guide future field research on employee substance use and productivity. Third, I outline several directions for future research on employee substance use and workplace injuries and accidents. Finally, I discuss the implications of past research and the proposed model for managers and workplace policy.

SUBSTANCE USE AND EMPLOYEE PRODUCTIVITY: A BRIEF OVERVIEW

Despite the widely held belief that the use of alcohol and other psychoactive drugs (i.e., marijuana; opioid or narcotic analgesics; depressants or sedatives; stimulants, including cocaine; hallucinogens or psychedelics; and inhalants) among employees may negatively affect employee productivity, past reviews of the literature suggest that this relation is neither consistent nor robust. For example, a panel convened by the National Research Council–Institute of Medicine (NRC-IOM) produced a detailed report entitled *Under the Influence? Drugs and the American Work Force* (Normand, Lempert, & O'Brien, 1994). In this report, when considering the relation of employee alcohol and drug use to productivity outcomes, the panel concluded: "Research support is most consistent . . . with absenteeism. For other types of outcomes, there is mixed or weak support, with very little support from better-designed studies" (Normand et al., 1994, p. 134). In the following sections I provide a brief overview of past research exploring the relation of employee substance use to (a) job accident and injury outcomes, (b) task performance and other on-the-job outcomes, and (c) attendance outcomes.

Accident and Injury Outcomes

To evaluate the extent to which alcohol and drugs play an etiologic role in workplace accidents and injuries, addressing two related questions is important (Zwerling, 1993). The first question is: What proportion of workplace injuries and accidents is associated with the use of alcohol and illicit drugs? The answer to this question does not tell us if alcohol and drug use are causally related to workplace injuries and accidents. However, it does provide an estimate of the maximum proportion of injuries and accidents that could be reduced if alcohol and drug use were causally related to these outcomes. The actual proportion of accidents and injuries due to employee substance use that could be reduced in the population of workers (i.e., population attributable fraction or risk) would further depend on the prevalence and pattern of employee substance use and the size of the true causal effect of employee substance use on accidents and injuries (e.g., Northridge, 1995; Rothman & Greenland, 1998). Nonetheless, if alcohol and drugs were causally implicated in only 1% or less of workplace injuries, determining the causal relation of employee substance use to injuries might not be viewed as terribly important. In contrast, if alcohol and drugs were causally implicated in 10% or more of workplace injuries, determining the causal relation of employee substance use to injuries might be an important research issue. Data to answer this first question generally come from coroner and medical examiner records and from emergency room visits. After reviewing the toxicology literature on blood-alcohol concentration in injured workers, Zwerling (1993) estimated that acute alcohol impairment is present in approximately 10% of fatal work injuries and 5% of nonfatal work injuries. However, it is important to point out that the percentage of accidents and injuries involving alcohol use may show substantial heterogeneity across occupations or workplace cultures. For example, using a sample of shipyard workers in a setting where alcohol was freely available during the workday, Moll van Charante and Mulder (1990) estimated that alcohol consumption was involved in 37% of nonfatal injuries.

Estimating the percentage of accidents and injuries involving the use of other psychoactive drugs is more difficult than estimating those involving alcohol for two reasons. First, although blood-alcohol concentration, assessed through either blood or Breathalyzer tests, is related to levels of impairment at the time of testing and could be used to estimate impairment at the time of an accident under favorable conditions, urine and blood tests for other psychoactive drugs generally provide evidence only for the amount of a drug or its metabolites present at the time of testing. Therefore, a positive drug test does not provide information regarding the dose used, time of administration (could have been days or weeks before the test), or level of impairment at the time of testing or at the time of an event (e.g., injury). Also, the presence of either alcohol or drugs in blood, urine, saliva, or hair does not provide sufficient information to differentiate between chronic and casual users (e.g., Kapur, 1993; Normand et al., 1994; Wolff et al., 1999). Second, less published toxicology research exists on the presence of drugs other than alcohol in workplace accidents and injuries. Nonetheless, Quest Diagnostics Incorporated, which is a large drug testing company (more than 10 million tests annually), publishes an annual drug testing index for employment-related drug tests. Quest Diagnostics Incorporated (2002) reported that of all postaccident tests conducted in 2001, 3.6% were positive for at least one drug in the federally mandated safety-sensitive workforce and 6.0% were positive in the general workforce.

The second question is: Is employee substance use related to workplace injuries and accidents? Answers to this question attempt to address the causal effect of alcohol and drug use on workplace injuries. Data to answer this question generally come from laboratory research examining the acute effect of substance use on cognitive and psychomotor performance and from epidemiologic field studies examining the relation of typical patterns of employee substance use to workplace accidents and injuries. Laboratory research on the acute effects of substance use generally shows that the use of alcohol, marijuana, opioid analgesics, and sedative drugs (e.g., benzodiazepines) either have no effect or impair performance on a variety of laboratory tasks, such as time estimation, divided attention, tracking, vigilance, postural stability, and complex reaction time (e.g., Adams & Martin, 1996; Coambs & McAndrews, 1994; Folton & Evans, 1993; Heishman, 1998; Maisto, Galizio, & Connors, 1999; C. S. Martin, 1998; Normand et al., 1994; Schwenk, 1998; Walker, Zacny, Galva, & Lichtor, 2001). In contrast, stimulants typically either have no effect or improve performance in laboratory settings (e.g., Coambs & McAndrews, 1994; Folton & Evans, 1993; Heishman, 1998; Maisto et al., 1999; Normand et al., 1994). Despite these laboratory findings, it is not yet clear whether and how strongly these cognitive and psychomotor effects will translate into actual changes in on-the-job behavior and performance (e.g., Heishman, 1998).

Although the postaccident prevalence rates reported earlier from Quest Diagnostics Incorporated (2002) cannot by themselves tell us whether impairment due to drug use played an important causal role in the occurrence of an accident, they can be compared with the prevalence rates obtained during random drug testing. If employee drug use played a major role in workplace accidents, one would expect the prevalence rates for postaccident drug tests (3.6% in the safety-sensitive workforce; 6.0% in the general workforce) to be substantially higher than the prevalence rates for random drug testing (2.2% in the safety-sensitive workforce; 7.0% in the general workforce). Thus, these comparisons do not support a strong causal role for employee drug use in workplace accidents.

Besides the prevalence rates for various types of drug testing, several qualitative literature reviews, conducted mostly during the early 1990s, summarize the results of epidemiologic field studies exploring the relation of employee substance use to workplace accidents and injuries. Each of these reviews found some studies that supported a relation and some studies that failed to support a relation between employee substance use and workplace injuries or accidents. Therefore, it is not surprising that past reviewers have unequivocally drawn the same conclusion. The consensus is that past research has failed to show that alcohol and drug use is a consistent and robust predictor of workplace accidents and injuries (e.g., Feinauer, 1990; Macdonald, 1997; Macdonald & Wells, 1994; Normand et al., 1994; Schwenk, 1998; Stallones & Kraus, 1993; Webb et al., 1994; Zwerling, 1993). However, because of various methodological weaknesses in past research, these reviewers also suggest that it is premature to conclude that employee substance use plays no causal role in the etiology of workplace injuries and accidents.

Task Performance and Other On-the-Job Outcomes

Several studies have explored the relation of employee substance use to on-the-job behaviors and outcomes other than injuries and accidents. Laboratory simulations by Price and colleagues (see Hahn & Price, 1994) and by others (e.g., Howland et al., 2001; Howland, Rohsenow, Cote, Siegel, & Mangione, 2000; Streufert et al., 1994) have shown that, even at low levels (e.g., blood-alcohol concentrations of .04% to .06%), acute exposure to alcohol can impair various dimensions of psychomotor performance used for many jobs and relates to impaired performance on actual occupational tasks, such as drill press operation, punch press operation, assembly tasks, welding, maintaining a power plant on a merchant ship, piloting a merchant ship, and managerial performance. In contrast, few occupational simulation studies have explored the acute effect of psychoactive drugs other than alcohol on task performance. An exception is Streufert and colleagues' work on the relation of the stimulant caffeine to dimensions of managerial performance (Streufert et al., 1995; Streufert et al., 1997). This research found that, on the one hand, increases in caffeine over an individual's usual daily dose had little overall effect—increasing performance on one outcome, decreasing performance on one outcome, and not affecting five outcomes. On the other hand, acute caffeine withdrawal had a consistently negative effect on performance.

Field research has explored a broader set of on-the-job outcomes. Several studies have reported that employee alcohol or drug use is related to poor job performance (e.g., Ames, Grube, & Moore, 1997; Blum, Roman, & Martin, 1993; Fisher, Hoffman, Austin-Lane, & Kao, 2000; Friedman, Granick, Utada, & Tomko, 1992), whereas other research has failed to support such a relation (e.g., S. Moore, Grunberg, & Greenberg, 2000; Parish, 1989). Research also has found a relation of alcohol and drug use to lower levels of positive contextual performance, such as working overtime, volunteering for additional work, or trying to improve the job (Lehman & Simpson, 1992), and to higher levels of counterproductive behavior, such as psychological and physical withdrawal at work and the perpetration of aggression and antagonistic behaviors at work (e.g., Ames et al., 1997; Greenberg & Barling, 1999; Lehman & Simpson, 1992; McFarlin, Fals-Stewart, Major, & Justice, 2001; S. Moore et al., 2000).

Attendance Outcomes

As for attendance outcomes, the general consensus across several older literature reviews is that absenteeism (tardiness has rarely been assessed) is the most consistently documented outcome related to employee alcohol and drug use (e.g., Blum et al., 1993; J. K. Martin, Kraft, & Roman, 1994; Normand et al., 1994; Zwerling, 1993). However, these conclusions were based on a small set of studies. In contrast, a closer examination of more recent research does show inconsistency, with some studies supporting a positive relation of employee alcohol and drug use to poor attendance outcomes (e.g., Bass et al., 1996; Department of Health and Human Services, 1999b; Fisher et al., 2000; Jones, Casswell, & Zhang, 1995; McFarlin & Fals-Stewart, 2002) and some studies failing to support this relation (e.g., French, Zarkin, & Dunlap, 1998; S. Moore et al., 2000; Peter & Siegrist, 1997; Vasse, Nijhus, & Kok, 1998).

A GENERAL MODEL OF EMPLOYEE SUBSTANCE USE AND PRODUCTIVITY

On the whole, much apparent inconsistency exists in research findings relating employee substance use to several types of productivity outcomes. Much of the inconsistency is found (a) across field studies exploring different outcomes (e.g., injuries, attendance, and task performance) and (b) across laboratory studies exploring cognitive and psychomotor impairment versus field studies of various productivity outcomes. However, inconsistencies in prior research findings may not be surprising given that most researchers have set out to test a simple and perhaps misguided hypothesis: Employee substance use is related to unfavorable productivity outcomes. Although its simplicity is appealing, this hypothesis is based on the often unrecognized assumption of causal homogeneity. That is, past researchers have implicitly assumed that the mere consumption of a psychoactive substance will have the same effect across all productivity outcomes for all employees despite the context of substance use. The general alcohol and drug literature, however, suggests that the underlying process linking employee substance use to workplace productivity is much more complicated. Failing to account for this complexity may explain much of the incongruity in past research findings.

In line with this argument, Figure 7.1 presents a general model of employee substance use and workplace productivity that incorporates several potentially important issues overlooked by past researchers. To date, a detailed conceptual model has not guided research on employee substance use and workplace productivity. Therefore, the proposed model provides an initial attempt to help identify past conceptual and measurement confounds that have undermined research in this area and to provide a framework to guide future field research on employee substance use and productivity. In the following sections I outline the important issues used in developing the proposed model.

Matching Context of Substance Use With Type of Productivity Outcome

One key issue is that past field research has failed to match the context of substance use with the type of productivity outcome under study. With very few exceptions, researchers have assessed employees' overall alcohol and drug use, which represents either (a) the mere use of a substance regardless of context or (b) the typical frequency or quantity of a substance consumed across all contexts. Thus, measures of overall alcohol and drug use largely reflect the assessment of substance use off the job (i.e., use away from work and outside an individual's normal work hours). I argue that distinguishing explicitly off-the-job substance use and impairment from on-the-job substance use and impairment is important. *Off-the-job substance use* represents the consumption of alcohol and other drugs at times that occur away from work and outside normal work hours. *Off-the-job substance impairment* represents impairment (i.e., intoxication and withdrawal, which are defined in the next section) due to alcohol and drug use experienced at times that occur away from work and outside normal work hours. In contrast, *on-the-job substance use* represents the consumption of alcohol and other drugs at times that occur just before or during one's work hours. Specifically, on-the-job substance use refers to the consumption of alcohol or drugs (a) within two hours of starting one's work shift, (b) during a lunch break, (c) during other work breaks, and (d) while performing one's job. *On-the-job substance impairment* represents impairment due to alcohol and drug use experienced during one's work hours.

Further inspection of past field research on employee substance use and productivity also reveals a general failure to distinguish between attendance outcomes and performance outcomes. Attendance outcomes represent the failure to come to work on time (tardiness) or the failure to come to work at all (absenteeism). Performance outcomes represent behaviors and outcomes that occur on the job, such as accidents and injuries, task performance, contextual performance, and counterproductive behaviors.

As shown in Figure 7.1, I propose that there is a correspondence between the context of employee substance use and impairment and the

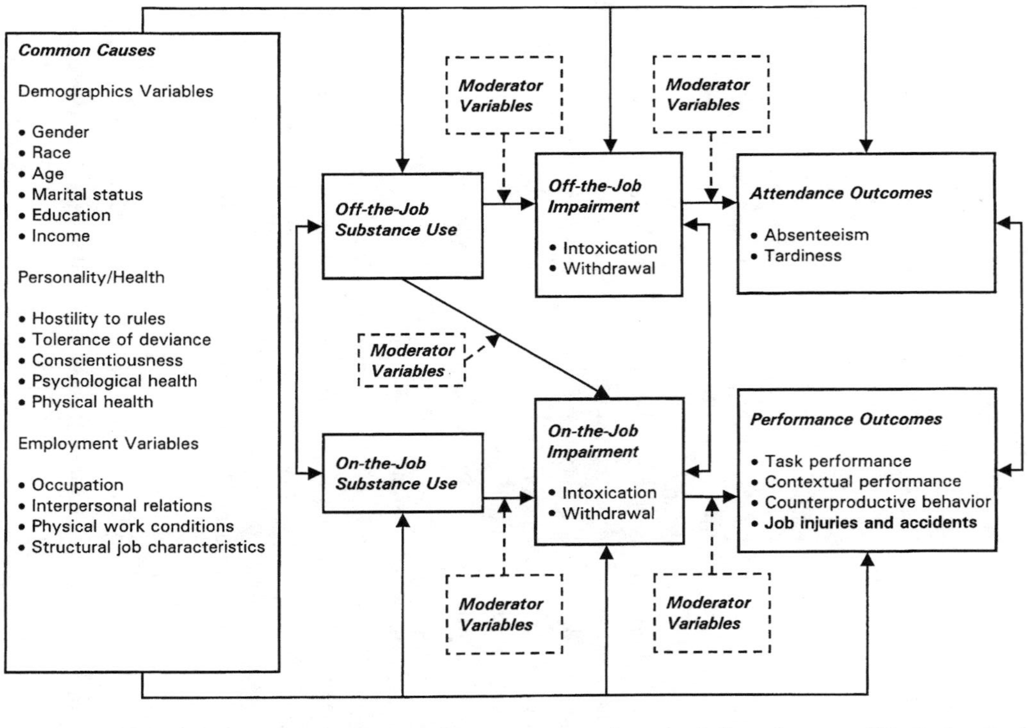

Figure 7.1. General model of employee substance use and productivity outcomes. (Single-headed arrows represent hypothesized causal effects. Double-headed arrows represent noncausal relations.)

type of productivity outcome affected. The first hypothesis derived from this general proposition is that off-the-job substance use and impairment predict attendance outcomes and that on-the-job substance use and impairment do not predict attendance outcomes. Because the measures of overall substance use used in past research primarily represent use and impairment outside the workplace, it is not surprising that early reviews of field research (e.g., Blum et al., 1993; J. K. Martin et al., 1994; Normand et al., 1994; Zwerling, 1993) concluded that there was a consistent positive relation between employee substance use and attendance outcomes. After all, if an employee is absent from work because of recent substance use, the use of alcohol or drugs must have occurred off the job.

The second hypothesis derived from the model is that employee performance outcomes are primarily the effect of substance use and impairment on the job. Nonetheless, off-the-job substance use may indirectly affect performance outcomes to the extent that it causes substance impairment during a person's work shift. Off-the-job impairment is not expected to relate to employee performance outcomes. The inability of many studies to document a relation between employee substance use and injuries or accidents at work, as well as task performance and other behaviors at work, may result from a failure to differentiate between off-the-job and on-the-job substance use and impairment. Most prior studies have relied on measures of overall substance use, which mainly assess the use of and impairment from psychoactive substances outside the workplace and outside an individual's hours of employment. Dawson (1994) noted that, by including extraneous (i.e., off-the-job) occasions of use, measures of overall alcohol use are likely to underestimate the size of the relation between alcohol use and job injuries. More generally, because poor performance at work is partly the result of psychomotor and cognitive impairment while an individual is at work, measures of on-the-job substance use and impairment should be more predictive of performance than measures of overall (i.e., off-the-job) substance use and impairment.

Indirect support for this contention comes from a study by Beaumont and Allsop (1983, cited in Gill, 1994), which found that 76% of alcohol-related accidents at work occurred within the times of 8–10 a.m. and 1–2 p.m. This suggests that work injuries may be the result of on-the-job impairment due to drinking just before work, heavy drinking the night before work, or drinking during one's work shift (i.e., during lunch breaks). More direct evidence comes from a study by Frone (1998), which found that when examined simultaneously a measure of on-the-job substance use and impairment was positively related to job injuries, whereas a measure of overall (i.e., off-the-job) substance use and impairment was unrelated to job injuries. Ames et al. (1997) also found that on-the-job alcohol use was positively related to several performance outcomes (problems with supervisor, sleeping on the job, problems with tasks and coworkers), whereas overall (i.e., off-the-job) alcohol use was unrelated to the performance outcomes.

Two exceptions exist to the pattern of findings discussed in the preceding paragraphs, but both studies suffer from important conceptual and methodological problems. For example, Mangione et al. (1999) reported that measures of both on-the-job alcohol use and overall heavy drinking were related to a composite measure of work performance. Interpreting these results is difficult, however, because the composite outcome measure confounded items assessing both attendance outcomes (e.g., absenteeism and tardiness) and performance outcomes (e.g., poor work quality and quantity, injured at work, argued with a coworker). According to the proposed conceptual model in Figure 7.1, the results reported by Mangione et al. using a confounded outcome measure are expected. However, they may not accurately capture the underlying process linking employees' alcohol use to both types of productivity outcomes. Therefore, their findings may result in misleading conclusions. Another exception was reported by Hingson, Lederman, and Walsh (1985), who found that overall heavy drinking was related to injuries at work requiring medical attention or hospitalization, whereas alcohol and drug use at work were not related to work injuries. However, this study suffered from a major methodological problem in that substance use was assessed for the preceding 30 days and injuries were reported for the preceding 12 months. It is plausible that a measure of overall heavy alcohol use during the preceding month captured a pattern of overall consumption that was stable during the preceding 12 months, whereas the substance use at work measures did not capture a stable pattern of substance use at work. This would make it more difficult for a measure of substance use at work during the preceding 30 days to predict a measure of work injuries during the preceding 12 months, because most of the injury episodes would have preceded the episodes of on-the-job substance use.

The distinction between off-the-job and on-the-job substance use and impairment may also explain prior inconsistencies when looking across field studies and laboratory studies. Past field studies have provided inconsistent findings regarding the relation between employee substance use and job performance. In contrast, more consistent evidence exists that acute intoxication from alcohol and many drugs can impair cognitive and psychomotor performance on laboratory tasks. A fundamental difference between field research and laboratory research is that the former has largely focused on substance use and impairment off the job by using measures of overall substance use, whereas the latter has focused on substance impairment while a person is performing a task. This underscores the importance of explicitly considering the context of substance use. Stronger and more consistent relations between employee substance use and workplace injuries and accidents, as well as other performance outcomes and behaviors at work, would be expected when field researchers begin to assess on-the-job substance use and impairment.

Distinguishing Between Substance Use and Substance Impairment

Another major issue that went into the development of the proposed model was the distinction between substance use and substance impairment. Measures of substance use reflect the mere use of a substance, the frequency of using a substance over some fixed period of time, or the quantity of a substance consumed on a typical occasion of use (see contextual definitions provided earlier in this chapter). Substance impairment has two dimensions: intoxication and withdrawal. In the medical and pharmacological literatures, *intoxication* simply refers to a state of being poisoned by some substance. Therefore, in this chapter, *substance intoxication* refers to reversible central nervous system impairment due to the direct pharmacological action of a substance that results in various behavioral, cognitive, and affective changes (e.g., American Psychiatric Association, 1994). With chronic heavy consumption of a substance, a process of physiological adaptation takes place whereby the central nervous system adjusts to the constant presence of the substance (see the following discussion of pharmacodynamic tolerance). Thus, *substance withdrawal* refers to central nervous system impairment due to a cessation of or reduction in substance use that results in various behavioral, cognitive, and affective changes (e.g., American Psychiatric Association, 1994; Saitz, 1998). Also included in the definition of withdrawal is the "hangover" syndrome most commonly associated with alcohol use. An alcohol hangover is believed to represent, at least partly, a mild form of acute alcohol withdrawal (e.g., R. S. Moore, 1998; Swift & Davidson, 1998; Wiese, Shlipak, & Browner, 2000). The precise signs and symptoms of substance intoxication and withdrawal may differ across types of psychoactive substances (e.g., American Psychiatric Association, 1994; Miller, Gold, & Smith, 1997).

Research generally shows that heavier use (i.e., increasing dose) of alcohol and drugs is associated with higher levels of intoxication and more severe withdrawal (e.g., Caetano, Clark, & Greenfield; 1998; Saitz, 1998; Swift & Davidson, 1998), though the relation of substance use to intoxication and withdrawal may not be strictly linear. Further, the laboratory research discussed earlier suggests that intoxication affects cognitive and psychomotor performance, and there is some evidence that substance withdrawal may have negative effects on various physical, cognitive, and affective outcomes as well (e.g., Saitz, 1998; Swift & Davidson, 1998; Wiese et al., 2000). However, the actual relevance of acute substance withdrawal for employee productivity is unclear at this point (e.g., R. S. Moore, 1998). In contrast to the laboratory and field research described earlier that explored the effects of substance use and intoxication on workplace productivity, very little research has explored the potential effect of substance withdrawal on workplace productivity.

Given the distinction between substance use and substance impairment, a general hypothesis is that substance impairment is a proximal cause of poor productivity outcomes and mediates the more distal effect of substance use on productivity. Specifically, increasing levels of off-the-job substance use are expected to cause higher levels of intoxication and more severe withdrawal (when substance use is decreased) off the job, which then causes poor attendance. Likewise, higher levels of off-the-job substance use may be related to intoxication at work because individuals may still have nonzero blood concentration when they arrive at work, and withdrawal symptoms may be more severe among chronic heavy off-the-job substance users if they do not consume the substance during their work shift. Also, higher levels of on-the-job substance use are expected to cause higher levels of on-the-job intoxication. In turn, on-the-job intoxication and withdrawal are expected to negatively affect performance outcomes.

The distinction between substance use and substance impairment may help explain inconsistencies in prior research. For example, as discussed earlier in this chapter, past reviewers concluded that employee substance use had the most consistent relation to attendance outcomes. However, I also pointed out that a broader review of more recent studies shows a fair amount of inconsistency regarding this relation. To explain this inconsistency, one must consider that some studies merely compare the attendance behaviors of substance users to nonusers. Also, even studies that used measures of overall frequency and quantity of use may have differed substantially in the proportion of subjects at the higher end of the frequency and quantity distributions. To the extent that past studies differed in the degree to which the measures of substance use captured meaningful substance impairment off the job, one would expect between-study variation in support for a relation between employee substance use and attendance outcomes. Parallel explanations also can be developed for inconsistencies in results involving the other productivity outcomes.

To date, however, no field research has directly tested the relation of off-the-job substance impairment to attendance outcomes. Likewise, little research has tested the relation of on-the-job substance impairment to performance outcomes. Finally, no research has tested whether context-specific substance impairment mediates the effect of context-specific substance use on the two categories of productivity outcomes. Clearly, future research needs to address these issues.

Control for Common Causes

The discussion until now has tried to explain why past research may have failed to document a consistent relation between employee substance use and productivity. Nonetheless, care must be taken in the interpretation of some prior studies that have supported such a relation. Much of this research has lacked adequate controls for common causes of substance use, substance impairment, and productivity. The conceptual model in Figure 7.1 shows that off-the-job substance use, off-the-job substance impairment, on-the-job substance use, on-the-job substance impairment, attendance outcomes, and performance outcomes are likely to share a set of common causes. For example, Normand et al. (1994) speculated that substance use and poor productivity may be related, at least in part, because both constructs reflect individuals' general predisposition to engage in deviant behaviors. Theories of general deviance (e.g., Gottfredson & Hirschi, 1990); problem behavior (e.g., Jessor, Donovan, & Costa, 1991); and employee unreliability (e.g., Hogan & Hogan, 1989) suggest that substance use and unreliable performance may be the result of personal characteristics representing behavioral undercontrol, such as hostility to rules and tolerance of deviance. Besides measures of behavioral undercontrol, Figure 7.1 shows that there are many other demographic, personality, health, and employment variables that need to be considered as potential common causes of employee substance use and productivity. For example, there is evidence that job stressors are related to both employee substance use (see Frone, 1999, for a review) and productivity (e.g., Jex, 1998).

Moderator Variables

As shown in Figure 7.1, variables may moderate the relations between employee substance use and productivity outcomes. These moderator variables represent various pharmacological, dispositional, motivational, and situational influences that may further account for past inconsistencies in the research literature. In the following subsections I summarize some variables that may function as important moderators of the substance-use/productivity relation. However, considering separately the relation between substance use and substance intoxication and between substance intoxication and productivity outcomes is helpful. Potential moderators of the relation between substance use and substance withdrawal and between substance withdrawal and productivity are not discussed because little research has explored this issue.

Relation Between Substance Use and Substance Intoxication

The moderating influence of other variables on the various relations between substance use and intoxication can occur at two different points (e.g., C. S. Martin, 1998). The first point of influence is the relation between the dose of a substance and its resulting concentration in blood. In other words, for a given dose of a substance, individual differences exist in the blood concentration achieved. Factors that affect blood concentration relate to the processes of absorption, distribution, and metabolism and elimination. Absorption refers to the rate and extent to which a drug leaves its site of administration and reaches its site of action in order to exert a pharmacological effect (Jenkins & Cone, 1998; Maisto et al., 1999). A major factor affecting absorption is mode of administration (e.g., oral, injection, inhalation, intranasal, sublingual). For example, compared with intravenous injection, inhalation, and sublingual (under-the-tongue) administration, oral ingestion is less efficient because the presence of food in the stomach and first-pass metabolism can impede absorption. First-pass metabolism refers to a process by which any ingested substance may be partially metabolized in the gastrointestinal tract or during the first pass of portal blood through the liver before it reaches systemic circulation (e.g., Jenkins & Cone, 1998; Jones & Pounder, 1998). Once alcohol and other drugs enter systemic circulation, they are distributed throughout the body. Factors affecting distribution include the biochemical properties of the drug (e.g., water or fat solubility and binding properties) and physiological factors (e.g., diffusibility of membranes and tissues, amount of body water, amount of body fat; e.g., Jenkins & Cone, 1998; Maisto et al., 1999). Finally, substances are either excreted directly from the body or metabolized into by-products called metabolites that are often pharmacologically inactive and then excreted. If the metabolites are pharmacologically active, they are either excreted directly or are further metabolized and then excreted (e.g., Jenkins & Cone, 1998; Maisto et al., 1999).

Because of individual differences in the processes underlying absorption, distribution, and metabolism and elimination, individual differences will be observed in blood concentration for a given dose of a substance (e.g., Maisto et al., 1999). For example, individual differences in mode of administration will lead to differences in blood concentration for a given dose of a specific substance. On average, women will have a higher blood concentration for a given dose of a substance than men because of their lower weight and their proportionately higher percentage of body fat, which leads to a corresponding lower amount of body water in which to distribute the substance. Elderly adults may experience higher blood concentration for a given dose than young adults because they have less body water and their ability to metabolize substances may become weakened. Also, after repeated exposure to a substance, pharmacokinetic tolerance (also known as dispositional tolerance or metabolic tolerance) may develop. This means that individuals who repeatedly use a drug may begin to experience an increase in the rate of metabolization and elimination of the drug (Maisto et al., 1999; C. S. Martin, 1998; Vogel-Sprott, 1992), which will lead to lower blood concentrations for a given dose compared with individuals who have not developed pharmacokinetic tolerance.

The second point of influence is the relation between the concentration of a drug in blood and the resulting intoxication (i.e., central nervous system impairment). In other words, for a given blood concentration of a drug, individual differences exist in the level of experienced intoxication. A potential moderating process is pharmacodynamic tolerance (also known as functional tolerance or cellular tolerance; e.g., Maisto et al., 1999; C. S. Martin, 1998; National Institute on Alcohol Abuse and Alcoholism, 1995; Vogel-Sprott, 1992). Pharmacodynamic tolerance refers to a process of physiological adaptation under which the central nervous system becomes less sensitive to the pharmacological effects of a given drug. Two forms of pharmacodynamic tolerance exist. Chronic tolerance develops after repeated exposure to a substance over time. However, for alcohol, researchers have also documented acute tolerance, which is reduced intoxication exhibited during a single episode of drinking. Specifically, it has been shown that the level of intoxication at a given blood-alcohol concentration (e.g., .08%) may be stronger while blood-alcohol concentration is rising (i.e., during the ascending limb of the blood-alcohol curve) compared with when blood-alcohol concentration is falling (i.e., during the descending limb of the blood-alcohol curve).

Another factor that affects degree of intoxication is the use of multiple substances. The use of multiple drugs can have either a synergistic effect that increases intoxication for a given dose of a substance or an antagonistic effect that reduces intoxication for a given dose of a substance (e.g., Maisto et al., 1999). For example, the impact of blood-alcohol concentration on intoxication may become stronger with the simultaneous use of sedative and depressant drugs. In contrast, the impact of blood-alcohol concentration on intoxication may be attenuated with the simultaneous use of stimulants, such as nicotine, caffeine, and cocaine (e.g., Farre et al., 1993; Kerr & Hindmarch, 1998).

Relation Between Substance Intoxication and Productivity Outcomes

A number of variables might moderate the relation between substance intoxication and employee productivity. A comprehensive set of variables is difficult to provide because of limited research on this issue and because the potential moderators are likely to vary by type of substance and by type of productivity outcome being assessed. One process that may moderate the relation of substance intoxication to job injuries and accidents, as well as to other performance and attendance outcomes, is behavioral (or learned) tolerance (e.g., Maisto et al., 1999; National Institute on Alcohol Abuse and Alcoholism, 1995; Vogel-Sprott, 1992). Behavioral tolerance refers to learned compensatory strategies to reduce the impact or expression of intoxication on cognitive and behavioral task performance. Tolerance to the effect of alcohol intoxication increases if a person explicitly practices a set of task behaviors or mentally rehearses the task behaviors while under acute intoxication (e.g., Vogel-Sprott, 1992). A number of factors or cues can enhance or impede the development or expression of behavioral tolerance. For example, when behavioral tolerance is developed in a specific environment, individuals performing a task under intoxication in a new setting will fail to show behavioral tolerance (e.g., Maisto et al., 1999). Behavioral tolerance also is more likely to be developed with increased experience using a substance (e.g., Fillmore & Vogel-Sprott, 1996). Finally, the development of behavioral tolerance is enhanced if individuals are rewarded for showing nonimpaired performance (e.g., Vogel-Sprott, 1992).

Another process that might reduce the impact or expression of substance intoxication on cognitive and behavioral task performance has been labeled resistance to impairment (George, Raynor, & Nochajski, 1992; Nochajski, 1993). Whereas behavioral tolerance develops over several occasions of performing a task while intoxicated, resistance to impairment refers to motivational aids that reduce the expression of intoxication in a single occasion. For example, past laboratory research shows that visual-motor performance decrements due to alcohol intoxication at blood-alcohol concentrations of .05% (George et al., 1992) and .10% (Nochajski, 1993) were eliminated in the first and subsequent postintoxication performance trials when individuals were instructed to "concentrate very hard" on the task at hand. Although speculative, these results suggest that intoxicated workers who engage in motivational self-talk that focuses their concentration on the task to be completed may be no more likely to perform more poorly or incur an injury at work than nonintoxicated workers. Moreover, intoxicated workers who do not use such motivational techniques may be more likely to incur an injury at work or perform poorly compared to nonintoxicated workers and intoxicated workers who use such motivational aids.

Taken together, research on behavioral tolerance and resistance to impairment suggests that the positive relation of on-the-job intoxication to job injuries and accidents, as well as to other dimensions of poor performance at work, may be attenuated (a) among workers who work in the same environment or at the same work station compared with workers who change work environments or work stations; (b) among experienced substance users compared with inexperienced users; (c) among individuals rewarded for exhibiting compensatory (i.e., nonimpaired) behavior compared with individuals who are not so rewarded; and (d) among individuals who use motivational strategies that focus their attention on the task to be accomplished compared with individuals who do not use such strategies. Turning to the positive relation of off-the-job intoxication to poor attendance outcomes, one might expect that behavioral tolerance would attenuate the magnitude of this relation as well. Specifically, experienced substance users and those who perceive an immediate cost to poor attendance (e.g., being disciplined or fired) may develop compensatory responses to residual intoxication the morning after a night of heavy drinking or drug use so that they show up for work on time compared with inexperienced user or users who do not perceive an immediate cost to poor attendance, respectively.

Although most studies of workplace injuries have assessed overall substance use and not the more proximal antecedent of on-the-job substance intoxication, some evidence exists that another potentially important moderator of the relation between on-the-job substance intoxication and job injuries and accidents may be the latent potential for injury and property damage accidents on a given job. In other words, injury liability due to increasing levels of intoxication may depend on the characteristics of and demands inherent in the jobs under study (e.g., Dawson, 1994; Holcom, Lehman, & Simpson, 1993). Finally, there is some indirect evidence that age may moderate the relation of on-the-job substance intoxication to work injuries. Wells and Macdonald (1999) reported that weekly alcohol consumption was positively related to work-related accidents among younger workers (ages 15–34) but was unrelated to work-related accidents among older employees (ages 35–64). This moderating effect of age is consistent with increased behavioral tolerance among older and perhaps more experienced drinkers or with increased risk-taking propensity among younger individuals.

Summary

The model outlined in Figure 7.1 presents a detailed guide for future researchers trying to gain a better and more comprehensive understanding of the impact of employee substance use on workplace productivity. To date, a full test of the proposed model has not been conducted because no single study has assessed even the set of six core variables (i.e., off-the job substance use, off-the-job substance impairment, on-the job substance use, on-the-job substance impairment, attendance outcomes, and performance outcomes). Nonetheless, several studies have provided suggestive partial support for the model.

Using data from the 1994 National Household Survey of Drug Abuse, Hoffman, Larison, and Sanderson (1997) examined the relation of overall drug use and heavy drinking to absenteeism and workplace accidents. They found that overall (i.e., off-the-job) drug use and heavy drinking were related to higher absenteeism and were unrelated to workplace accidents. Similarly, an evaluation study of preemployment drug testing, which can only assess recent use that most likely occurs off the job, found that positive test results were related to higher levels of absenteeism and were unrelated to the occurrence of workplace injuries and accidents (Normand, Salyards,

& Mahoney, 1990). Frone (1998) examined the relation of overall (i.e., off-the-job) and on-the-job substance use (combined use of alcohol and marijuana) to work injuries in a sample of young employees. At the bivariate level, both overall and on-the-job substance use were related to an elevated rate of work injuries. However, when both substance use measures were forced into a regression analysis simultaneously, only on-the-job substance use predicted injuries. This pattern of results was observed after controlling for 18 potential common causes of employee substance use and work injuries: demographics (gender, age); personality (negative affectivity, rebelliousness, impulsivity); employment (job tenure, hours worked per week, exposure to physical hazards, supervisor monitoring, workload, job boredom, role ambiguity, supervisor conflict, coworker conflict, work–school conflict, job dissatisfaction); and health (depression, somatic symptoms). Ames et al. (1997) examined the simultaneous relation of on-the-job alcohol use and overall alcohol use to several dimensions of work performance in a sample of manufacturing employees. In a multiple regression analysis that controlled for several demographic and employment covariates, only on-the-job alcohol use was related to poor performance.

Collectively, very little research has explored the main issues underlying the proposed general model of employee substance use and productivity. Much more focused research on these issues needs to be undertaken.

FUTURE RESEARCH ON EMPLOYEE SUBSTANCE USE AND WORKPLACE ACCIDENTS AND INJURIES

Regarding the various components of the model presented in Figure 7.1, I have outlined many important directions for future research throughout this review. Nonetheless, I would like to highlight several other general issues in need of additional research. Although my focus is primarily on workplace injuries and accidents, some of this discussion will have relevance to other productivity outcomes. Future research should explore multiple dimensions of negative safety outcomes. For example, is on-the-job substance impairment more likely to be related to the simple experience of having an injury during a fixed period of time or to the total number of injuries during that same period of time? Will on-the-job substance impairment lead to the same liability when looking across injuries experienced by the impaired employee, injuries experienced by nonimpaired coworkers, or noninjury (i.e., property damage) accidents? The safety- and morale-related costs associated with employee substance use would likely increase if, in addition to the impaired worker, coworkers are likely to be injured and company property is likely to be destroyed. Is on-the-job substance impairment more likely to result in severe or minor injuries? Finally, is on-the-job substance impairment more likely to be related to injuries that are not reported to the employer? If this is true, the relation of on-the-job substance impairment to worker injuries might be underestimated in studies using company records to assess injuries and in self-report surveys that implicitly confound reported and unreported injury incidents.

In addition to characteristics of the negative safety outcomes, future field research may need to differentiate more strongly between the separate and joint effects of various types of psychoactive substances on workplace injuries and accidents. Because of the low overall prevalence rates for the use of and impairment from many psychoactive substances, and the even lower prevalence rates for on-the-job use and impairment, aggregate measures of off-the-job and on-the-job substance may need to be used. However, given the results of past laboratory research, researchers may want to differentiate minimally between alcohol, cannabis, central nervous system depressants (e.g., narcotic analgesics, depressants, or sedatives), and central nervous system stimulants (e.g., nicotine, caffeine, cocaine) when sample sizes and statistical power allow.

Finally, as suggested earlier in this chapter, there is a need for better research on the pattern and context of employee substance use, as well as on the relation of context-specific assessments of employee substance use to workplace safety. For example, it is likely that the workforce contains "hot spots" representing specific combinations of demographic and occupational characteristics, where employee substance use may be both more prevalent and exert a larger and more consistent effect on workplace safety. To date, little empirical research in the general workforce has explicitly focused on this issue. Identification of such hot spots is important because it can ensure that the limited resources (time and funding) for research on employee substance use and safety are progressively used in samples of employees that will maximize statistical power and sound policy recommendations. Therefore, research needs to begin (a) documenting actual patterns of both off-the-job and on-the-job substance use and impairment in specific populations or subpopulations of workers and (b) exploring the causal relation of these actual substance use and impairment patterns to employee safety in those populations or subpopulations of employees.

ORGANIZATIONAL, MANAGERIAL, AND POLICY IMPLICATIONS REGARDING EMPLOYEE SUBSTANCE USE AND WORKPLACE SAFETY

As noted earlier in this chapter, organizational researchers, managers, human resource professionals, and policymakers have asked an overly simplistic question about employee substance use: Is employee substance use related to unfavorable productivity (including workplace safety) outcomes?

In asking this simple question, they have assumed that the mere consumption of a psychoactive substance will have the same effect across all productivity outcomes for all employees despite the context of substance use. However, regarding workplace safety, the research reviewed earlier suggests that the mere consumption of a psychoactive substance is unlikely to affect workplace safety. For substance use to affect workplace safety, the dose consumed, the timing of use, and the location of use must coincide in a manner that leads to impairment (intoxication or withdrawal) on the job. Moreover, it must be emphasized that the impact of on-the-job impairment on safety outcomes may be moderated by a number of dispositional, pharmacological, and environmental circumstances. Thus, the answer to the question above is "It depends." The relation between substance use and safety outcomes may depend on all of the key dimensions reviewed in detail earlier in this chapter.

If the relation between employee substance use and workplace safety is more complicated than generally assumed, what are the practical implications for managers, human resource professionals, and policymakers? The first practical implication is that given the likely prevalence rates and patterns of off-the-job and on-the-job substance use and impairment for most employees, coupled with their actual work environments and other moderating influences, one might expect employee substance use to be at best a modest causal antecedent of workplace safety outcomes (as well as other performance outcomes) in most workplaces. In fact, no credible scientific evidence currently exists to suggest that employee substance use is among the major and consistent causes of workplace injuries and accidents (or any other productivity outcome). Therefore, it is of little scientific and practical value to make general, exaggerated, and largely unsubstantiated claims regarding the outcomes of alcohol and drug use in the workforce and in the workplace. Nevertheless, given the vested interests (e.g., financial, political, moral) of various stakeholders (e.g., employers, policymakers, drug testing industry) in the relation of employee substance use to workplace safety and productivity, we will continue to see questionable claims based on inadequate data in both the scientific literature and popular press until a new generation of more sophisticated research begins to tackle this issue. For example, in a recent trade article, Chen and Kleiner (2001) exclaim that "it is obvious that drug abuse in the workplace leads to decreased productivity. Also, employees' use of illegal drugs significantly affects profitability and workplace safety" (p. 145). Freeing oneself of the burden of proof and ignoring all of the issues and complexities reviewed earlier, this statement has an air of face validity. However, one might ask, "Is it obvious that drug abuse in the workplace (on-the-job impairment) is prevalent?" Chen and Kleiner fail to provide any evidence that on-the-job impairment is prevalent or that any use of illicit drugs has any general and practically important relation to profitability and workplace safety. I am not suggesting that employee substance use does not affect workplace productivity or that workplace policies forbidding on-the-job use and impairment should not implemented. However, I am suggesting that the effects of employee substance use on workplace productivity may be more limited than commonly claimed. If policy decisions are to be based on the most accurate information, additional research, as described earlier in this chapter, will be required to determine which dimensions of substance use and impairment affect which productivity outcomes for which subpopulations of workers (i.e., hot spots).

A second practical implication is that the findings reviewed earlier in this chapter raise doubts about the overall usefulness of employee drug testing for improving workplace safety and productivity. One of the central arguments in defense of drug testing is that it can reduce workplace accidents and injuries by reducing the likelihood that employees will engage in illicit substance use. However, there is little sound scientific data evaluating the impact of drug testing on workplace safety outcomes. And, of the research that does exist, there is no evidence that drug testing programs reduce workplace accidents and injuries (e.g., Macdonald, 1997; Macdonald & Wells, 1994; Normand et al., 1994). Given the present review of the literature, the failure of past research to support a relation between drug testing programs and lower rates of workplace injuries and accidents is not surprising. The relation between employee substance use and workplace injuries and accidents is not as simple and robust as typically assumed by proponents of drug testing. Although a number of factors account for the complexity of this relation, three fundamental elements are (a) whether impairment occurs, (b) the context of the impairment, and (c) the productivity outcome under consideration. As noted earlier in this chapter, a positive preemployment or random drug test is not sufficient evidence that an individual has ever been impaired while working and a positive postaccident drug test is not sufficient evidence that an individual was impaired at the time of accident or injury.

One might argue that drug testing, by increasing the odds of discovery, may reduce illicit drug use across the board and therefore should reduce accidents and injuries and improve other productivity outcomes. However, such an argument is based on the faulty, or at least unsubstantiated, assumption that most individuals who use illicit drugs do so in such a manner that they are impaired while working. To date, however, no convincing evidence exists to suggest that on-the-job impairment due to illicit drug use is generally prevalent or that drug testing reduces the use of illicit drugs. And, even if drug testing did reduce illicit drug use, the reduction is likely to occur primarily among casual drug users who do not present at work impaired. Among the smaller group of problem (i.e., dependent) drug users who are more likely to present at work impaired, drug testing programs may have little deterrence value.

Another problem with drug testing programs is that they generally do not test for the use of alcohol, which is likely to play larger role in workplace accidents and injuries than the use of illicit drugs. Even though alcohol testing does not suffer from the same weaknesses as drug testing in terms of determining impairment and context of use, its overall value is still dependent on the absolute prevalence and frequency of on-the-job impairment due to alcohol. At present, no credible prevalence and frequency data exist on alcohol use and impairment while on the job to support the wide-scale use of alcohol testing in most occupations. Finally, and more fundamentally, drug testing does not directly address the underlying causes of alcohol and drug misuse in the workforce (Bennett & Lehman, 2003).

It must be emphasized that even if a positive relation were to be found between positive drug test results and the prevalence of workplace injuries, the typical study design does not allow for causal conclusions. Because a positive drug test merely identifies someone who has used drugs in the recent past—it does not provide evidence of on-the-job impairment—there is no way to causally link drug use to workplace injuries in most studies designed to evaluate drug testing. An alternative explanation is that some other variable, such as the personality characteristic of being intolerant of rules, might be a cause of both drug use (which is detected in a drug test) and being injured at work. Thus, although drug use and workplace injuries may be empirically related, drug use as detected by a workplace drug testing program may not be the actual cause of the increased risk of injuries among individual who use drugs. Drug testing, therefore, may lead to an incorrect course of action regarding the reduction of workplace injuries and accidents.

The limitations of alcohol and drug testing discussed to this point do not mean that testing should never be used. Rather, the goal of this discussion is to promote a more thoughtful consideration of the virtues and limitations of testing with regard to workplace accidents and injuries, and other productivity outcomes. Unfortunately, much of the rhetoric around workplace alcohol and drug testing, especially as it relates to workplace safety outcomes, has been motivated by economic considerations, political and moral agendas, and issues regarding compliance (e.g., Bennett, 2003; Cavanaugh & Prasad, 1994; Draper, 1998).

But when is drug testing a legitimate course of action with regard to minimizing workplace injuries and accidents? A report by the Canadian Human Rights Commission (CHRC, 2002) provides several well-reasoned recommendations regarding workplace alcohol and drug testing that are generally consistent with the large literature reviewed in this chapter. The CHRC states that because they are not a bona fide occupational requirement and because they cannot indicate actual on-the-job impairment or the likelihood of being at work impaired, the following types of testing are not acceptable: (a) preemployment drug testing, (b) preemployment alcohol testing, (c) random drug testing, and (d) random alcohol testing of employees in non-safety-sensitive positions. In contrast, the CHRC suggests that, if they can be shown to be bona fide occupational requirements, the following types of workplace alcohol and drug testing may be acceptable: (1) random alcohol testing of employees in safety-sensitive positions; (2) drug or alcohol testing for reasonable cause or postaccident; and (3) periodic or random alcohol or drug testing following disclosure of a current drug or alcohol dependency or abuse problem, if part of a broader program of monitoring and support.

A final practical implication is that employers need to move beyond drug testing. Although drug testing may have its place, for most employees and for most occupations it represents an intrusive management technology. The other chapters in this volume outline many other issues that need to be addressed, such as working conditions, compensation systems, safety culture and leadership, and safety training. In most circumstances, because these factors have an effect on all employees, giving attention to them is likely to pay bigger dividends than drug testing will in reducing workplace accidents and injuries. Also, attention to reducing employee alcohol and drug misuse inside and outside the workplace needs to shift from a single-minded focus on alcohol and drug testing to a multidimensional and proactive emphasis on workplace substance use prevention interventions and employee education (Bennett & Lehman, 2003). However, because of the long-standing interest in alcohol and drug testing, the development of workplace prevention interventions is still in its infancy (see Bennett & Lehman, 2003, for a detailed discussion).

CONCLUSIONS

The potential effect of employee substance use on workplace safety, and other productivity outcomes, represents an important social policy issue. Thus, the lack of rigorous conceptual and empirical attention to this issue is surprising. In contrast to the simple hypothesis tested in past research, this chapter shows that the relation of employee substance use to workplace productivity outcomes is anything but simple and straightforward. The proposed general model of employee substance use and productivity represents a first step toward building a stronger empirical database to inform researchers, employers, treatment providers, and workplace policymakers.

REFERENCES

Adams, I. B., & Martin, B. (1996). Cannabis: Pharmacology and toxicology in animals and humans. *Addiction, 91*, 1585–1614.

American Psychiatric Association. (1994). *Diagnostic and statistical manual of mental disorders* (4th ed.). Washington, DC: Author.

Ames, G. M. (1993). Research strategies for the primary prevention of workplace alcohol problems. *Alcohol Health and Research World, 17*, 19–27.

Ames, G. M., Grube, J. W., & Moore, R. S. (1997). The relationship of drinking and hangovers to workplace problems: An empirical study. *Journal of Studies on Alcohol, 58*, 37–47.

Bass, A. R., Bharucha-Reid, R., Delaplane-Harris, K., Schork, M. A., Kaufmann, R., McCann, D., et al. (1996). Employee drug use, demographic characteristics, work reactions, and absenteeism. *Journal of Occupational Health Psychology, 1*, 92–99.

Beaumont, P. B., & Alsop, S. J. (1983). Beverage report. *Occupational Safety and Health, 13*, 25–27.

Bennett, J. B. (2003). Introduction. In J. B. Bennett & W. E. K. Lehman (Eds.), *Preventing workplace substance abuse: Beyond drug testing to wellness* (pp. 3–28). Washington, DC: American Psychological Association.

Bennett, J. B., & Lehman, W. E. K. (Eds.). (2003). *Preventing workplace substance abuse: Beyond drug testing to wellness*. Washington, DC: American Psychological Association.

Blum, T. C., Roman, P. M., & Martin, J. K. (1993). Alcohol consumption and work performance. *Journal of Studies on Alcohol, 54*, 61–70.

Caetano, R., Clark, C. L., & Greenfield, T. K. (1998). Prevalence, trends, and incidence of alcohol withdrawal symptoms. *Alcohol Health and Research World, 22*, 73–80.

Canadian Human Rights Commission. (2002). *Canadian human rights commission policy on alcohol and drug testing*. Ottawa, Canada: Author. Retrieved November 15, 2002, from http://www.chrc-ccdp.ca/Legis&Poli/DrgTPol_PolSLDrg/PolDrgAlcEng.pdf

Cavanaugh, J. M., & Prasad, P. (1994). Drug testing as symbolic managerial action: In response to "A case against workplace drug testing." *Organizational Science, 5*, 267–271.

Chen, C.-L., & Kleiner, B. H. (2001). How to manage personnel with positive drug test results. *Management Research News, 24*, 145–148.

Coambs, R. B., & McAndrews, M. P. (1994). The effects of psychoactive substances on workplace performance. In S. Macdonald & P. M. Roman (Eds.), *Research advances in alcohol and drug problems: Vol. 4. Drug testing in the workplace*. New York: Plenum.

Dawson, D. A. (1994). Heavy drinking and the risk of occupational injury. *Accident Analysis and Prevention, 26*, 655–665.

Department of Health and Human Services. (1999a). *Substance use and mental health characteristics by employment status* (DHHS Publication No. SMA 99-3311). Washington, DC: U.S. Government Printing Office.

Department of Health and Human Services. (1999b). *Worker drug use and workplace policies and programs: Results from the 1994 and 1997 NHSDA* (DHHS Publication No. SMA 99-3352). Washington, DC: U.S. Government Printing Office.

Draper, E. (1998). Drug testing in the workplace: The allure of management technologies. *International Journal of Sociology and Social Policy, 18*, 62–103.

Farre, M., De La Torre, R., Llorente, X., Ugena, B., Segura, J., & Cami, J. (1993). Alcohol and cocaine interactions in humans. *Journal of Pharmacology and Experimental Therapeutics, 266*, 1364–1373.

Feinauer, D. M. (1990). The relationship between workplace accident rates and drug and alcohol abuse: The unproven hypothesis. *Labor Studies Journal, 15*, 3–15.

Fillmore, M. T., & Vogel-Sprott, M. (1996). Social drinking history, behavioral tolerance and the expectation of alcohol. *Psychopharmacology, 127*, 359–364.

Fisher, C. A., Hoffman, K. J., Austin-Lane, J., & Kao, T.-C. (2000). The relationship between heavy alcohol use and work productivity loss in active duty military personnel: A secondary analysis of the 1995 Department of Defense worldwide survey. *Military Medicine, 165*, 355–361.

Folton, R. W., & Evans, S. M. (1993). Performance effects of drugs of abuse: A methodological survey. *Human Psychopharmacology, 8*, 9–19.

French, M. T., Zarkin, G. A., & Dunlap, L. J. (1998). Illicit drug use, absenteeism, and earnings at six U.S. worksites. *Contemporary Economic Policy, 16*, 334–346.

Friedman, A. S., Granick, S., Utada, A., & Tomko, L. A. (1992). Drug use/abuse and supermarket workers' job performance. *Employee Assistance Quarterly, 7*, 17–34.

Frone, M. R. (1998). Predictors of work injuries among employed adolescents. *Journal of Applied Psychology, 83*, 565–576.

Frone, M. R. (1999). Work stress and alcohol use. *Alcohol Research and Health, 23*, 284–291.

George, W. H., Raynor, J. O., & Nochajski, T. H. (1992). Resistance to alcohol impairment of visual-motor performance. *Journal of Studies on Alcohol, 53*, 507–513.

Gill, J. (1994). Alcohol problems in employment: Epidemiology and responses. *Alcohol and Alcoholism, 29*, 233–248.

Gottfredson, M., & Hirschi, T. (1990). *A general theory of crime*. Palo Alto, CA: Stanford University Press.

Greenberg, L., & Barling, J. (1999). Predicting employee aggression against coworkers, subordinates and supervisors: The roles of person behaviors and perceived workplace factors. *Journal of Organizational Behavior, 20*, 897–913.

Hahn, H. A., & Price, D. L. (1994). Assessment of the relative effects of alcohol on different types of job behavior. *Ergonomics, 37*, 435–448.

Harwood, H., Fountain, D., & Livermore, G. (1998). *The economic costs of alcohol and drug abuse in the United States, 1992* (NIH Publication No. 98-4327). Washington, DC: U.S. Government Printing Office.

Heishman, S. J. (1998). Effects of abused drugs on human performance: Laboratory assessment. In S. B. Karach (Ed.), *Drug abuse handbook* (pp. 206–235). New York: CRC Press.

Hingson, R., Lederman, R. I., & Walsh, D. C. (1985). Employee drinking patterns and accidental injury: A study of four New England states. *Journal of Studies on Alcohol, 46*, 298–303.

Hoffman, J. P., Larison, C., & Sanderson, A. (1997). *An analysis of worker drug use and workplace policies and programs*. (DHHS Publication No. SMA 97-3142). Washington, DC: U.S. Government Printing Office.

Hogan, J., & Hogan, R. (1989). How to measure employee reliability. *Journal of Applied Psychology, 74*, 273–279.

Holcom, M. L., Lehman, W. E. K., & Simpson, D. D. (1993). Employee accidents: Influences of personal characteristics, job characteristics, and substance use in jobs differing in accident potential. *Journal of Safety Research, 24*, 205–221.

Howland, J., Rohsenow, D. J., Cote, J., Gomez, B., Mangione, T. W., & Laramie, A. K. (2001). Effects of low-dose alcohol exposure on simulated merchant ship piloting by maritime cadets. *Accident Analysis and Prevention, 33*, 257–265.

Howland, J., Rohsenow, D. J., Cote, J., Siegel, M., & Mangione, T. W. (2000). Effects of low-dose alcohol exposure on simulated merchant ship handling power plant operations by maritime cadets. *Addiction, 95*, 719–726.

Jenkins, A. J., & Cone, E. J. (1998). Pharmacokinetics: Drug absorption, distribution, and elimination. In S. B. Karach (Ed.), *Drug abuse handbook* (pp. 151–201). New York: CRC Press.

Jessor, R., Donovan, J. E., & Costa, F. M. (1991). *Beyond adolescence: Problem behavior and young adult development*. New York: Cambridge University Press.

Jex, S. M. (1998). *Stress and job performance: Theory, research, and implications for managerial practice*. Thousand Oaks, CA: Sage.

Jones, A. W., & Pounder, D. J. (1998). Measuring blood-alcohol concentration for clinical and forensic purposes. In S. B. Karach (Ed.), *Drug abuse handbook* (pp. 327–356). New York: CRC Press.

Jones, S., Casswell, S., & Zhang, J.-F. (1995). The economic costs of alcohol-related absenteeism and reduced productivity among the working population of New Zealand. *Addiction, 90*, 1455–1461.

Kapur, B. M. (1993). Drug testing methods and clinical interpretations of test results. *Bulletin on Narcotics, 45*, 115–154.

Kerr, J. S., & Hindmarch, I. (1998). Effects of alcohol alone or in combination with other drugs on information processing, task performance, and subjective responses. *Human Psychopharmacology, 13*, 1–9.

Lehman, W. E. K., & Simpson, D. D. (1992). Employee substance use and on-the-job behaviors. *Journal of Applied Psychology, 77,* 308–321.

Leigh, J. P., Markowitz, S. B., Fahs, M., Shin, C., & Landrigan, P. J. (1997). Occupational injury and illness in the United States: Estimates of costs, morbidity, and mortality. *Archives of Internal Medicine, 157,* 1557–1568.

Macdonald, S. (1997). Work-place alcohol and drug testing: A review of the scientific evidence. *Drug and Alcohol Review, 16,* 251–259.

Macdonald, S., & Wells, S. (1994). The impact and effectiveness of drug testing programs in the workplace. In S. Macdonald & P. M. Roman (Eds.), *Research advances in alcohol and drug problems: Vol. 2. Drug testing in the workplace.* New York: Plenum.

Maisto, S. A., Galizio, M., & Connors, G. J. (1999). *Drug use and abuse* (3rd ed.). New York: Harcourt Brace.

Mangione, T. W., Howland, J., Amick, B., Cote, J., Lee, M., Bell, N., et al. (1999). Employee drinking practices and work performance. *Journal of Studies on Alcohol, 60,* 261–271.

Martin, C. S. (1998). Measuring acute alcohol impairment. In S. B. Karach (Ed.), *Drug abuse handbook* (pp. 309–326). New York: CRC Press.

Martin, J. K., Kraft, J. M., & Roman, P. M. (1994). Extent and impact of alcohol and drug use problems in the workplace: A review of empirical evidence. In S. Macdonald & P. M. Roman (Eds.), *Research advances in alcohol and drug problems: Vol. 2. Drug testing in the workplace.* New York: Plenum.

McFarlin, S. K., & Fals-Stewart, W. (2002). Workplace absenteeism and alcohol use: A sequential analysis. *Psychology of Addictive Behaviors, 16,* 17–21.

McFarlin, S. K., Fals-Stewart, W., Major, D. A., & Justice, E. M. (2001). Alcohol use and workplace aggression: An examination of perpetration and victimization. *Journal of Substance Abuse, 13,* 303–321.

Miller, N. S., Gold, M. S., & Smith, D. E. (Eds.). (1997). *Manual of therapeutics for addictions.* New York: Wiley-Liss.

Moll van Charante, A. W., & Mulder, P. G. H. (1990). Perceptual acuity and the risk of industrial accidents. *American Journal of Epidemiology, 131,* 652–663.

Moore, R. S. (1998). The hangover: An ambiguous concept in workplace alcohol policy. *Contemporary Drug Problems, 25,* 49–63.

Moore, S., Grunberg, L., & Greenberg, E. (2000). The relationships between alcohol problems and well-being, work attitudes, and performance: Are they monotonic? *Journal of Substance Abuse, 11,* 183–204.

National Institute on Alcohol Abuse and Alcoholism. (1995). *Alcohol and tolerance* (Alcohol Alert No. 28). Rockville, MD: Author.

Newcomb, M. D. (1994). Prevalence of alcohol and other drug use on the job: Cause for concern or irrational hysteria? *Journal of Drug Issues, 24,* 403–416.

Nochajski, T. H. (1993). Instructional set and visual-motor performance. In H. D. Utzelmann, G. Berghaus, & G. Kroj (Eds.), *Alcohol, drugs, and traffic safety—T92.* Cologne: Verlag TUV Rheinland.

Normand, J., Lempert, R. O., & O'Brien, C. P. (1994). *Under the influence? Drugs and the American work force*. Washington, DC: National Academy Press.

Normand, J., Salyards, S. D., & Mahoney, J. J. (1990). An evaluation of preemployment drug testing. *Journal of Applied Psychology, 75*, 629–639.

Northridge, M. E. (1995). Annotation: Public health methods—attributable risk as a link between causality and public health action. *American Journal of Public Health, 85*, 1202–1204.

Parish, D. C. (1989). Relation of the pre-employment drug testing results to employment status: A one year follow-up. *Journal of General Internal Medicine, 4*, 44–47.

Peter, R., & Siegrist, J. (1997). Chronic work stress, sickness absence, and hypertension in middle managers: General or specific sociological explanations? *Social Science and Medicine, 45*, 1111–1120.

Quest Diagnostics Incorporated. (2002). *The drug testing index*. Teterboro, NJ: Author.

Roman, P. M., & Blum, T. C. (1995). Employers. In R. H. Coombs & D. M. Ziedonis (Eds.), *Handbook on drug prevention*. Boston: Allyn & Bacon.

Rothman, K. J., & Greenland, S. (1998). *Modern epidemiology* (2nd ed.). New York: Lippincott, Williams, & Wilkins.

Saitz, R. (1998). Introduction to alcohol withdrawal. *Alcohol Health and Research World, 22*, 5–12.

Schwenk, C. R. (1998). Marijuana and job performance: Comparing the major streams of research. *Journal of Drug Issues, 28*, 941–970.

Sindelar, J. (1998). Social costs of alcohol. *Journal of Drug Issues, 28*, 763–780.

Stallones, L., & Kraus, J. F. (1993). The occurrence and epidemiologic features of alcohol-related occupational injuries. *Addiction, 88*, 945–951.

Streufert, S., Pogash, R., Miller, J., Gingrich, D., Landis, R., Lonardi, L., et al. (1995). Effects of caffeine deprivation on complex human functioning. *Psychopharmacology, 118*, 377–384.

Streufert, S., Pogash, R., Roache, J., Severs, W., Gingrich, D., Landis, R., et al. (1994). Alcohol and managerial performance. *Journal of Studies on Alcohol, 55*, 230–238.

Streufert, S., Satish, U., Pogash, R., Gingrich, D., Landis, R., Roache, J., et al. (1997). Excess coffee consumption in simulated complex work settings: Detriment or facilitation of performance? *Journal of Applied Psychology, 82*, 774–782.

Swift, R., & Davidson, D. (1998). Alcohol hangover: Mechanisms and mediators. *Alcohol Health and Research World, 22*, 73–80.

U.S. Department of Labor (2001). *Workplace injuries and illnesses in 2000* (USDL Publication No. 01-472). Washington, DC: Author.

Vasse, R. M., Nijhus, F. J. N., & Kok, G. (1998). Associations between work stress, alcohol consumption and sickness absence. *Addiction, 93*, 231–241.

Vogel-Sprott, M. (1992). *Alcohol tolerance and social drinking*. New York: Guilford.

Walker, D. J., Zacny, J. P., Galva, K. E., & Lichtor, J. L. (2001). Subjective, psychomotor, and physiological effects of cumulative doses of mixed-action opioids in healthy volunteers. *Psychopharmacology, 155*, 362–371.

Webb, G. R., Redman, S., Hennrikus, D. J., Kelman, G. R., Gibberd, R. W., & Sanson-Fisher, R. W. (1994). The relationships between high-risk and problem drinking and the occurrence of work injuries and related absences. *Journal of Studies on Alcohol, 55,* 434–446.

Wells, S., & Macdonald, S. (1999). The relationship between alcohol consumption patterns and car, work, sports and home accidents for different age groups. *Accident Analysis and Prevention, 31,* 663–665.

Wiese, J. G., Shlipak, M. G., & Browner, W. S. (2000). The alcohol hangover. *Annals of Internal Medicine, 132,* 897–902.

Wolff, K., Farrell, M., Marsden, J., Monteiro, M. G., Ali, R., Welch, S., et al. (1999). A review of biological indicators of illicit drug use, practical considerations and clinical usefulness. *Addiction, 94,* 1279–1298.

Zwerling, C. (1993). Current practice and experience in drug and alcohol testing. *Bulletin on Narcotics, 45,* 155–196.

III

MANAGING SAFETY AND THE RETURN TO WORK

8

THE ROLE OF LEADERSHIP IN SAFETY

DAVID A. HOFMANN AND FREDERICK P. MORGESON

Leadership has often been cited as playing a critical role in safety performance, particularly in the practitioner literature (Cooper, 2001; Geller, 2000; Grubbs, 1999). Although safety climate and culture research often reference leadership in an indirect way—for example, by tapping into employee perceptions of leaders' commitment to safety—there has been proportionally very little research investigating leadership per se. As such, the purpose of this chapter is threefold: (a) to provide a brief overview of the leadership-based safety research conducted to date; (b) to integrate this safety research into a general overview of selective leadership theories that we believe are relevant for safety performance; and (c) to discuss the implications and future research opportunities for the various leadership theories, particularly those that have been underresearched to date.

SAFETY-ORIENTED LEADERSHIP RESEARCH

We should mention at the outset that, with a few exceptions, we do not review safety climate research in this chapter even though much of this research highlights the importance of management's commitment to safety.

Chapter 2 of this volume deals more directly with safety climate research. Instead, we focus on the relatively small body of research investigating leadership and the role it plays in safety performance.

The vast majority of the research linking leadership to safety has adopted a behavioral perspective. In other words, the focus is on actions leaders take that impact safety performance. The types of leader actions investigated have ranged from operant feedback studies investigating the effects of positive feedback to the generalized style of leadership adopted. In addition, this research has investigated both safety-specific leadership behavior and more general leader actions.

Safety-Specific Leader Behaviors

Although there are a number of studies that document the effectiveness of providing feedback and incentives on safety performance and accidents (e.g., Komaki, Barwick, & Scott, 1978; Nasanen & Saari, 1987; Saari & Nasanen, 1989), the vast majority of this research has provided feedback and reinforcement directly to the groups themselves. In other words, they did not rely on the leader to deliver the feedback or incentives. Zohar (2001), however, focused his investigation directly on supervisor actions and their impact on employee safety. Adopting a transactional leadership perspective, Zohar (2001) developed and implemented a training program designed to increase leaders' safety reward and monitoring behavior. Overall, the intervention was effective, with the employees of the trained leaders reporting higher perceptions of safety climate, improved earplug use, and fewer accidents. In sum, and not surprisingly, the results suggested that when supervisors actively monitor and reward safety performance their subordinates engage in more safety-related behavior and experience safer overall performance than when they are not actively monitored.

Other researchers who have investigated the influence of supervisor safety-related behavior and safety outcomes have used either self-report or employee perceptions of supervisor behavior instead of observational data. Simard and Marchand (1994) found that supervisors' self-reported involvement was significantly higher in safe versus unsafe plants—although this effect was found not to be significant after the researchers controlled for the development of safety programs. Even though the authors did not fully investigate the possibility of a mediated relationship, the pattern of results does suggest that supervisor participation in safety activities may very well lead to the development of safety programs, which, in turn, is related to fewer accidents.

Similarly, both Zohar (2000a) and Tomás, Meliá, and Oliver (1999) investigated employee perceptions of supervisor safety-related behavior. In both cases, supervisor safety-related behavior was significantly related to safety outcomes. For example, Zohar (2000a) investigated employee perceptions of supervisor reactions to subordinate safety-related conduct as well as supervisors' relative emphasis on safety versus production. He found that supervisor behavior—as perceived by the employees—was significantly related to subsequent accidents experienced by the work group. Tomás et al. investigated employee perceptions of supervisors' safety response (i.e., supervisor attitude, positive–negative contingencies associated with safety, and supervisor safety performance). They found that supervisors' attitudes toward safety played a critical role in an explanatory chain linking safety climate to coworkers' responses as well as safety-related outcomes (i.e., attitudes, behavior, risk, and accidents).

Along these lines of focusing on leader safety-specific behavior, Barling, Loughlin, and Kelloway (2002) investigated the influence of safety-specific transformational leadership on safety climate, safety events, and occupational injuries. The final structural model revealed that safety-specific transformational leadership and role overload (Hofmann & Stetzer, 1996) influenced occupational safety through their relationships with perceived safety climate, safety consciousness, and safety-related events.

The studies investigating supervisor safety-related behavior—through either self-reports, observational data, or employee responses—all indicate that when supervisors emphasize, discuss, reward, and encourage safe performance, safe performance ensues. Although these results are as expected, it is important that empirical research has confirmed this relationship. In addition to these findings, the results of Zohar's (2001) leadership intervention suggest that organizations can implement training sessions to increase the safety-related behavior of supervisors and, as a result, experience safer performance within their work group. This type of leadership intervention is quite advantageous from an organizational perspective. First, it is less costly because the intervention is done at the leader level instead of the employee level. Second, given the relationship between supervisor behavior and safety climate within work groups (Zohar, 2000a), these interventions serve to change the climate within the group and, as a result, may be more effective in the long run than individual interventions that attempt to change individual behavior without addressing the context (i.e., supervisor attitudes and behavior) within which these behaviors occur.

General Leader Behavior

In contrast to research focusing on safety-specific leader behavior, a number of studies have investigated the relationship between more general leader behavior and safety outcomes. On the operant conditioning front, Mattila, Hyttinen, and Rantanen (1994) investigated the behavior of leaders using the Operant Supervisory Taxonomy and Index (OSTI; Komaki, Zlotnick, & Jensen, 1986). The results suggested that supervisory behaviors that predicted effective financial performance (i.e., completing a construction project on budget) overlapped significantly with the behaviors that resulted in effective safety performance. In general, those supervisors who spent more time on the building site—and more frequently gave feedback, monitored performance, communicated about non-work-related issues, and engaged in a participatory style of leadership—had more safely performing units. Many of these same behaviors were associated with effective financial performance as well.

Moving away from operant conditioning toward a focus on the relationship between leaders and subordinates, Simard and Marchand (1997) investigated both micro- and macro-organizational influences on compliance with safety rules. The results suggested that micro factors were more predictive of compliance. In particular, cooperative supervisor–employee relationships and supervisory involvement in decisions regarding safety issues were significant predictors of safety compliance (as rated by the supervisor). Of all the factors investigated, cooperative supervisor–employee relationships were the best predictor of safety compliance.

Hofmann and Morgeson (1999) have also investigated the notion of cooperative relationships from the theoretical vantage point of social exchange theory (Blau, 1964; Gouldner, 1960). Drawing on research suggesting that high-quality leader–member exchanges (i.e., LMX) between leaders and subordinates result in more open communication and increased value congruence, the authors hypothesized that these relationships occurring in high-risk environments would result in increased upward communication of safety issues, increased safety commitment, and ultimately fewer accidents. The authors found support for these relationships.

Also focusing on relationships between supervisors and employees, Thompson, Hilton, and Witt (1998) investigated supervisor fairness, which they operationalized using interactional justice items (which are similar in content to LMX items). They found that supervisor fairness was significantly related to safety compliance and that this relationship was mediated by supervisor support for safety. In other words, supervisor fairness was related to support for safety, which in turn was associated with safety compliance.

These investigations focusing on general leader behavior and styles raise a question regarding the independence of leadership behavior or style and leader emphasis on safety. As noted by Zohar (2000b), leadership and safety orientation are likely to be independent constructs. In other words, one leader could have high-quality LMX and strongly emphasize safety, whereas another leader may have similarly high LMX and emphasize safety to a lesser degree. The independence of these constructs suggests the possibility of an interaction. In fact, several recent investigations have focused on the possibility of this interaction.

Zohar (2000b), for example, found support for an interaction between leadership (i.e., transactional or transformational) and the leaders' safety priority in the prediction of safety climate. As part of a larger study, Zohar (2000b) hypothesized that transactional leadership would be positively related to safety only when the supervisor values safety, whereas the relationship would be negative when the leader does not value safety (i.e., the leader provides reward contingencies for other behaviors). He did find support for this interaction.

Hofmann, Morgeson, and Gerras (2003), extending their previous work on the relationship between LMX and safety outcomes (Hofmann & Morgeson, 1999), investigated safety climate as a moderator of the relationship between LMX and subordinate safety-related role definitions and behavior. Of particular interest was when employees defined citizenship-type safety behaviors (i.e., extra-role behaviors focusing on safety) as part of their expected role. Again adopting a social exchange theoretical foundation and drawing on previous LMX research, the researchers hypothesized that employees would reciprocate high-quality LMX relationships with safety-oriented citizenship behaviors only when the leaders' behavior engendered a climate within the group that highly valued safety. The authors found support for the interaction. Under positive safety climates, there was a significant positive relationship between LMX and subordinate role definitions; that is, employees were more likely to view safety citizenship behaviors as in-role when a positive safety climate and high-quality LMX relationships existed. Alternatively, under poor safety climates, the relationship between LMX and safety role definitions was nonsignificant. In other words, when the surrounding context did not signal a strong commitment to safety, employees did not reciprocate high-quality LMX relationships by expanding their safety-related role definitions, nor did they engage in these behaviors more frequently.

Conclusion and Discussion of Safety-Oriented Leadership Research

Although the summary in the preceding paragraphs provides a general overview of the various leadership dimensions that have been investigated, Table 8.1 provides a more in-depth summary of these investigations. Review of this table allows for several conclusions regarding the current state of

TABLE 8.1
Summary of Leadership and Safety Research

Authors	Leadership dimensions	Level of analysis	Research design	Results
Barling et al. (2002)	Safety-oriented transformational leadership	Individual level of analysis	Investigated a structural model linking leadership to safety consciousness, safety climate, safety-related events, and occupational injuries.	Safety-oriented transformational leadership significantly influenced perceived safety climate and safety consciousness. Perceived safety climate predicted both safety-related events and occupational injuries.
Hofmann & Morgeson (1999)	Leader–member exchange	Individual level of analysis	Investigated bivariate correlations and structural model linking LMX to safety communication, safety commitment, and subsequent accidents.	LMX was significantly related to safety communication, safety commitment, and subsequent accidents. Relationship between LMX and accidents was mediated by safety communication and commitment.
Hofmann, Morgeson, & Gerras (2003)	Leader–member exchange	Multilevel analysis—both individual and group	Investigated the moderating effect of safety climate on the relationship between LMX and safety-related role definitions. Also investigated the relationship between safety-related role definitions and safety behavior.	Found the safety climate significantly moderated the relationship between LMX and safety-related role definitions. Safety-related role definitions were significantly related to safety behavior.

Mattila et al. (1994)	Operant Supervisory Taxonomy and Index: • Consequences • Monitoring • Antecedents • Referring to his/her own performance • Work-related communication • Non-work-related communication • Solitary	Site-level analysis with multiple observations within each site	Categorized construction projects into those that were financially successful (on budget) versus those that were not. Investigated differences in supervisory behavior. Similar to financial performance, categorized safe versus less safe projects, and investigated differences in supervisory behavior.	Supervisors of safety projects spent more time at work site, more frequently gave feedback, spent more time monitoring performance, spent less time discussing antecedents, used more time for communication on non-work-related issues, more frequently used neutral or positive communication, gave incentives more frequently, and made use of a more participatory style of leadership. Many of the same supervisory behaviors were liked to effective financial performance.
Simard & Marchand (1994)	Supervisor behavior toward safety: • Frequency of personal involvement in safety • Frequency of employee participation in safety Senior management commitment: • Involvement in safety • Attitudes toward safety	Plant level	Categorized plants as effective (safe) or not effective (less safe). Investigated differences across plants on leadership and other factors.	Supervisor involvement in safety was related to effectiveness but not after controlling for the development of safety programs. Senior management commitment not significant.

(continues)

TABLE 8.1
Summary of Leadership and Safety Research (Continued)

Authors	Leadership dimensions	Level of analysis	Research design	Results
Simard & Marchand (1997)	Workgroup-level leadership variables: • Cooperative relationships among workgroup and with supervisor • Supervisor participative management of safety issues • Plant level leadership variables: • Structural safety commitment (develop of safety programs and joint regulatory committees) • Safety leadership (regular involvement of upper management in safety issues)	Workgroup- and plant-level analysis using Hierarchical Linear Modeling (HLM; Bryk & Raudenbush, 1992)	Multilevel analysis investigated group and organizational predictors of compliance with safety rules (supervisors' ratings of their employees' compliance behavior)	With respect to the leadership variables, the results revealed that cooperative relationships between supervisors and work group, and supervisory participative management of safety were significant predictors of compliance. Overall conclusion was that employee–supervisor cooperative relationships were the best predictors of compliance behavior.
Thompson, Hilton, & Witt (1998)	Supervisor fairness (i.e., interactional justice), management support for safety, and supervisor support for safety	Individual level of analysis	As part of a larger model, the authors investigated the influence of supervisor fairness (i.e., interactional justice) on support for safety and safety conditions/compliance.	Supervisor fairness was significantly related to safety compliance, and this relationship was mediated by supervisor support for safety. Manager support for safety was shown to be related to safety conditions and safety compliance. The relationship between manager support for safety and safety compliance was mediated by supervisor support for safety.

Tomás, Meliá, & Oliver (1999)	Supervisors' safety response (i.e., supervisors' attitudes towards safety, positive or negative contingencies provided by supervisor, and supervisors' safety performance)	Individual level of analysis	Investigated structural model including safety climate, supervisors' response, coworkers' response, worker attitude, safety behavior, actual risk, and accidents.	Found evidence for supervisors' response playing a critical role in an explanatory chain linking safety climate to coworkers' response as well as safety-related outcomes (i.e., attitudes, behavior, risk, and accidents).
Williams, Turner, & Parker (2000)	Assessed transformational leadership (Bass & Avolio, 1995)	Individual level of analysis	Investigated the relationship between transformation leadership and safety commitment, self-managing orientation, safety compliance, and safety proactivity.	Transformation leadership was positively related to safety compliance and safety proactivity, but was not significantly related to either safety commitment or self-managing orientation. Some evidence for safety commitment as a mediator of the relationship between transformation leadership and safety compliance. Evidence of an interaction between transformational leadership and employee safety commitment predicting safety compliance.
Zohar (2000a)	Operationalized safety climate as leader safety behavior (2 factors): Overt supervisory reaction to subordinates conduct Supervisory expectations regarding relative importance of safety vs. production	Workgroup level of analysis	Aggregated supervisory behaviors to form climate score for each group. Investigated relationship between climate and microaccidents 5 months later.	Safety climate (supervisor behavior) significantly predicted microaccidents 5 months later.

(continues)

TABLE 8.1
Summary of Leadership and Safety Research (Continued)

Authors	Leadership dimensions	Level of analysis	Research design	Results
Zohar (2000b)	Measured transactional and transformational leadership (Bass & Avolio, 1997)	Workgroup level of analysis	Investigated the interaction between leadership and leaders' safety priority in the prediction of safety climate and microaccidents.	Found evidence for an interaction between leadership and leaders' safety priority in the prediction of employee perceptions of safety climate. In addition, a structural model indicated that leadership influences climate, which in turn predicts microaccidents.
Zohar (2001)	Focused on safety-related transactional leadership (i.e., safety-related rewarding and monitoring interactions)	Individual level of analysis with accidents measured at subunit level	Developed leadership-based intervention designed to increase safety-related rewarding and monitoring. Investigated effects of intervention on safety climate, earplug use, and accidents.	Leadership-based intervention was effective. Experimental group had significantly higher perceptions of safety climate, improved earplug use, and fewer accidents.

affairs of safety-related leadership research. With respect to the independent variable (i.e., leadership), there are essentially two perspectives represented in the literature: those that have investigated safety-specific leader behavior (e.g., Barling et al., 2002; Simard & Marchand, 1994; Tomás et al., 1999; Zohar, 2000a, 2001) and those that have investigated the relationship between general leadership constructs and safety-related outcomes (e.g., Hofmann & Morgeson, 1999; Hofmann et al., 2003; Thompson et al., 1998; Williams, Turner, & Parker, 2000; Zohar, 2000a). The investigations that have focused on safety-specific leader behavior typically have used as their theoretical foundation an operant conditioning model or an implied organizational climate perspective. Both of these perspectives suggest that leader behavior establishes a context in which certain behavior is valued, rewarded, and expected. Not surprisingly, in contexts in which safety is valued, employees are more likely to engage in safe behavior.

Alternatively, investigations linking more general leadership constructs to safety outcomes are typically grounded in current leadership paradigms and their respective theoretical foundations. But linking general leadership constructs to safety outcomes raises a number of questions about the relationship between a leader's general leadership behavior or style and the emphasis the leader places on safety. Although the research suggests that there is likely to be a significant relationship between a leadership style that places a high value on employee well-being and the leader's concern for safety, recent research suggests that this relationship may be interactive as well (e.g., Hofmann et al., 2003; Zohar, 2000b).

Viewed together, the studies involving facet-specific (i.e., safety-related) and facet-free leadership constructs seem complementary. In other words, it seems as though there is some mounting evidence suggesting that general leadership behaviors or styles that emphasize employee well-being may be necessary, but not sufficient, for developing employee commitment to safety and safety behavior. The leader must also create a context that signals a high value for safety. One way the leader can do this is by engaging in the safety-specific behaviors being investigated under the purview of facet-specific leadership research (i.e., Simard & Marchand, 1994; Tomás et al., 1999; Zohar, 2000a, 2001). In fact, Hofmann et al. (2003) and Zohar (2000b) have investigated the interrelationships among facet-free leadership constructs; safety-specific leader behavior (e.g., safety climate); and safety-related outcomes. This integrative approach seems like a fruitful avenue for future research.

Table 8.1 also reveals a number of different dependent variables and levels of analysis that have been investigated in this research. Several researchers (e.g., Mattila et al., 1994; Simard & Marchand, 1994) have categorized plants as safe versus unsafe and have investigated differences in leadership at the plant or work-site level. Others (Zohar, 2000a, 2000b) have investigated accident rates at the group level of analysis, and some (Hofmann et al., 2003; Simard & Marchand, 1997) have investigated the effects of independent variables at two levels of analysis on individual-level outcomes. Finally, there have been a number of individual-level investigations of employee safety outcomes (Hofmann & Morgeson, 1999; Thompson et al., 1998; Tomás et al., 1999; Williams et al., 2000; Zohar, 2001). With respect to dependent variables and level of analysis, there is no right or wrong answer. But clearly, the level of analysis needs to be congruent with the level of the theory (Klein, Dansereau, & Hall, 1994; Morgeson & Hofmann, 1999). We encourage scholars to justify their choice of levels and link it to their theoretical model.

Finally, we encourage researchers to continue the recent trend of investigating the structural relationship among these variables as well as potential moderators, yet we feel that research should move away from situations where all the variables are measured from the same source in order to avoid potential problems with common method variance.

LEADERSHIP THEORIES

It is important, when either reviewing or investigating the relationship between leadership and safety, to move beyond the safety-specific literature to consider the broader leadership literature. This is important because it may yield additional insight into how leadership can impact safety. Although several of the studies reviewed in the preceding section focused on leadership, they did not cite and seemed to be relatively uninfluenced by the broader research into leadership. In order to understand the influence of leadership on safety—or any other organizational outcome of interest for that matter—it is necessary to look at the various approaches to leadership and consider the theoretical linkages to the outcome of interest. In the following section, we highlight several existing leadership theories. Where possible, we integrate the safety research that has been conducted. Where no research has been conducted, we provide some thoughts about how the respective theory might influence safety outcomes.

Leadership has been defined in many different ways. Yukl and Van Fleet (1992) offered a comprehensive account: "Leadership is viewed as a process that includes influencing the task objectives and strategies of a group or organization, influencing people in the organization to implement the strategies and achieve the objectives, influencing group maintenance and identification, and influencing the culture of the organization" (p. 149). Leaders enact many different strategies to effectively influence others in their environment, be they subordinates, peers, or superiors. Likewise, scholars have studied organizational leadership from numerous perspectives. For our purposes, the leadership literature can be divided into three major approaches: (a) behavioral, (b) situational, and (c) power and influence.

Behavioral Approaches

Behavioral approaches to the study of leadership focus on the range of behaviors leaders exhibit. Researchers at Ohio State University conducted early research into leader behaviors. They developed extensive lists of leader behaviors, collected questionnaire data on the performance of those behaviors, and subjected the data to factor analysis. This research revealed two primary factors: consideration and initiating structure (Fleishman, 1953). Consideration involved people-oriented behaviors, whereas initiating structure involved task-oriented behaviors.

Although important historically, this two-factor conceptualization has been criticized for being overly general. Consequently, various more complex behavioral taxonomies have been proposed. For example, Mintzberg (1973) intensively studied a small number of senior managers, concluding that their jobs could be described in terms of 10 discrete roles, and Yukl (1989) proposed an integrating taxonomy of leader behavior consisting of 14 generic categories: planning and organizing, problem solving, clarifying, informing, monitoring, motivating, consulting, recognizing, supporting, managing conflict and team building, networking, delegating, developing and mentoring, and rewarding.

Two additional lines of research that fall under the behavioral category deserve mention. First, the increased use of teams in organizations has called into question the need for leadership. As a consequence, scholars have begun to investigate what leaders actually do in these more participative situations. Manz and Sims (1987) found that a key leader activity in these settings is supporting the team's self-management, thus facilitating the team's ability to manage itself. Second, Quinn (1988) outlined a set of managerial roles and suggested that a manager's job is inherently paradoxical and that reconciling the paradoxes would lead to better managerial performance. Denison, Hooijberg, and Quinn (1995) empirically tested this proposition and found that executives who evidenced greater behavioral complexity (i.e., engaged in a greater variety of leadership roles) were more effective than executives who engaged in a more limited variety of leadership roles.

Implications of Behavioral Approaches for Safety

With respect to safety, the majority of the research has focused on safety-specific actions. Evidence has clearly shown that when leaders monitor and reward safe performance, employees perform more safely. Evidence for this effect has been found with both safety-specific behaviors (Zohar, 2001) and more general monitoring, rewarding, and communicating behaviors (Mattila et al., 1994). These relationships, particularly with the general leader behavior, however, would be considered moderate only to small effect sizes. This seems to open the door for potential moderators.

One explanation for the relationship between general leader behavior and safety outcomes is that this research has been conducted in environments where safety is highly valued. Thus, one might assume that, on average, leaders to at least some degree emphasize safety. But we know from the climate literature (e.g., Hofmann & Stetzer, 1996, 1998; Hofmann et al., 2003; Zohar, 2000a) that even leaders within the same organization working under similar conditions can significantly vary in their emphasis on safety. If leaders vary in their emphasis of safety, then it seems that the relationship between generalized leader behavior and safety outcomes would be moderated by safety commitment or climate (e.g., Hofmann et al., 2003; Zohar, 2000b). In other words, leaders who more frequently monitor and reward their subordinates *and* are committed to safety (and create climates that emphasize safety) will likely monitor and reward safety behavior. It is certainly possible, however, that some leaders who frequently monitor and reward performance do not particularly care about safety. In this case, we would not expect to find a significant relationship between leader behavior and safety outcomes.

This notion of the moderation of the relationship between leader behavior and safety outcomes by leader values or organizational climate is related indirectly to Quinn's (1988) emphasis on the inherent paradoxical nature of leadership. Specifically, leaders are confronted with a number of often competing goals and objectives. The key to effective leadership is to manage these different goals and roles and do so such that the appropriate goals and roles are matched to the appropriate situation. Some leaders will view safety and production or efficiency as trade-offs and choose to emphasize (i.e., establish a climate the emphasizes) production to a greater extent than safety. We would, as a result, expect their behaviors—such as rewarding, monitoring, initiation of structure, and motivating—to be focused on production. Other managers, perhaps those who have been better able to master the inherent paradoxes of leadership, will be able to emphasize both safety and production and, as a result, will engender both effective and safe performance (see Mattila et al., 1994). This suggests that leadership development focused on managing seemingly contradictory goals may be worthwhile.

Situational Approaches

Situational approaches to the study of leadership suggest that only under certain circumstances will leadership be effective. Although situational research peaked in the 1960s and 1970s, the implications of a situational approach might hold some relevance for the role leaders play in safety at work.

Normative decision theory (Vroom & Yetton, 1973) offered a process for deciding what type of decision procedures are most likely to result in effective decisions. The five types of decision procedures range from completely autocratic to a group decision. Seven situational variables determine how a particular decision procedure will affect decision outcomes: (a) whether decision quality is important, (b) whether the decision problem is structured, (c) whether the leader already has sufficient information to make a good decision, (d) whether subordinate acceptance is important for effective implementation, (e) whether subordinate acceptance is likely with an autocratic decision, (f) whether subordinates share the organizational objectives sought by the leader, and (g) whether conflict exists among the subordinates.

Building off valence-instrumentality-expectancy (VIE) theory (i.e., expectancy theory; Vroom, 1964), the path–goal theory of leadership (House, 1971) suggested that leadership is centered on showing subordinates how they can achieve valued goals. In particular, leaders can motivate higher performance by behaving in ways that influence subordinates to believe that valued outcomes can be achieved if they exert effort toward those outcomes. House suggested that different leadership styles (supportive, directive, participative, and achievement-oriented) are differentially effective depending on subordinate, task, and organizational characteristics.

Finally, substitutes for leadership theory (Kerr & Jermier, 1978) have indicated that certain subordinate, task, and organization characteristics reduce the importance of leaders. The four subordinate characteristics were (a) ability, experience, training, and knowledge; (b) need for independence; (c) "professional" orientation; and (d) indifference toward organizational rewards. The four task characteristics were (a) unambiguous and routine; (b) methodologically invariant; (c) provides its own feedback concerning accomplishment; and (d) intrinsically satisfying. The six organizational characteristics were (a) formalization (explicit plans, goals, and areas of responsibility); (b) inflexibility (rigid, unbending rules and procedures); (c) highly specified and active advisory and staff functions; (d) close-knit, cohesive work groups; (e) organizational rewards not within the leader's control; and (f) spatial distance between superior and subordinates.

Implications of Situational Approaches for Safety

The situational approaches to leadership raise a number of potentially interesting questions when considered from a safety vantage point. For example, the literature on substitutes for leadership identifies a number of potentially interesting constraints on the opportunity for leaders to influence subordinate safety. It may be the case that when employees have a great deal of experience and training, the task context is very routine, a large number of safety policies and procedures have been developed (i.e., high formalization), and organizational rewards are linked to safety, then leadership will have less of an impact than it would otherwise have. In other words, when individuals know what to do, know how to do it, know when to do it, and are rewarded to do it, the importance of leadership may be decreased.

The trend toward high-performance work practices (e.g., Arthur, 1992; Huselid, 1995) and more flexible manufacturing systems (Weick, 1990), however, renders obsolete many of the conditions noted in the preceding paragraph. These types of work systems typically involve many more technical exceptions that require real-time decision making in which the decisions have both safety and performance implications. When this is the case, the organization wants to establish a context in which both safety and performance concerns factor into the decision premises used by employees. In other words, the organization wants to create a culture that encourages employees—as they go about making decisions—to consider both safety and performance implications of different potential actions. It is only through considering both implications that employees will be able to identify the course of action that jointly optimizes both outcomes.

This raises the question of how organizations go about creating such a context. This is where, we would argue, that leadership becomes critical. The path–goal theory of leadership offers some initial guidance. Specifically, the leader needs to convey expectations regarding safe performance and provide appropriate rewards and recognition. Coupling this with the behavioral perspective described earlier in this section suggests that after establishing these expectations leaders need to effectively monitor and reward decisions and actions that take into consideration both safety and performance. In other words, leaders need to establish an organizational climate that stresses the importance of safety and puts in place rewards and accountabilities that further encourage the consideration of the safety implications of difference actions (Frink & Klimoski, 1998).

We believe that the implications of normative decision theory (Vroom & Yetton, 1973) are most significant for the identification and implementation of safety programs. In general, this theory suggests that the leader should engage in different decision-making processes depending on the characteristics of the situation. Seemingly all too often in the context of safety-related organizational learning, a culture of blame develops in which the organization attempts to find a convenient scapegoat (Hofmann & Stetzer, 1998; Reason, 1994). This is a maladaptive response for the organization because it short-circuits organizational learning and improvement (Hofmann & Stetzer, 1998; Reason, 1994). A better approach is a problem-oriented one whereby the organization attempts to identify the true underlying causes and implement potential remedies. This, it seems, returns us to normative decision theory with respect to how the organization should go about diagnosing the root causes and developing remedies. Different situations may require different approaches. When employees have a great deal of information and when the acceptance of that information is critical to effective implementation (e.g., changes in work practices), then normative decision theory would recommend a more participative process. Other types of changes (e.g., government-required procedures) may be effectively implemented with a less participative approach. At any rate, thinking more critically about the process used to diagnose root causes and initiate safety-related organizational learning might benefit from a consideration of both the theory and research findings underlying normative decision theory (see Vroom & Jago, 1988).

Power and Influence Approaches

Power and influence approaches to the study of leadership focus on how leaders are able to influence their subordinates. This ability to influence includes the sources of power leaders can tap into when influencing subordinates, the different influence tactics that can be used, and the relationship leaders have with their subordinates. Research in this area is most directly related to the core of the definition of *leadership* offered above, namely, influencing others in an organizational system.

The classic work of French and Raven (1959), who identified several different types of leadership power, represents the beginning of research on the sources of a leader's power. Yukl and Falbe (1990) extended this work by suggesting that there are two primary sources of power, with several more specific types of power for each primary source. *Position power* is given to a leader by the organization and consists of legitimate, reward, coercive, and informational sources. *Personal power* is given to a leader by his or her subordinates and includes expert, referent, persuasive, and charismatic sources.

Just having certain sources of power, however, does not mean that the leader will use them. As a consequence, research has been conducted on the types of influence tactics leaders use (Kipnis & Schmidt, 1988; Kipnis, Schmidt, & Wilkinson, 1980). Yukl and Van Fleet (1992) summarized the range of influence tactics that have been studied. These include legitimating tactics, rational persuasion, inspirational appeals, consultation, exchange, pressure, ingratiation, personal appeals, coalition tactics, and upward appeals.

Other researchers have focused on the reciprocal influence leaders and subordinates can have on each other. Foremost among these approaches is the leader–member exchange (LMX) approach developed by Dansereau, Graen, and Haga (1975). The hundreds of studies of this approach have been meta-analytically summarized by Gerstner and Day (1997). At its most basic level, LMX suggests that leaders develop different quality exchange relationships with individual subordinates. There is considerable evidence that the quality of exchange relationship is related to a host of affective and behavioral outcomes.

The final approach within power and influence is transformational leadership. Although transformational leadership could be placed under the behavioral approach category, we have placed it here because transformational leadership concerns itself with "influencing major changes in attitudes and assumptions of organizational members . . . and building commitment for major changes in the organization's objectives and strategies" (Yukl & Van Fleet, 1992, p. 174). Leaders empower and elevate their subordinates by espousing a compelling vision and convincing them to transcend their self-interest. These subordinates then become change agents and leaders in their own right. Bass (1985) has suggested that transformational leadership contains three components: charisma, intellectual stimulation, and individualized consideration.

Implications of Power and Influence Approaches for Safety

There has been no research into how a leader's sources of power or use of influence tactics impacts safety behavior. It is likely that personal sources of power will better enable leaders to influence the safety behavior of subordinates. But, as noted earlier in this chapter, this influence will impact safety behavior only if safety is a core value of the leader.

The work of Hofmann and Morgeson (1999; also Hofmann et al., 2003) has shown the influence relationships between leaders and subordinates (operationalized with LMX) can have on such things as safety communication, commitment, role definitions, and accidents. Future work can investigate how leader–member relationships impact the adoption of other kinds of role behavior and subordinate values.

With few exceptions (Barling et al., 2002; Zohar, 2000b), researchers have left the area of transformational leadership untapped. It may be the case that if leaders can articulate a compelling vision of safety and persuade subordinates to transcend their self-interest, this will motivate greater safety behavior, increase helping behavior among subordinates, and serve as a substitute for leadership. One challenge in conducting research into transformational leadership, however, is the need for samples where there is variance in transformational leadership. This may require research that spans multiple organizations.

CONCLUSIONS

Although much has been done with respect to understanding the link between leadership and safety, much more needs to be done. In addition to reviewing existing safety-related research, we have highlighted some relatively untapped areas of leadership theory that hold promise for increasing our understanding of how leaders can improve safety in organizations.

Looking across the findings of the relatively small body of leadership-based safety research, we are encouraged. Overall, the findings suggest that perhaps the age-old assumption that safety and production are inherent trade-offs is not as certain as once believed. Although Mattila et al. (1994) bring some direct evidence to the table, there is much indirect evidence as well. Specifically, research has linked LMX and transformational leadership to safety outcomes. These two leadership constructs are also associated with a number of other positive outcomes in organizations such as increased organizational commitment, productivity via the performance of both in-role and citizenship behaviors, and reduced absenteeism and turnover (e.g., see Gerstner & Day, 1997). Perhaps it is the case that it is possible for leaders to develop organizational cultures that are healthy in the sense that the organization is effective and safe. We believe that for safety research to continue to make significant contributions to mainstream organizational research, establishing this link between safety performance and organizational performance is critical. We hope that future research will further explore these relationships.

REFERENCES

Arthur, J. B. (1992). The link between business strategy and industrial relations systems in American steel mini-mills. *Industrial and Labor Relations Review, 45*, 488–507.

Barling, J., Loughlin, C., & Kelloway, E. K. (2002). Development and test of a model linking safety-specific transformational leadership and occupational safety. *Journal of Applied Psychology, 87*, 488–496.

Bass, B. M. (1985). *Leadership and performance beyond expectations.* New York: Free Press.

Bass, B. M., & Avolio, B. J. (1995). *MLQ multifactor leadership questionnaire.* Palo Alto, CA: Mind Garden.

Blau, P. (1964). *Exchange and power in social life.* New York: Wiley.

Bryk, A. S., & Raudenbush, S. W. (1992). *Hierarchical linear models: Application and data analysis methods.* Newbury Park, CA: Sage.

Cooper, D. (2001). Treating safety as a value. *Professional Safety, 46*, 17–21.

Dansereau, F., Graen, G., & Haga, W. J. (1975). A vertical dyad linkage approach to leadership within formal organizations: A longitudinal investigation of the role making process. *Organizational Behavior and Human Performance, 13*, 46–78.

Denison, D. R., Hooijberg, R., & Quinn, R. E. (1995). Paradox and performance: Toward a theory of behavioral complexity in managerial leadership. *Organization Science*, 6, 524–540.

Fleishman, E. A. (1953). The description of supervisory behavior. *Personnel Psychology, 37*, 1–6.

French, J., & Raven, B. H. (1959). The bases of social power. In D. Cartwright (Ed.), *Studies of social power.* Ann Arbor, MI: Institute for Social Research.

Frink, D. D., & Klimoski, R. J. (1998). Toward a theory of accountability in organizations and human resource management. In G. R. Ferris (Ed.), *Research in personnel and human resource management* (Vol. 16, pp. 1–51). Stamford, CT: JAI Press.

Geller, E. S. (2000). 10 leadership qualities for a total safety culture. *Professional Safety, 45*, 38–41.

Gerstner, C. R., & Day, D. V. (1997). Meta-analytic review of leader–member exchange theory: Correlates and construct issues. *Journal of Applied Psychology, 82*, 827–844.

Gouldner, A. W. (1960). The norm of reciprocity. *American Sociological Review, 25*, 165–167.

Grubbs, J. R. (1999). A transformational leader. *Occupational Health and Safety, 68*, 22–26.

Hofmann, D. A., & Morgeson, F. P. (1999). Safety-related behavior as a social exchange: The role of perceived organizational support and leader–member exchange. *Journal of Applied Psychology, 84*, 286–296.

Hofmann, D. A., Morgeson, F. P., & Gerras, S. J. (2003). Climate as a moderator of the relationship between LMX and content specific citizenship behavior: Safety climate as an exemplar. *Journal of Applied Psychology, 88*, 170–178.

Hofmann, D. A., & Stetzer, A. (1996). A cross-level investigation of factors influencing unsafe behaviors and accidents. *Personnel Psychology, 49*, 307–339.

Hofmann, D. A., & Stetzer, A. (1998). The role of safety climate and communication in accident interpretation: Implications for learning from negative events. *Academy of Management Journal, 41*, 644–657.

House, R. J. (1971). A path–goal theory of leader effectiveness. *Administrative Science Quarterly, 16*, 321–339.

Huselid, M. A. (1995). The impact of human resource management practices on turnover, productivity, and corporate financial performance. *Academy of Management Journal, 38*, 635–672.

Kerr, S., & Jermier, J. M. (1978). Substitutes for leadership: Their meaning and measurement. *Organizational Behavior and Human Performance, 22*, 375–403.

Kipnis, D., & Schmidt, S. M. (1988). Upward-influence styles: Relationship with performance evaluations, salary, and stress. *Administrative Science Quarterly, 33*, 528–542.

Kipnis, D., Schmidt, S. M., & Wilkinson, I. (1980). Intra-organizational influence tactics: Explorations in getting one's way. *Journal of Applied Psychology, 65*, 440–452.

Klein, K. J., Dansereau, F., & Hall, R. J. (1994). Levels issues in theory development, data collection, and analysis. *Academy of Management Review, 19*, 105–229.

Komaki, J., Barwick, K. D., & Scott, L. R. (1978). A behavioral approach to occupational safety: Pinpointing and reinforcing safe performance in a food manufacturing plant. *Journal of Applied Psychology, 63*, 434–445.

Komaki, J. L., Zlotnick, S., & Jensen, M. (1986). Development of an operant-based taxonomy and observational index of supervisory behavior. *Journal of Applied Psychology, 71*, 260–269.

Manz, C. C., & Sims, H. P., Jr. (1987). Leading workers to lead themselves: The external leadership of self-managing work teams. *Administrative Science Quarterly, 32*, 106–128.

Mattila, M., Hyttinen, M., & Rantanen, E. (1994). Effective supervisory behaviour and safety at the building site. *International Journal of Industrial Ergonomics, 13*, 85–93.

Mintzberg. H. (1973). *The nature of managerial work*. New York: Harper & Row.

Morgeson, F. P., & Hofmann, D. A. (1999). The structure and function of collective constructs: Implications for multilevel research and theory development. *Academy of Management Review, 24*, 249–265.

Nasanen, M., & Saari, J. (1987). The effects of positive feedback on housekeeping and accidents at a shipyard. *Journal of Occupational Accidents, 8*, 237–250.

Quinn, R. E. (1988). *Beyond rational management: Mastering the paradoxes and competing demands of high performance*. San Francisco: Jossey-Bass.

Reason, J. T. (1994). Forward. In M. S. Bogner (Ed.), *Human error in medicine* (pp. vii–xv). Hillsdale, NJ: Erlbaum.

Saari, J., & Nasanen, M. (1989). The effect of positive feedback on industrial housekeeping and accidents: A long-term study at a shipyard. *International Journal of Industrial Ergonomics, 4*, 201–211.

Simard, M., & Marchand, A. (1994). The behavior of first-line supervisors in accident prevention and effectiveness in occupational safety. *Safety Science, 17*, 169–185.

Simard, M., & Marchand, A. (1997). Workgroups' propensity to comply with safety rules: The influence of micro–macro organisational factors. *Ergonomics, 40*, 172–188.

Thompson, R. C., Hilton, T. F., & Witt, L. A. (1998). Where the safety rubber meets the shop floor: A confirmatory model of management influence on workplace safety. *Journal of Safety Research, 29*, 15–24.

Tomás, J. M., Meliá, J. L., & Oliver, A. (1999). A cross-validation of a structural equation model of accidents: Organizational and psychological variables as predictors of work safety. *Work and Stress, 13*, 49–58.

Vroom, V. H. (1964). *Work and motivation*. New York: Wiley.

Vroom, V. H., & Jago, A. G. (1988). *The new leadership: Managing participation in organizations*. Englewood Cliffs, NJ: Prentice Hall.

Vroom, V. H., & Yetton, P. W. (1973). *Leadership and decision-making*. Pittsburgh, PA: University of Pittsburgh Press.

Weick, K. E. (1990). Technology as equivoque: Sensemaking in new technologies. In P. S. Goodman, & L. S. Sproull (Eds.), *Technology and organizations* (pp. 1–41). San Francisco: Jossey-Bass.

Williams, H., Turner, N., & Parker, S. K. (2000, August). *The compensatory role of transformational leadership in promoting safety behaviors*. Paper presented at the Academy of Management Annual Meeting.

Yukl, G. (1989). *Leadership in organizations* (2nd ed.). Englewood Cliffs, NJ: Prentice Hall.

Yukl, G., & Falbe, C. M. (1990). Influence tactics and objectives in upward, downward, and lateral influence attempts. *Journal of Applied Psychology, 75*, 132–140.

Yukl, G., & Van Fleet, D. D. (1992). Theory and research on leadership in organizations. In M. D. Dunnette & L. M. Hough (Eds.), *Handbook of industrial and organizational psychology* (Vol. 3, pp. 147–197). Palo Alto, CA: Consulting Psychologists Press.

Zohar, D. (2000a). A group-level model of safety climate: Testing the effects of group climate on microaccidents in manufacturing jobs. *Journal of Applied Psychology, 85*, 587–596.

Zohar, D. (2000b, August). *Safety climate and leadership factors as predictors of injury records in work-groups*. Paper presented at the Academy of Management Annual Meeting.

9

PAY AND BENEFITS: THE ROLE OF COMPENSATION SYSTEMS IN WORKPLACE SAFETY

ROBERT R. SINCLAIR AND LOIS E. TETRICK

Employment typically is defined as the exchange of labor for money. This definition fit most employee–employment relationships in the United States until the 1920s and 1930s. At this time, wage controls encouraged employers to develop alternative forms of wages such as paid health insurance and stock options (Milkovich & Stevens, 2000). The primary goal of pay and monetary benefits was to attract and retain employees (Lawler, 1990). This goal continues today, although employment often includes a vast array of nonmonetary benefits such as valet services, referrals for child care services, and financial planning services. Therefore, the compensation system has blossomed to include an array of possible pay schedules and benefits. There also is a growing awareness that the compensation system may be useful in influencing employees' behaviors other than for attraction and retention.

This research was partially supported by a summer faculty development grant to the first author from Portland State University. We thank Dianne Burt-Green, Amy Galbreath, Tammy Gibbons, Michael Leo, Paul Paris, Amy Sinclair, and Jennifer Sommers for their contributions to this chapter.

As separate fields of study, compensation systems and occupational health surely are among the most heavily studied topics in a wide array of academic disciplines. An electronic abstract search on terms such as *compensation* and *benefits* yields hundreds of citations. Much of this research documents efforts to change behavior through reward systems and other compensation strategies. Similarly, psychology is one of many fields with a central interest in protecting and promoting workers' health. Taken together, these interests suggest that promoting safe behavior at work through the effective application of compensation and benefits programs would be an obvious topic of interest to researchers and practitioners alike. However, relatively few studies have examined the effective use of compensation systems to promote safety and health. For example, reviews of research on compensation systems reveal scant mention of occupational health and safety issues (e.g., Bartol & Durham, 2000; Bartol & Locke, 2000; Gerhart & Milkovich, 1992; Lawler, 1990; Lawler & Jenkins, 1992; Milkovich, Milkovich, & Newman, 1996). Similar reviews of occupational health and safety literature reveal little mention of compensation and benefits practices (Schabracq, Winnubst, & Cooper, 1996; Stellman, 1998). In response to these gaps, this chapter describes a framework for understanding the effects of compensation systems on occupational safety and health issues, describes specific strategies for encouraging safety, reviews compensation research addressing health issues, and presents recommendations for further research and effective practice.

THE NEED FOR A SYSTEMS PERSPECTIVE

Systems concepts are ubiquitous in organizational psychology and can be traced back at least to the early-1960s depiction of open systems theory by Katz and Kahn (see Katz & Kahn, 1978). A full discussion of the nuances of the historical and current variations of systems theory is far beyond the scope of this chapter. However, three features of systems perspectives are particularly relevant to understanding compensation systems.

Holistic Thinking About Organizations

Systems thinking encourages psychologists to recognize that all components of organizations are interconnected and, as a result, changes to one aspect of a system are likely to produce organization-wide consequences (Sterman, 2000). Thus, a systems perspective requires holistic thinking—understanding problems as the consequences of complex interactions between various organizational subsystems that differ in power, goals and values, ideologies, and assumptions about organizational life.

Focusing on connections among organizational subsystems enables researchers and practitioners to understand how events in one aspect of an organization affect, and are affected by, other subsystems. This perspective encourages researchers to view safety or compensation issues as part of a much larger organizational landscape composed of, for example, connections between human resources, production systems, top management, and labor unions. Workers are influenced by the goals, values, and functions of all of these systems. In the safety and health domain, an individual's unsafe behavior may be simultaneously influenced by compensation policies, team norms, and management practices—some of which may encourage safety and others of which encourage competing goals, such as profitability. Moreover, systems thinking encourages decision makers to recognize that the ascribed source of an organization's problem may simply be the part of a system to which it is easiest to blame. As one example, managers often attribute injuries to negative characteristics of employees rather than to work design or other organizational factors (Kenny, 1995a, 1995b).

Connections Between Systems

A second important contribution of a systemic perspective on compensation systems is the recognition that organizational issues such as health and safety are located in a complex organizational context. This context is composed of several broad social systems that affect safety and health concerns. For instance, Balgopal and Nofz (1989) advocate for an ecological perspective on workplace injuries that recognizes the legal, medical, family, and work environment aspects of recovery from work injuries. Similarly, Walker (1992) indicates that employees' responses to the workers' compensation system are shaped by at least six critical social influences: (a) employers, (b) physicians, (c) insurance carriers, (d) families, (e) rehabilitation specialists, and (f) attorneys. Accounting for the complex relationships among these systems can be complicated. For example, Walker points out that the workers' compensation system provides incentives for learned helplessness as workers typically continue to receive benefits whether or not they actively attempt to return to work. Such complexities are not easily handled by traditional research designs in psychology or by individual-level organizational interventions. Moreover, recognizing the complex fabric of the organizational context also implies a need for holistic or system-level interventions (e.g., Balgopal & Nofz, 1989; Kenny, 1995a, 1995b; Walker, 1992)—efforts to change institutions and/or relationships among institutions—rather than behavioral interventions.

Systemic Feedback

One of the most important concepts in systems thinking is that the structure of a system generates its behavior (Sterman, 2000). The most important structural features of a system are the cycles of feedback between different actors or subsystems. The implication is that, to understand these cycles, researchers must forgo the traditional view of organizational events as sets of relatively isolated chains of causes and effects. Several cycles may be operating simultaneously, some encouraging change or growth and others operating to counteract change. The joint effects of these cycles can lead to a wide array of behavioral patterns that range from seemingly stable patterns that mask complex underlying dynamics to seemingly random behavior with no apparent explanation (Sterman, 2000). Failure to understand these feedback cycles leads to cases in which the "logical" efforts to solve organizational problems produce unintended consequences and may even worsen the problems they were intended to solve.

One of the most important feedback cycles for occupational safety concerns reactions of organizational subsystems to safety events (e.g., injuries, accidents, and violence). Within the organization, examples of these cycles include the causal attributions of employees, teams, and organizations about the causes of unsafe behavior and their resulting responses. These reactions have multiple effects on the system, including establishing future supply and demand parameters for wage premiums, influencing social exchange processes that guide safety motivation, and introducing new challenges for workers attempting to attend to safety issues.

A FRAMEWORK FOR COMPENSATION SYSTEMS AND SAFE BEHAVIOR

Companies design compensation systems to meet legal, financial, fairness, and productivity criteria (Bartol & Durham, 2000). However, with respect to safety, we assume that compensation systems mostly influence safety and health through their effects on individual-level motivation. Figure 9.1 depicts the processes through which compensation systems might influence safe behavior and highlights several contextual influences on safe behavior. Space restrictions prevent a full discussion of each motivational theory related to this model (see Bartol & Durham, 2000; Bartol & Locke, 2000; and Rousseau & Ho, 2000, for excellent summaries of these models). Therefore, Table 9.1 presents a set of basic theoretical propositions about safety motivational process implied by general work motivation theories.

The core of our model concerns the four points at which compensation systems could plausibly affect organizational safety issues: job choice, safety motivation, on-task behavior, and outcomes of safety events. *Job*

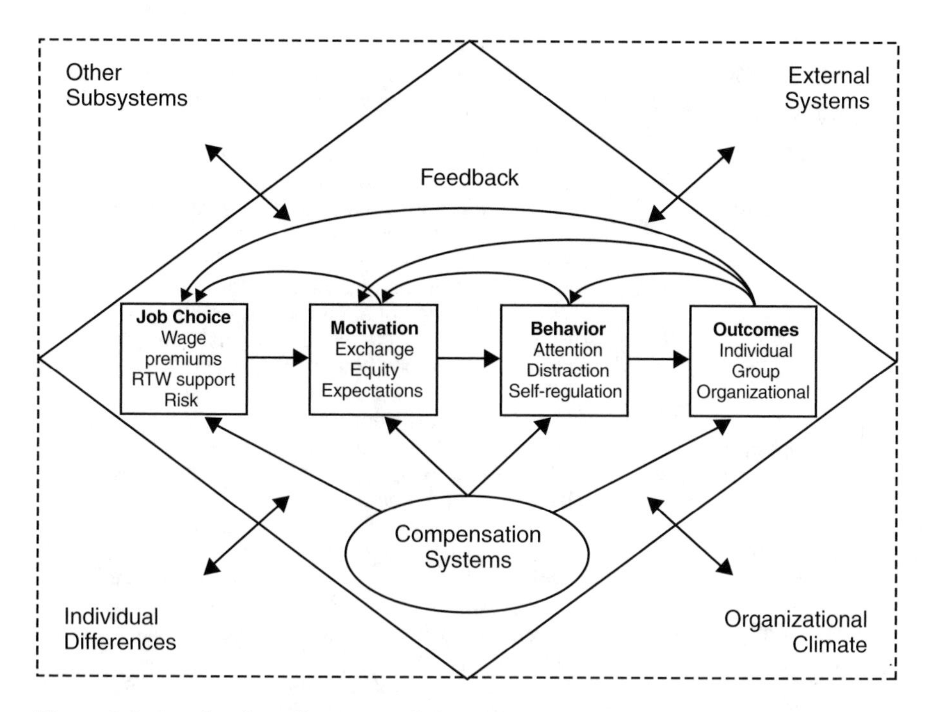

Figure 9.1. A safety benefit systems framework.

choice refers to compensation system attributes that encourage workers to choose unsafe jobs and system attributes that encourage workers to return to work following an injury. Potential influences on job choice include premiums for hazardous work or undesirable shifts, death benefits, and workers' compensation support for injured workers.

Safety motivation refers to the constellation of motivational processes that encourage choices to engage in safe behaviors rather than unsafe behaviors, levels of effort toward safety, and persistence in safe work. These motivational processes include most of the models that have dominated industrial psychology research for the past several decades, such as equity and fairness models, expectancy-value theories, intrinsic motivation, and social exchange theories. Relevant compensation programs include performance bonuses for safe behavior, the reciprocity and expectancy motives associated with management and work team reactions to past safety incidents, and workers' referents for determining equitable compensation for safety.

On-task behavior refers to motivational processes that operate during actual task engagement. These include allocation of attentional resources toward safety (rather than other issues) and self-regulation processes that operate to correct deviations from safety. These processes are effectively captured by goal-setting theory, which suggests that goals serve both as cues

TABLE 9.1
Basic Propositions About Safety From Several Motivational Theories

Theory	Safety motivation is increased . . .
Distributive justice	When the ratio of safety outcomes and inputs is proportional to a referent other.
Procedural fairness	When the process used to determine safety outcomes is perceived as fair. When key decision makers implement safety compensation policies in a fair manner.
Expectancy	When outcomes of safety are desirable. When there is a perceived relationship between safety behavior and receiving outcomes. When there is a perceived relationship between effort and safety behavior.
Reinforcement	When desirable outcomes follow safe behavior.
Intrinsic/extrinsic motivation	When safety rewards enhance self-efficacy without being perceived as controlling.
Agency theory	When safety interests of principles (workers) match those of agents (e.g., employers).
Prospect theory	When safe behavior is framed as beneficial for employees.
Goal setting	When employees are committed to specific and difficult safety goals. Because safety goals focus attention on safe behavior, facilitate the development of safety strategies, and encourage and persistence of efforts to be safe.
Social exchange theory	Because safe behavior is an act of reciprocation for good treatment by the organization.

to allocate cognitive resources toward a particular problem and toward self-regulation processes that occur during task engagement. Applied to safety, compensation policies (e.g., incentive levels) contribute to safety by increasing the attractiveness of safety goals and encouraging persistence in the face of potential challenges to safety, such as fatigue.

The outcomes component includes effects on other aspects of organizational functioning and effects on the likelihood of future safe behavior. Much of the motivational literature deals with individual-level analyses of workers' expectations about whether safe behavior is rewarded. This literature suggests that workers who do not receive potential rewards when they are unsafe should redouble their efforts to be safe employees. However, the pay-for-performance literature points out that workers who do not receive safety-contingent pay rewards may make external attributions for their behavior, blaming supervisors, team members, or the performance management system for their outcomes (e.g., Taylor & Pierce, 1999). These attributions may lead to future problems with both safety performance and general performance and satisfaction. Thus, safety outcomes are dynamically related to future motivation.

Other outcomes include safety-related consequences for group and organizational processes. An individual's safe behavior influences productivity and well-being goals for the group and for the organization at large. Moreover, group and organizational outcomes are reciprocally related to safety motivational processes. For example, a reported injury most likely stimulates appraisals of (a) the injured worker by the team, (b) the work climate by the team, (c) the team (based on its responses) by the worker, and (d) the work climate by the worker. These appraisals and accompanying reactions should affect all workers' willingness and capability of returning to work following an injury and their motivation to work safely in the future. Incorporating managers' behavior as well as larger organizational norms, values, and policies introduces further complexity to the situation. The key question for compensation research concerns how compensation policies affect these responses.

Several scholars have acknowledged that organizational climate perceptions play an important role in motivating safe behavior. Griffin and Neal (2000) found evidence for a second-order global safety climate factor reflecting four first-order factors: management values, beliefs about the efficacy of safety inspections, perceptions of the adequacy of training procedures, and safety communication. However, they concluded that there is little agreement about the specific content of safety climate. They also acknowledged that other researchers have studied a wide array of factors and concluded that it was "premature to propose a definitive structure of first-order factors" (p. 348).

Presumably, safety climate has several motivational effects, including the establishment of expectations about whether safe behavior will be rewarded and how the organization will respond to unsafe behavior. At a minimum, this idea highlights the notion that intervention effectiveness depends on the organization's climate. This idea also has been acknowledged in the compensation literature (Wisdom & Patzig, 1987). For instance, efforts to develop safety reward systems may be undermined if workers and/or managers perceive that other organizational goals conflict with safety goals and that safety is not actually valued. Thus, compensation-related safety interventions should be based on a careful assessment of both the safety and compensation climates in the organization. Goals, values, and perceptions in each of these domains need to be carefully aligned or interventions may have little effect or even be met with active resistance.

Finally, it is important to consider the role of individual differences. A quick skim of most bodies of applied psychological research will reveal ample interest in individual differences and, more specifically, individual differences in health outcomes. This research could help practitioners understand why organization-wide compensation systems are not likely to be equally effective at promoting safe behavior. For example, workers in different occupations may have quite different perceptions of the value of a given compensation policy. However, there is regrettably little empirical research to guide safety programs in this area (Ilgen, 1990).

COMPENSATION FOR HIGH-RISK JOBS

According to economic theory, firms pay a premium for jobs that are dangerous. The underlying rationale for this practice is that organizations should compensate people for a possible loss of livelihood at some future point from work-related injuries or illness. Firms essentially pay employees for the risks they assume by accepting dangerous jobs. However, passing the risk along to employees does not encourage firms to improve workplace safety.

Studies of safety-related compensating differentials (i.e., higher wage rates for dangerous work) raise questions about the economic perspective on hazard compensation. For instance, Dorman and Hagstrom (1998), using data from the 1982 Panel Study of Income Dynamics, concluded that safety-related compensating differentials disappear once industry and demographic characteristics are controlled. They also found that unionized workers generally were better paid and faced less risk of fatality than nonunionized workers. Dorman and Hagstrom suggest that unionized workers are, because of their stronger bargaining power, treated *as if* they had a stronger value for safe work. In contrast, non-unionized workers in dangerous jobs may have restricted labor market mobility and cannot simply leave their unsafe jobs in search of work with a better balance of compensation and hazard levels. Similary, Loomis and Richardson (1998), using data from medical examiner cases of occupational fatalities from 1977 to 1991 in North Carolina, also concluded that labor market segmentation and discrimination in employment accounted for racial disparities in occupational injury mortality. Differential compensation for dangerous jobs did not address the safety issues, and Loomis and Richardson called for safety standards and ensuring compliance with safety standards as necessary to reduce the risk. Similarly, the report by Greenhouse (2001) supported Dorman and Hagstrom's conclusions, as well as Loomis and Richardson's conclusions, suggesting that organizations may not be adequately compensating individuals for dangerous work. Many work environments remain unsafe, and labor market segmentation is still an issue.

Regardless of whether hazardous duty pay adequately compensates individuals for dangerous work, it could be argued that informing individuals of the risks associated with the hazardous duty pay should increase their vigilance regarding hazards. However, Jermier, Gaines, and McIntosh's (1989) study of physically dangerous work suggested otherwise. These researchers found that physically dangerous work produced fear and related affective distress among members of a city police department. Such distress is inconsistent with increased vigilance and attention paid to hazards in the workplace. However, it is recognized that police officers face dangers very different from those that workers in other segments of the economy (e.g., construction, transportation, education) face. Therefore, although it could be argued that hazardous duty pay directs workers' attention to unsafe work conditions and increases their vigilance, existing data do not appear to support this proposition.

Another compensation practice that attempts to provide for injured workers is the workers' compensation system. Workers' compensation systems differ across states in the United States as well as around the world (Ison, 1998; Rey & Lesage, 1998). Nonetheless, most of these systems essentially represent no-fault insurance systems that enable employers to avoid litigation concerning negligence or responsibility for injuries (Rey, 1998). Ison (1998), however, suggests that the costs of workers' compensation are not fully borne by employers, but are instead passed on to workers in the form of lower wage rates and reduced benefits.

Workers' compensation initially was concerned with injuries that occurred in the workplace but has been extended to cover occupational diseases. However, as Rey (1998) notes, more than 50% of occupational disease claims are rejected whereas only about 2% of injury claims are rejected. It is not clear how the implementation of the Americans with Disabilities Act (ADA) of 1990 for physical and mental disabilities may impact these rates. Although some research is beginning to address compensation claims for occupational stress (Dollard, Winefield, & Winefield, 1999), this research is complicated by differences in legal criteria for occupational stress claims across different jurisdictions (Brogmus, 1996; Ison, 1998). Under workers' compensation, occupational diseases that include mental disorders must be shown to have a work-related origin. Dealing with workers' compensation claims for psychological injuries or illnesses raises measurement issues as well as the need for better understanding of the etiology of occupational diseases. For example, further research is needed to establish the relationship between events such as harassment, witnessing violence at work, or chronic job stress and outcomes such as posttraumatic stress disorder or depression. These challenges create ambiguities for employees, employers, insurance providers, and legislators accustomed to dealing with traditional injuries (Dollard et al., 1999; Lippel, 1989). However, under the ADA, the etiology of a disability is not an issue. The concern is whether the mental disability qualifies as a disability covered under the ADA and whether the individual can perform the essential functions of the job.

Despite the potential implications of the ADA, the literature examining the effects of workers' compensation suggests that the system actually discourages workers from reporting injuries. This may be a function of the bureaucracy involved, lack of recognition of work-related injuries or illnesses among physicians, or workers' insufficient knowledge of their rights under their workers' compensation system. Walsh and Dumitru (1998; as cited in Rey, 1998) concluded that liberal compensation under workers' compensation systems results in increased claims, longer periods of recovery, and slower returns to work. However, Curington (1994) reported data indicating that increasing the value of workers' compensation benefits after individuals returned to work reduced the time before workers returned to work. Thus, the incentive value of benefits after returning to work encouraged workers to return to work sooner after severe impairments. The complexity of most workers' compensation systems, coupled with the complexity of most workplaces, makes it difficult to determine the effects of workers' compensation systems on injuries. Moreover, past research does not adequately address the effects of the system on individuals who returned to work earlier as compared with those who stayed out of work longer or the effects of workers' return to work on their coworkers.

Rey (1998) suggested that the workers' compensation system may signal to management the degree of risk in the work environment, with higher experience rates reflecting more hazardous work. Habeck, Leahy, Hunt, Chan, and Welch (1991), using workers' compensation claims as indicators of workplace safety, found that organizations in the same industry with fewer claims systematically differed from those with more workers' compensation claims. Organizations with fewer claims reported that they (a) did more monitoring and correcting of unsafe behaviors, (b) provided more safety training for new and transferred employees, (c) invested more resources in employee health promotion, (d) used modified duty assignments to facilitate workers' return to work, and (e) had procedures to ensure that supervisors were involved in workers' return-to-work programs. Although workers' compensation does not appear to directly provide a safer work environment or necessarily to affect health outcomes, compensation claims could signal the need to improve workplace safety. Future research also could analyze health care and disability insurance claims in a similar fashion.

Previous literature suggests that hazardous duty pay and workers' compensation may be related to workers' job choice or their perceptions of risk, as shown in Figure 9.1. However, the evidence is weak. Surprisingly little research has been conducted on perceptions of occupational hazards and their relation to occupational safety and health. It appears that the effects of hazardous duty pay and workers' compensation systems on worker safety and health are more indirect, being more distal from their immediate work behaviors.

FINANCIAL INCENTIVES FOR SAFE BEHAVIOR

The use of financial incentives to promote safety continues to be a controversial subject, with different bodies of literature drawing different conclusions. Some commentators firmly conclude that incentive programs are ineffective or even counterproductive (Kohn, 1993). Moreover, research shows that performance-contingent rewards are associated with increased levels of turnover (Guthrie, 2000); increased depression and somatic complaints (Shirom, Westman, & Melamed, 1999); and increased blood pressure (Schleifer & Okogbaa, 1990).

Practitioner reports still note that financial incentives play an important role in safety management (Tompkins, 1996). Moreover, several studies have addressed the use of incentives to promote a wide array of safe behaviors, including reduced vehicle accidents; increased usage of personal protective equipment (e.g., earplugs and seat belts); improved "housekeeping"; and lower injury rates. McAfee and Winn (1989) reviewed this literature and concluded that "every study, without exception, found that incentives or feedback enhanced safety and/or reduced accidents" (p. 15). More recent research has reported similar findings (e.g., LaMere, Dickinson, Henry, Henry, & Poling, 1996), and other research has described safety benefits of general compensation strategies such as gainsharing (Doherty, Nord, & McAdams, 1989).

Most of the propositions about safety motivation in Table 9.1 apply to incentives for safety. However, in practice, designing effective incentive programs can be challenging. First, one must carefully consider the parameters of the incentive itself. The nature and level of the incentive being offered must be proportional to the intended behavior change. Simply, one should not expect wholesale behavioral change from the application of a small incentive. Moreover, nonfinancial incentives such as safety feedback ultimately may be as effective as financial incentives without some of the potential negative consequences. Finally, it is important to note that many criticisms of the incentive literature have not been explicitly examined in the safety and health context. Thus, generalizations from compensation studies to safety management should be viewed with caution because the incentive literature includes many examples of contingency factors, intervening variables, and unintended consequences.

Potential Problems With Safety Incentives

Kerr's (1995) analysis of the frequent misalignment of intended and actual consequences of outcome-contingent goals can be readily applied to the safety incentive literature. Kerr pointed out that reward systems often foster the wrong behavior by using rewards incorrectly. In fact, reviews of performance-contingent pay research (e.g., Lawler & Jenkins, 1992) and the general incentive literature (Bartol & Durham, 2000; Bartol & Locke, 2000) have described several unintended consequences of reward systems, at least some of which are related to safety concerns. For example, Kaminski (2001) found that firms with pay-for-performance plans also reported higher injury rates than firms with no pay-for-performance plans. She suggested that performance-contingent pay systems lead to increased work pace and reduced adherence to equipment maintenance schedules, each of which increases chances of injuries. These effects probably do not apply universally. In fact, Kaminski suggested (but did not directly test) the hypothesis that performance-contingent pay is more likely to lead to injuries when other hazard control systems are poorly implemented. Thus, poorly conceived and poorly implemented incentive programs are likely to create more problems than they solve. Although most of the literature has not addressed compensation systems specifically designed to increase safe behavior, we have identified several potential difficulties associated specifically with safety incentives.

Money at Risk

One argument in favor of safety incentives is that incentive programs place an increased amount of income directly under worker control. This is particularly true with safety bonuses that increase the potential pay that can be derived from a job. Moreover, variable pay plans, as they are sometimes called, increase organizational flexibility and control over labor costs. However, such programs also increase the proportion of income at risk for employees, reducing the amount of guaranteed income from the job (Holmstrom, 1987). These problems often are compounded by poor system design, including poorly specified goals and managers' unwillingness to differentiate among employees (Bartol & Durham, 2000). Not surprisingly, employees often react negatively to these plans.

Work Climate

Some studies of safety incentive programs have described informal competitions arising across different work teams (Komaki, Barwick, & Scott, 1978). Although informal competitions can be beneficial, Lawler and Jenkins (1992) suggested that incentive systems can split workers into a we-versus-they mentality when some workers lose income because of other workers' performance; these incentive systems work "against creating a climate of openness, trust, joint problem solving, and commitment to organizational objectives" (p. 1028). The potential for divisiveness may be attenuated for safety incentives because the amount of pay at risk typically is lower for safety incentives than for performance incentives. Thus, it is unclear whether the merits of competition outweigh the potential for creating divisiveness.

Underreporting

Perhaps the best example of unintended consequences of reward systems concerns underreporting of occupational illnesses and injuries. In an investigation of manufacturing firms' Occupational Safety and Health Administration (OSHA) injury and illness logs, Eisenberg and McDonald (1988) found evidence of underreporting in 20% of nearly 200 firms (overreporting was noted in 15%). Pransky, Snyder, Dember, and Himmelstein (1999) cited several reasons for underreporting, including difficulties in diagnosing injuries and poor organizational record keeping. However, incentive systems may promote underreporting if they encourage employees to attempt to beat the system by not reporting minor events that would affect their compensation (Pransky et al., 1999). Group-based incentive systems also may encourage employees to socially pressure each other not to report injuries. Over time, untreated minor injuries can become serious and ultimately result in greater costs for both employees and employers (Gallagher & Myers, 1996).

Other researchers have studied "presenteeism," which is typically defined as perceived pressures from coworkers or managers to be at work when one is sick. For example, Dodier (1985) described cases in which employees negatively evaluated coworkers who took sick leave. Similarly, McKevitt, Morgan, Dundas, and Holland (1997) found that many professional workers described cultural and organizational pressures not to take sick leave, with more than 80% indicating that they continued to work when they felt they should have taken sick leave. Grinyer and Singleton (2000) described negative emotions generated in an organization in which, after taking a fourth period of sick leave, employees were interviewed and required to explain why they had been using sick leave. Grinyer and Singleton even described one of their interview sites as having instituted a "clamp down on short-term 'casual sick' employees" (p. 14). The researchers also pointed out that the fears generated by such pressures often lengthen illness recovery time rather than shortening it. Thus, it is not surprising that presenteeism has been associated with higher rates of various psychosomatic symptoms, including back or neck pain, fatigue, and sleep disturbances (Aronsson, Gustafsson, & Dallner, 2000). Although presenteeism has not directly been linked to incentive systems, it certainly seems to be a likely consequence of incentive pressures toward safety, particularly when one worker's safe behavior affects the financial rewards of other workers.

Are Productivity and Safety Compatible Goals?

Careful consideration of the safety–productivity relationship raises a troubling set of issues. Safety incentive programs are designed to address goals that may be perceived as conflicting with firms' financial and productivity goals (Kenny, 1995b). Particularly in difficult economic circumstances, organizations may face real or perceived costs associated with careful attention to safety policy enforcement. For example, labor unions have long recognized that few organizations can effectively function if workers follow all work rules to the letter of the law. In fact, the work-to-rules bargaining strategy involves devout attention to policies and procedures in unions that either do not wish to or cannot go on strike (Fossum, 1999, pp. 363–364). The practical consequence of this strategy is essentially the same as a work stoppage—productivity decreases substantially.

Kaminski (2001) pointed out that, in the United States, socioeconomic structures contribute to the undervaluing of workplace safety, as the joint cost of workers' compensation and fines imposed for OSHA violations may be small enough that it makes financial sense for organizations not to be concerned with safety. The combination of this socioeconomic context and the practical challenges of following detailed safety regulations suggests that strict adherence to safety regulations may be impractical. And even well-designed programs may be ineffective if they are perceived as interfering with other goals. Thus, behavioral scientists face considerable challenges in their efforts to identify programs that balance organizational productivity goals against the health needs of workers.

EMERGING ISSUES AND RESEARCH DIRECTIONS

What Compensation Programs Work and How?

Perhaps the most important topic for further research is to understand how various compensation strategies and systems affect occupational health and safety. Throughout this chapter we emphasize the importance of considering previously unanticipated consequences of compensation systems. One way to avoid such consequences is to clarify the criterion domain—that is, to be clear about the goals of a particular program prior to its implementation. By distinguishing different levels of analysis and stages of safety motivation, we hope we have taken a first step toward that clarity. However, further research is needed to document which programs work in practice. Given the complexity of occupational safety and health challenges, we expect researchers interested in this area to be pressed to consider the use of alternative research strategies, such as qualitative research, and alternate theoretical perspectives, such as systems theory, to best identify these interventions.

Program evaluation research is needed to establish both the overt and hidden costs and benefits of safety compensation systems. For workers, perhaps the most important outcomes include workers' preferences for and reactions to different types of safety programs. Both individual-level and group-level analyses of the safety climate will be critical to these efforts. For example, although researchers laud the importance of compensation system flexibility (e.g., Lawler, 1990), it is unclear whether such flexibility should be applied to safety benefits.

Cost–benefit analysis research might be able to borrow from traditional approaches to utility analysis in organizational psychology (see Boudreau, 1991). However, the complexities of compensation systems raise several questions: How much of a particular benefit is needed to shape behavior? Are multifaceted compensation strategies more effective? How high must premiums be set before they influence job choice? Should companies use limited benefits systems for temporary and contingent workers to promote safe behavior? Are there low-cost alternatives to benefit compensation such as employee participation systems? What can a company do to mitigate some of the negative effects of incentives?

For Whom Do Safety Compensation Programs Work?

Individual difference studies can help explain the theoretical mechanisms linking compensation systems to behavior and can provide practitioners with guidelines for designing compensation systems to match the characteristics of a target workforce. However, in our view, individual differences research on safety compensation has lacked integration or unifying purpose. Our proposed model suggests several points at which individual difference concepts might usefully contribute to understanding benefit systems. These concepts include direct effects on each stage in the proposed model (e.g., differences in how people weigh safety compensation when making job choices; dispositional determinants of safety motivation and behavior) and potential moderators of the reactions to benefits programs on each of those stages.

One particularly promising area for further individual differences research is examination of the relationship between relatively deep dispositions (e.g., personal values, vocational interests, and personality traits) and reactions to benefit programs. For example, we could speculate that conscientious people (who prefer structured situations, order, and predictability) might respond better to safety incentives than people who react negatively to having their safety behavior linked to financial outcomes. Finally, although the jury still is out with respect to demographic differences in reactions to compensation systems, we believe that the ever-changing demographic characteristics of the workforce necessitate continued attention to the potentially varied needs of different groups.

New Methodological Strategies and Theoretical Perspectives

A systems perspective requires new strategies for looking at old problems. For example, organizational psychologists are increasingly recognizing the need for research strategies that simultaneously address multiple levels of analysis (Klein & Kozlowski, 2000), whereas business systems researchers are developing sophisticated quantitative models of dynamic organizational feedback cycles (Sterman, 2000). At the same time, researchers in occupational rehabilitation, industrial sociology, and social work are increasingly applying qualitative methods to systems issues in health and compensation (Kenny, 1995a, 1995b). Finally, compensation research should benefit from adapting systemic perspectives such as institutional theory (see Brass, 2000; Meyerson, 1994) to properly locate individual behavior in its context. Each of these developments holds ample promise for future safety research.

RECOMMENDATIONS TO PRACTITIONERS

The review given in this chapter clearly indicates that there are considerable gaps in current knowledge of the relationship between the various components of compensation systems and worker safety. However, there appear to be some consistent findings with important practical implications. Therefore, we end the chapter with our recommendations for occupational health psychologists, human resources practitioners, safety engineers, managers, union representatives, and other professionals involved with workplace safety.

We concluded that hazardous duty pay and workers' compensation systems are unlikely to directly affect workers' safety. However, claims experienced under an organization's workers' compensation system may function as a barometer of workplace safety. Therefore, we recommend that organizations track workers' compensation claims, health care insurance usage, and disability claims to assess occupational safety issues. These data would only identify potential needs for improving the work environment and changing workers' behavior; they would not specifically identify underlying causes of safety issues. Moreover, some previous research indicates that organizations have difficulties accurately tracking and reporting claims—with respect to both under- and overreporting (Eisenberg & McDonald, 1988). Thus, claims should not be relied on as the sole source of information about safety.

Kaminski (2001) describes four levels of hazard controls, in order of proposed effectiveness: elimination of hazards, engineering controls, administrative controls, and personal protective equipment. As she points out, these controls imply that organizations should regularly gather data about (a) the presence or absence of hazards, (b) the functioning of engineering controls such as ventilation systems, (c) the effectiveness of administrative controls such as training, and (d) the use of personal protective equipment. Compensation systems are a form of administrative control and, as such, are not likely to be effective if other types of controls are not in place.

One prevailing theme of this chapter is that because safety-oriented incentive programs have several pitfalls, implementation of these programs should proceed with caution. One of the biggest challenges practitioners face is aligning the actual and intended goals of safety incentive programs. These programs often reward safety outcomes (e.g., reporting behavior, financial costs) rather than safe behaviors and, as a result, may yield as many negative consequences as benefits. Prior to implementing a program, practitioners should carefully consider what rewards it implies. One useful strategy might be to use focus groups or other employee involvement strategies to decide whether a proposed program will increase safe behavior or motivate employees to affect safety outcomes without actually increasing safety.

Rather than focusing on safety outcomes such as reports of injuries, one might base a safety incentive program, at least in part, on pay for safety skill development. This practice would reward employees for demonstrating safety and health-related competencies such as knowledge of safety procedures, healthy work strategies, or even stress management skills. Although we know of no safety-related research in this area, studies have indicated that substantial gains may be associated with skill-based pay for productivity (e.g., Murray & Gerhart, 1998). Other research has indicated that skill-based pay programs are associated with lower turnover than group incentive programs are (Guthrie, 2000). It would be interesting to determine whether these results generalize to safety training programs.

Another possibility to improve safety would be to engage supervisors in directly providing rewards on the basis of their observations of safe behavior (Zohar, 2000b). This is consistent with Habeck et al.'s (1991) conclusions regarding workers' compensation claims. For example, Smith, Cohen, Cohen, and Cleveland (1978) found that greater management commitment and involvement in safety differentiated high-accident plants from low-accident plants. More recently, the literature on safety climate (Griffin & Neal, 2000; Zohar, 2000a) has reiterated the importance of management commitment to safety as a key factor in safety performance. Therefore, we conclude that compensation systems that cultivate strong safety climates and management commitment to safety are more likely than other systems to result in intended improvement in safety.

REFERENCES

Aronsson, G., Gustafsson, K., & Dallner, M. (2000). Sick but yet at work: An empirical study of sickness presenteeism. *Journal of Epidemiology and Community Health*, *54*, 502–509.

Balgopal, P. R., & Nofz, M. P. (1989). Injured workers: From statutory compensation to holistic work services. *Journal of Sociology and Social Welfare*, *16*, 147–149.

Bartol, K. M., & Durham, C. C. (2000). Incentives: Theory and practice. In C. L. Cooper & E. A. Locke (Eds.), *Industrial and organizational psychology: Linking theory with practice* (pp. 1–33). Oxford: Blackwell.

Bartol, K. M., & Locke, E. A. (2000). Incentives and motivation. In S. L. Rynes & B. Gerhart (Eds.), *Compensation in organizations: Current research and practice* (pp. 104–147). San Francisco: Jossey-Bass.

Boudreau, J. W. (1991). Utility analysis for decisions in human resource management. In M. D. Dunnette & L. M. Hough (Eds.), *Handbook of industrial and organizational psychology* (2nd ed., Vol. 2, pp. 621–745). Palo Alto, CA: Consulting Psychologists Press.

Brass, D. J. (2000). Networks and frog ponds: Trends in multilevel research. In K. Klein & S. W. J. Kozlowski (Eds.), *Multilevel theory, research, and methods in organizations: Foundations, extensions, and new directions* (pp. 557–571). San Francisco: Jossey-Bass.

Brogmus, G. E. (1996). The rise and fall? of mental stress claims in the USA. *Work and Stress, 10,* 24–35.

Curington, W. P. (1994). Compensation for permanent impairment and the duration of work absence: Evidence from four natural experiments. *Journal of Human Resources, 29,* 888–910.

Dodier, N. (1985). Social uses of illness at the workplace: Sick leave and moral evaluation. *Social Science and Medicine, 20,* 123–128.

Doherty, E. M., Nord, W. R., & McAdams, J. L. (1989). Gainsharing and organization development: A productive synergy. *Journal of Applied Behavioral Science, 25,* 209–229.

Dollard, M. F., Winefield, H. R., & Winefield, A. H. (1999). Predicting work-stress compensation claims and return to work in welfare workers. *Journal of Occupational Health Psychology, 4,* 279–287.

Dorman, P., & Hagstrom, P. A. (1998). Wage compensation for dangerous work revisited. *Industrial and Labor Relations Review, 52,* 116–135.

Eisenberg, W. M., & McDonald, H. (1988). Evaluating workplace injury and illness records: Testing a procedure. *Monthly Labor Review, 111,* 58–60.

Fossum, J. A. (1999). *Labor Relations: Development, structure, process* (7th ed.). Irwin/McGraw-Hill: New York.

Gallagher, R. M., & Myers, P. (1996). Referral delay in back pain patients on worker's compensation: Costs and policy implications. *Psychosomatics, 37,* 270–284.

Gerhart, B., & Milkovich, G. T. (1992). Employee compensation: Research and practice. In M. D. Dunnette & L. M. Hough (Eds.), *Handbook of industrial and organizational psychology* (2nd ed., Vol. 3, pp. 481–569). Palo Alto, CA: Consulting Psychologists Press.

Greenhouse, S. (2001, July 16). Hispanic workers die at higher rate: More likely than others to do the dangerous, low-end jobs. *New York Times,* p. A11.

Griffin, M. A., & Neal, A. (2000). Perceptions of safety at work: A framework for linking safety climate to safety performance, knowledge, and motivation. *Journal of Occupational Health Psychology, 5*, 347–358.

Grinyer, A. & Singleton, V. (2000). Sickness absence as risk-taking behaviour: A study of organisational and cultural factors in the public sector. *Health, Risk, and Society, 2*, 7–21.

Guthrie, J. P. (2000). Alternative pay practices and employee turnover: An organization economics perspective. *Group and Organization Management, 25*, 419–439.

Habeck, R. V., Leahy, M. J., Hunt, H. A., Chan, F., & Welch, E. M. (1991). Employer factors related to workers' compensation claims and disability management. *Rehabilitation Counseling Bulletin, 34*, 210–226.

Holmstrom, B. (1987). Incentive compensation: Practical design from a theory point of view. In H. R. Nalbantian (Ed.), *Incentives, cooperation, and risk sharing: Economic and psychological perspectives on employment contracts* (pp. 176–185). Savage, MD: Rowman & Littlefield.

Ilgen, D. R. (1990). Health issues at work: Opportunities for industrial/organizational psychology. *American Psychologist, 45*, 273–283.

Ison, T. G. (1998). Workers' compensation systems. In J. M. Stellman (Editor-in-Chief), *Encyclopaedia of occupational health and safety* (4th ed., pp. 25.2–25.24). Geneva: International Labour Office.

Jermier, J. M., Gaines, J., & McIntosh, N. J. (1989). Reactions to physically dangerous work: A conceptual and empirical analysis. *Journal of Organizational Behavior, 10*, 15–33.

Kaminski, M. (2001). Unintended consequences: Organizational practices and their impact on workplace safety and productivity. *Journal of Occupational Health Psychology*, 6, 127–138.

Katz, D., & Kahn, R. L. (1978). *The social psychology of organizations* (2nd ed.) New York: Wiley.

Kenny, D. T. (1995a). Common themes, different perspectives: A systemic analysis of employer–employee experiences in occupational rehabilitation. *Rehabilitation Counseling Bulletin, 39*, 54–77.

Kenny, D. T. (1995b). Stressed organizations and organizational stressors: A systemic analysis of workplace injury. *International Journal of Stress Management, 2*, 181–196.

Kerr, S. (1995). On the folly of rewarding A, while hoping for B. *Academy of Management Executive, 9*, 7–14.

Klein, K. J., & Kozlowski, S. W. J. (Eds.). (2000). *Multilevel theory, research, and methods in organizations: Foundations, extensions, and new directions.* San Francisco: Jossey-Bass.

Kohn, A. (1993). Why incentive plans cannot work. *Harvard Business Review, 71*, 54–60.

Komaki, J., Barwick, K. D., & Scott, L. R. (1978). A behavioral approach to occupational safety: Pinpointing and reinforcing safe performance in a food manufacturing plant. *Journal of Applied Psychology, 63*, 434–445.

LaMere, J. M., Dickinson, A. M., Henry, M., Henry, G., & Poling, A. (1996). Effects of a multicomponent monetary incentive program on the performance of truck drivers: A longitudinal study. *Behavior Modification, 20*, 385–405.

Lawler, E. E., III. (1990). *Strategic pay: Aligning organizational strategies and pay systems*. San Francisco: Jossey-Bass.

Lawler, E. E., III, & Jenkins, G. D., Jr. (1992). Strategic reward systems. In M. D. Dunnette & L. M. Hough (Eds), *Handbook of industrial and organizational psychology* (2nd ed., Vol. 3, pp. 1009–1055). Palo Alto, CA: Consulting Psychologists Press.

Lippel, K. (1989). Workers' compensation and psychological stress claims in North American law: A microcosmic model of systemic discrimination. *International Journal of Law and Psychiatry, 12*, 41–70.

Loomis, D., & Richardson, D. (1998). Race and the risk of fatal injury at work. *American Journal of Public Health, 88*, 40–44.

McAfee, R. B., & Winn, A. R. (1989). The use of incentives/feedback to enhance workplace safety: A critique of the literature. *Journal of Safety Research, 20*, 7–19.

McKevitt, C., Morgan, M., Dundas, D., & Holland W. W. (1997). Sickness absence and "working through" illness: A comparison of two professional groups. *Journal of Public Health Medicine, 19*, 295–300.

Meyerson, D. E. (1994). Interpretations of stress in institutions: The cultural production of ambiguity and burnout. *Administrative Science Quarterly, 39*, 628–653.

Milkovich, G. T., Milkovich, C., & Newman, J. M. (1996). *Compensation* (5th ed.). Chicago: Irwin.

Milkovich, G. T., & Stevens, J. (2000). From pay to rewards: 100 years of change. *ACA Journal, 9*, 6–18.

Murray, B., & Gerhart, B. (1998). An empirical analysis of a skill-based pay program and plant performance outcomes. *Academy of Management Journal, 41*, 68–78.

Pransky, G., Snyder, T., Dembe, A., & Himmelstein, J. (1999). Under-reporting of work-related disorders in the workplace: A case study and review of the literature. *Ergonomics, 42*, 171–182.

Rey, P. (1998). Workers' compensation: Trends and perspectives. In J. M. Stellman (Editor-in-Chief) and P. Rey & M. Lesage (Chapter Eds.), *Encyclopaedia of occupational health and safety* (4th ed., pp. 26.6–26.14). Geneva: International Labour Office.

Rey, P. & Lesage, M. (1998). Topics in workers' compensation systems. In J. M. Stellman (Editor-in-Chief), *Encyclopaedia of occupational health and safety* (4th ed., pp. 26.1–26.28). Geneva: International Labour Office.

Rousseau, D. M., & Ho, V. T. (2000). Psychological contract issues in compensation. In S. L. Rynes, & B. Gerhart (Eds.), *Compensation in organizations: Current research and practice* (pp. 273–310). San Francisco: Jossey-Bass.

Schabracq, M. J., Winnubst, J. A. M., & Cooper, C. L. (Eds.). (1996). *Handbook of work and health psychology.* Chichester, England: Wiley.

Schleifer, L. M., & Okogbaa, O. G. (1990). System response time and method of pay: Cardiovascular stress effects in computer-based tasks. *Ergonomics, 33,* 1495–1509.

Shirom, A., Westman, M., & Melamed, S. (1999). The effects of pay systems on blue-collar employees' emotional distress: The mediating effects of objective and subjective work monotony. *Human Relations, 52,* 1077–1097.

Smith, M. J., Cohen, H. H., Cohen, A., & Cleveland, R. J. (1978). Characteristics of successful safety programs. *Journal of Safety Research, 10,* 5–15.

Stellman, J. M. (Editor-in-Chief). (1998). *Encyclopaedia of occupational health and safety* (4th ed., Vols. 1–4). Geneva: International Labour Office.

Sterman, J. (2000). *Business dynamics: Systems thinking and modeling for a complex world.* New York: McGraw-Hill.

Taylor, P. J., & Pierce, J. L. (1999). Effects of introducing a performance management system on employees' subsequent attitudes and effort. *Public Personnel Management, 28,* 423–452.

Tompkins, N. C. (1996). *Safety incentive and award programs.* Society for Human Resource Management White Paper.

Walker, J. M. (1992). Injured worker helplessness: Critical relationships and systems level approach for intervention. *Journal of Occupational Rehabilitation, 2,* 201–209.

Walsh, N. E., & Dumitru, D. (1988). The influence of compensation on recovery from low back pain. *Occupational Medicine State of the Art Reviews, 3,* 109–121.

Wisdom, B., & Patzig, D. (1987). Does your organization have the right climate for merit? *Public Personnel Management, 16,* 127–133.

Zohar, D. (2000a). A group-level model of safety climate: Testing the effect of group climate on microaccidents in manufacturing jobs. *Journal of Applied Psychology, 85,* 587–596.

Zohar, D. (2000b, August). *Safety climate and leadership factors as predictors of injury records in work groups.* Paper presented at the Academy of Management meetings, Toronto.

10

HIGH-PERFORMANCE WORK SYSTEMS AND OCCUPATIONAL SAFETY

ANTHEA ZACHARATOS AND JULIAN BARLING

Organizations have long focused on the human resource function. Most recently, attention has been focused on how human resource functions can add value to the organization. The potential benefits of an integrated human resource management system have been noted (O'Reilly & Pfeffer, 2000; Pfeffer, 1998a, 1998c), and initial research has supported these ideas (e.g., Ichniowski, Shaw, & Prennushi, 1997; Huselid, 1995). In this chapter, we focus on the extent to which "modern" human resource management practices, which focus on the recruitment, development, and management of employees (Wood & Wall, 2002), might affect occupational safety with the aim of stimulating thinking, encouraging research, and considering potential practical implications of this topic.

Traditionally, occupational safety has been managed by taking a control-oriented approach to human resources (Barling & Hutchinson, 2000), what Wood and Wall (2002) see as the polar opposite of a high-functioning approach to human resource management. The control orientation is based on the assumption that workers will exert only as much effort as it takes to get the job done. Thus, for employees to work effectively, it becomes necessary for management to use control and coercion to ensure desirable behaviors (Walton, 1985), and punishment to reduce undesirable behaviors. With respect to occupational safety, the control orientation emphasizes the use of rules to enforce behaviors and ensure compliance with government regulations and the provisions in a collective agreement (Barling & Hutchinson, 2000).

More recently, greater attention has been paid to managing human resources by way of a commitment-oriented or high-performance work system approach. In contrast to the control orientation, this approach assumes that workers are capable of performing at high levels if encouraged and allowed to do so. It is argued that workers will be more committed to the organization and more trusting of management if given respect and treated as capable and intelligent individuals—and that organizations that employ such human resource approaches will reap the benefits in terms of improved performance (Walton, 1985). A number of studies now provide support for a relationship between high-performance work systems and employee- and organizational-level performance (Arthur, 1992, 1994; Huselid, 1995; Ichniowski et al., 1997; MacDuffie, 1995; Patterson, West, & Wall, 2001).

With respect to workplace safety, the high-performance work approach emphasizes the role of management in promoting safe work. This is in stark contrast to the control orientation, essentially a victim-blaming approach in which employee behaviors are deemed to be the primary cause of workplace injuries and fatalities. We argue that in order to promote workplace safety, management must adopt a set of high-performance work practices that would serve to improve workplace safety by increasing employee trust in management, commitment to the organization, and positive perceptions of safety climate.

Following the commitment-oriented strategy described by Walton (1985) and the framework proposed by Jeffrey Pfeffer (1998c), we propose 10 human resource practices that would promote workplace safety. These comprise the seven practices described by Pfeffer (1998c) with the addition of three practices we deem equally important.

The seven practices Pfeffer (1998c) described are the following:

1. Organizations must ensure employment security for their employees.
2. Organizations must subject all new personnel to a selective hiring process.
3. Employees should be provided with extensive training.
4. The design of organizations should emphasize decentralized decision making and self-managed teams.
5. The organization must make an effort to reduce status symbols that separate employees into different hierarchical levels.
6. Information sharing throughout the organization should be encouraged.
7. Employees should receive relatively high compensation dependent on the organization's performance.

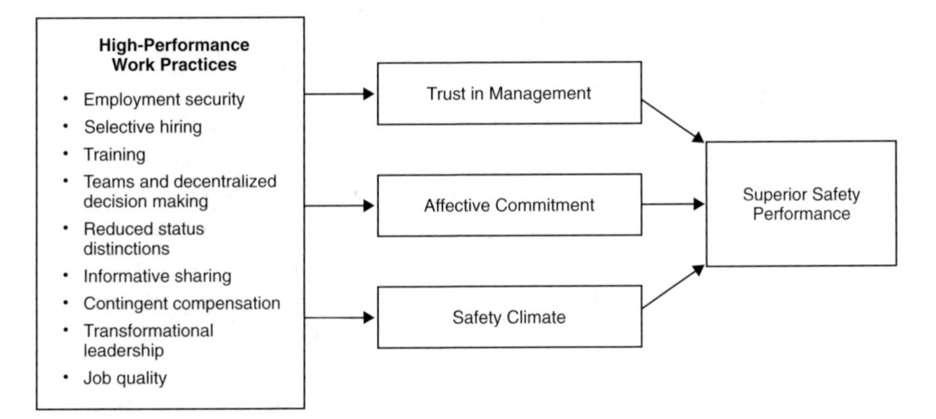

Figure 10.1. High-performance work practices and their impact on workplace safety.

We believe these seven practices can be extended by considering three additional factors: the role of transformational leadership, job quality, and the measurement of variables critical to organizational success.

Although Pfeffer (1998b) argued that the implementation of these high-performance work practices results in improved performance at the individual and firm levels, he also claimed that such success extends beyond financial reporting measures; there are now numerous studies supporting this general notion. Following this argument, we predict that occupational safety would be an additional outcome of a high-performance work system. The purpose of this chapter, therefore, is to describe a model of how a high-performance work system can affect occupational safety (see Figure 10.1). We first describe each of the 10 practices that comprise such a system and then discuss the mechanisms that mediate the relationship between these practices and occupational safety—namely, trust in management, affective commitment to the organization, and positive perceptions of safety climate.

TEN HUMAN RESOURCE PRACTICES THAT AFFECT OCCUPATIONAL SAFETY

Employment Security

Employment security encourages a long-term outlook within organizations, promoting trust and organizational commitment. As such, employment security is one means by which firms can improve their performance (Pfeffer, 1998c). We believe that employment security benefits workplace safety for many of the same reasons it benefits a firm's economic performance. First, increased turnover encourages the existence of an inexperienced and less trained workforce, and consequently increased involvement in safety incidents (Kincaid, 1996). For instance, recent work on the relationship between contingent work and worker safety has found that contract workers typically are less trained, are less experienced, and receive less supervision than their permanent peers (Collinson, 1999; Hofmann, Jacobs, & Landy, 1995; Kochan, Smith, Wells, & Rebitzer, 1994). In addition, within-group instability that arises from absenteeism has been associated with occupational accidents (Goodman & Garber, 1988).

As well, employment security has been shown to increase trust in management (Pfeffer, 1998c; Walton, 1985) and affective commitment (Barling, Weber, & Kelloway, 1996; Meyer & Allen, 1997). In a study of nuclear power plant employees, workers less committed to the organization reported the probability of an accident to be greater than those more highly committed to the organization (Kivimäki & Kalimo, 1993). Similarly, employees' perceptions of nuclear accident risk increases as their trust in management decreases (Kivimäki, Kalimo, & Salminen, 1995).

Finally, employment security encourages a long-term perspective that would benefit workplace safety. It is in the best interest of organizations to protect the safety of employees in whom time and financial resources have been invested. When managers lack a long-term focus, employees are seen as dispensable and short-term profits override concerns for safety (Jackall, 1988; Sells, 1994). Given this, it is not surprising that employment or job security differentiates between low-injury and high-injury companies (Smith, Cohen, Cohen, & Cleveland, 1978; Zohar 1980). In addition, Grunberg, Moore, and Greenberg (1996) found that employees reporting high employment insecurity reported more injuries and resulting days lost in a study of the wood products industry. Similarly, employees suffering from feelings of job insecurity in a study of a food processing plant reported lower safety motivation and safety compliance and greater numbers of injuries than those who felt their jobs were secure (Probst & Brubaker, 2001).

Selective Hiring

If anything, the way in which employees are selected with the intention of promoting workplace safety is consistent with the control orientation. Most studies evaluate the effects of personality screening inventories to exclude potentially unsafe (or accident-prone) employees based on criteria such as drug addiction, alcoholism, emotional maturity, and trustworthiness (see, e.g., Borofsky, Bielema, & Hoffman, 1993; Jones, 1991; Jones & Wuebker, 1988). This approach, however, has serious ethical, scientific, and practical problems. First, from an ethical perspective, this approach focuses on the "thin skull" approach, blaming the victim for his or her own misfortune. Second, scientific problems emerge because of the flawed experimental designs on which such studies are based. Third, practical problems emerge because the statistical power of the findings from such studies simply does not justify practical decisions within organizations of the magnitude of excluding individuals from the organization.

Of all the 10 aspects of the high-performance work system, selective hiring has received the least support and is an area in need of greater research scrutiny. Nonetheless, selective hiring could play a critical role in ensuring occupational safety. For example, involving teams in the selection of future members may prove beneficial, as would requiring applicants to participate in several rounds of interviews. The values of the organization more generally, and specifically with respect to safety, could be emphasized during the interview process. In support of this argument, A. Cohen (1977) and Smith et al. (1978) both found that companies with low injury and accident rates had more elaborate selection procedures than did high-injury-rate companies. Further understanding and empirical research are obviously necessary with respect to the link between selective hiring and occupational safety.

Extensive Training

Training is a crucial aspect of any human resource system (Pfeffer, 1998a, 1998c; Whitefield, 2000; Wood & Wall, 2002) and is probably the most frequently used method for ensuring adequate levels of occupational safety in organizations. A review of the literature examining the role of training in workplace safety is well beyond the scope of this chapter. Overall, it has been found when considering well-designed studies that employees who have undergone safety training experience fewer injuries than their untrained counterparts (see chapter 11 of this volume; Hale, 1984). Safety training is especially salient in those instances in which work is inherently dangerous, given the high cost of an error and the inability to learn by trial and error (Weick, 1987). However, the potential benefits go beyond the training itself. It is critical not only that employees are well trained but also that management is seen to be committed to safety training (Zohar, 1980) and that training goes beyond that required by government regulations or the provisions of a collective agreements.

Beyond providing employees with the knowledge and skills to do their jobs to the best of their abilities or to complete the tasks safely, training has the added benefit of increasing organizational commitment. In a study of naval trainees, both the extent to which training met participants' expectations and how satisfied the trainees were with the training they received predicted their subsequent commitment to the organization (Tannenbaum, Mathieu, Salas, & Cannon-Bowers, 1991). This is important because Parker, Axtell, and Turner (2001) found that organizational commitment predicted the safe working of employees. Overall, therefore, extensive training is an integral part of any high-performance work system targeted at occupational safety.

Self-Managed Teams and Decentralized Decision Making

There is an expanding body of research showing that teamwork and the decentralization of decision making benefit employee performance (Pfeffer, 1998c) and should affect occupational safety as well. For instance, in a study of 1,061 work groups, it was found that the cohesion among members of a work group was a significant predictor of workers' propensity to comply with safety rules (Simard & Marchand, 1997). Earlier, Trist, Susman, and Brown (1977) found that miners working in autonomous teams experienced fewer injuries than their peers working individually, whereas Goodman and Garber (1988) found that as the familiarity between miners working in pairs decreased, safety infractions increased.

One explanation for these findings is that employees working in high-functioning teams tend to feel more accountable for safety in general and for each other's safety in particular. For instance, Geller, Roberts, and Gilmore (1996) found that workers' propensity to actively care for their coworkers' safety was predicted by a sense of belongingness to the group as well as by personal control. Teamwork should also improve workplace safety because teams provide their members with the most knowledge and familiarity with the situation, and more opportunities for control. In fact, Hechanova-Alampay and Beehr (2001) found that the safest teams in a chemical company were those with the greatest control over varied aspects of their work.

Furthermore, and again supporting Wood and Wall's (2002) notion that the elements of a high-performance work system are mutually reinforcing, teams promote the sharing of ideas and this promotion would have a significant impact on worker safety. For instance, Tjosvold (1990) found that flight crews faced with dangerous situations performed more effectively when organized as a group than when organized in a hierarchy with the captain in command. Under such conditions, employees are encouraged to share ideas and develop the best possible solutions to problems. Similarly, in the Trist et al. (1977) study, it was found that sharing ideas and experiencing common goals explained why miners in autonomous groups demonstrated improved safety performance.

To summarize, the implementation of teams has the potential to enhance occupational safety because positive team dynamics encourage team members to assume responsibility and to provide each other with feedback and encouragement for working safely. Furthermore, teamwork ensures the sharing of ideas and greater control over work—factors that will encourage employee safety.

Reduce Status Distinctions

> The biggest problem this plant has is that anybody with a degree thinks they're above the men on the bottom rung. There is no communication whatsoever. They think we're a bunch of dummies because we don't have a degree. (Statement by a chemical operator, Nelkin & Brown, 1984, p. 54)

Status distinctions that convey the message that some members of the organization are more important to its functioning than are others create unwanted barriers between organizational members and breed resentment and harm motivation (Pfeffer, 1998a). Within a high-performance work system, employees from all levels should feel able to contribute their knowledge and energy to benefit diverse aspects of the organization (Pfeffer, 1998c). We argue that reduced status distinctions, which would mutually reinforce the effectiveness of teams and decentralized decision making, would also benefit workplace safety.

In organizations where status distinctions between employees are evident, it would be difficult for management to appreciate the risks encountered by frontline employees. In addition, the objective and perceived gap would make it less likely that employees would perceive the extent to which management is concerned with their well-being. In a study of British Rail employees, Clarke (1999) found that both workers and management failed to appreciate the value that each group placed on the other's workplace safety. Furthermore, when hierarchical status distinctions are evident, each group is more likely to lay blame for safety incidents on the other group (DeJoy, 1994). In contrast, increasing exposure between the members of hierarchically distinct groups increases the likelihood that both parties will realize their groups are interdependent with regard to safety and will feel jointly responsible for workplace safety.

There is one empirical study that is relevant to status distinctions and workplace safety. Milanovich, Driskell, Stout, and Salas (1998) note that in airline cockpits, individuals of lower status (namely, first officers) were often too compliant and obedient in the presence of captains (who enjoyed higher status), whereas the captains often missed opportunities to listen to their first officers. Milanovich et al. further showed that individuals held higher general and specific expectations of pilots than of copilots. This important finding highlights the danger of status distinctions because it suggests that copilots holding higher expectations of pilots may change their own behavior by, for example, becoming more subservient in their presence. This could have serious consequences in terms of both performance in general and safety performance in particular. Thus, much as a reduction in status distinctions would serve to improve employee and organization performance within a high-performance work system described by Pfeffer

(1998c), it would similarly serve to improve workplace safety. By encouraging communication, the sharing of ideas, and greater mutual concern and trust among workers, reduced status distinctions play an important role in enhancing occupational safety.

Share Information

Information is one of the most important organizational resources (Pfeffer, 1998c), and providing employees with information allows them to best understand the operation and its goals and thereby increase organizational functioning. The critical role of information sharing is also evident within occupational health and safety legislation, which is typically based on the assumption that workers have access to complete information about their jobs and can therefore refuse unsafe work. However, the potential benefits go beyond the sharing of information per se: Organizations that share information with their employees signal to those employees that they are trusted.

The role of information sharing in workplace safety is illustrated by a number of empirical studies. Both A. Cohen (1977) and Smith et al. (1978) found that companies in which there was more contact and more open discussion between management and shop floor employees reported fewer safety incidents. Similarly, Hofmann and Morgeson (1999) found that to the extent that employees felt comfortable discussing safety-related issues with their supervisors they also more closely followed safety procedures and practices and, in turn, experienced fewer injuries.

Other than simply giving employees the information they need to work safely, information sharing may also impact worker safety by ensuring that employees feel they are an important part of the organization, and this has positive consequences. Sjöberg and Drottz-Sjöberg (1991) found that to the extent that nuclear power plant workers lacked knowledge regarding safety issues, they also experienced greater perceived risk. This finding is important because perceived risk had been found to be negatively correlated with organizational commitment (Kivimäki et al., 1995) and positively correlated with task distraction (McLain, 1995), both of which are potentially important factors in workplace safety.

Information sharing is also particularly important in the safety context, for it allows workers to learn vicariously about their work. In an environment in which mistakes are costly both financially and personally, and in which learning from mistakes is undesirable, information sharing is critical to learning and to incident prevention (Weick, 1987). In sum, the sharing of information between organizational members is an important part of any high-performance work system. It is also has an integral role in the reduction of safety incidents such as near misses, injuries, and fatalities.

Compensation Contingent on Safe Performance

Within the high-performance work system described by Pfeffer (1998c), paying people well is argued to enhance organizational performance. Paying employees well expands the pool of potential applicants for any available jobs in the organization and signals to current employees that they are valued by the organization. The same holds true with respect to workplace safety. As one North Sea oil worker stated, "If you're getting paid a wage that you're happy with, then you're happy at your work so you're switched on and alert. You don't mind doing your bit" (Collinson, 1999, p. 591).

Even more important is the opportunity that contingent compensation provides to organizations to convey unambiguously which behaviors are most valued. Under such a compensation system, employees are also motivated to contribute to the organization to the extent to which their interests are aligned with those of the organization. In terms of enhancing worker safety, compensation that is dependent on occupational safety (rather than on the number of accident reports) would signal to employees what the organization really values.

There are data demonstrating that rewarding employees on the basis of their safety performance is effective in reducing workplace injuries. The use of tokens and other reinforcers has been examined in a number of studies and has been shown to reduce injury rates (see, e.g., Austin, Kessler, Riccobono, & Bailey, 1996; Haynes, Pine, & Fitch, 1982). Nevertheless, this approach is undesirable from a high-performance work system perspective. Beyond the facts that many of the behaviors learned are highly specific and do not generalize to other contexts and that the long-term effects are ill understood (McAfee & Winn, 1989), this method of encouraging occupational safety also suffers because of its inherent reliance on the use of control to encourage behaviors. This is undesirable to the extent that it constitutes manipulation; consequently, these methods fail to be endorsed by unions (Walker, 1998).

More consistent with Pfeffer's (1998c) proposition are (a) that employees be compensated for organizational performance at the group level and (b) that compensation must be provided for behaviors beyond individual-level safety. For instance, it was found in an open-pit mine that employees who were rewarded with tokens for working safely not only on an individual basis but also as a group, and who were rewarded for making suggestions of ways to improve safety and for taking unusual action to prevent injuries, experienced significantly fewer injuries than their unrewarded peers (Fox, Hopkins, & Anger, 1987). Although the approach applied in this study is more in line with a high-performance work system model than other compensation approaches are, it is necessary for future research to more thoroughly examine the role of fair and contingent compensation on workplace safety.

Nonetheless, for conceptual, measurement, and empirical reasons, we conclude our comments on contingent compensation with a word of caution. Conceptually, it should be noted that not all researchers view contingent compensation as part of a high-performance work system. As Wood and Wall (2002) remind us, Arthur (1994) initially conceptualized compensation as part of a control orientation. Measurement issues also loom large because of reports showing that focusing employees' attention on a reduction in the number of accidents rather than an improvement in occupational safety can result in less safety (Collinson, 1999).

Transformational Leadership

The focus in the literature on the role of leadership in enhancing workplace safety is by no means new (e.g., Butler & Jones, 1979; Dunbar, 1975). For instance, both A. Cohen (1977) and Smith et al. (1978) found that strong management commitment to worker safety characterized low-injury-rate companies, and Hofmann et al. (1995) showed recently that management commitment to safety impacted how motivated employees were to work safely. We believe that transformational leadership provides an appropriate leadership model for demonstrating a commitment to safety and in turn enhancing workplace safety.

Transformational leaders are able to act as role models to their followers. They are highly respected because they do what is right and not necessarily what is easy or personally profitable. They are able to inspire their followers to go beyond their individual needs and work toward the collective good of the group, and they are able to encourage followers to challenge assumptions and to examine problems from new angles. Finally, transformational leaders are concerned with the needs and interests of their followers (Bass, 1998). With respect to workplace safety, transformational leaders would be able to convey to followers the value they place on occupational safety and would be able to persuade employees to work as safely as possible. They would also be capable of encouraging their followers to look at safety problems from different angles and would demonstrate concern for each individual's safety.

Indeed, transformational leadership has been found to be associated with improved safety. In one study of the offshore oil and gas industry, managers' transformational leadership predicted the willingness of workers to take initiative in safety matters (O'Dea & Flin, 2000). Similarly, to the extent that supervisors exhibited transformational leadership, restaurant workers experienced fewer occupational injuries (Barling, Loughlin, & Kelloway, 2002).

Similar results were found within the team context. Williams, Turner, and Parker (2000) found that transformational leadership predicted the extent to which individuals working in teams followed safety procedures and participated in safety behaviors beyond those required. Zohar (2002) showed that transformational leadership was negatively associated with the number of minor injuries experienced by workers in a metal processing plant. Collectively, therefore, the findings from these studies provide evidence for the role of transformational leadership in reducing the number of safety incidents experienced, and injuries suffered, by workers.

High-Quality Work

A job that is "more fulfilling and effective" (Parker & Wall, 1998, p. ix) will ensure that workers are focused, attentive, and emotionally engaged. In Wheatley's (1997) words, "You can't direct people into perfection; you can only engage them enough so they want to do perfect work" (p. 25), and job quality will also promote safe working. Although there are a number of dimensions of job quality (task significance, feedback), we will consider only three: appropriate workload, employee control, and role clarity.

With respect to appropriate workload (i.e., work that is neither overly taxing nor boring), both work overload and underload are associated with worker safety. In a study of petrochemical workers, high job demands were associated with greater perceived risk (Baugher & Roberts, 1999). Eyssen, Hoffman, and Spengler (1980) found that safety was compromised when managers suffered from unusually heavy workloads. Similarly, Hofmann and Stetzer (1996) demonstrated that a greater push for production was associated with diminished workplace safety in a chemical processing plant. Work underload can have similarly negative effects on worker safety. In a study of employed adolescents (Frone, 1998), on-the-job boredom and work overload were both positively related to work-related injuries.

The provision of greater autonomy is a further aspect of job quality that has been found to benefit workplace safety (Barling, Kelloway & Iverson, 2003). Parker et al. (2001) reported that job autonomy increases employee commitment to the organization and, in turn, employee safety compliance. Similarly, Simard and Marchand (1995) found that a participative approach to managing safety predicted the extent to which employees were proactively involved in their own safety; autonomy also predicted group cohesion, a further factor in individuals' propensity to work safely.

Finally, role clarity—a further component of job quality—is also important to workplace safety. In a study of junior medical doctors, Houston and Allt (1997) showed that assuming a new role and the associated distress of not knowing the role requirements were associated with a greater number of medical errors. The potential for the generalizability of these findings to occupational safety is strong but remains to be tested empirically. Similarly, Hemingway and Smith (1999) found that role ambiguity among nurses was significantly associated with the number of injuries they experienced. Overall, therefore, high-quality work characterized by an appropriate amount of work, role clarity, and employee control has been associated in the literature with workplace safety.

Measurement of Variables Critical to Organizational Success

Measurement is an important human resource practice, given that "organizations get the kinds of behaviors they measure and reward" (Lawler, 1996, p. 232) and should therefore be a part of current conceptualizations of human resource management systems. In terms of workplace safety, there are a number of measurement issues that deserve attention. First, most organizations are preoccupied with complying with government regulations and provisions in collective agreements that regulate occupational safety. This is problematic in the extent to which this directs attention on the number of accidents and events, and as a result measures behaviors consistent with a control orientation. Second, as Wheatley (1997, p. 25) notes, focusing on compliance with government regulations encourages the identification and correction of safety-related infractions while neglecting those opportunities to go beyond regulations and enhance safety through, for instance, the sharing of ideas. Third, government fines for safety infractions, increasing insurance premiums, and collective agreements provide numerous incentives to organizations to underreport safety infractions (Collinson, 1999; Conway & Svenson, 1998). As such, safety levels measured in terms of the number of incidents may provide organizations with data that are unreliable (Collinson, 1999; Donald, 1995). Furthermore, this type of record keeping does not provide the organization with any information regarding the number of near misses or less severe injuries that occur.

In contrast, optimal measures will provide information that is useful for interventions. The usefulness of financial reporting systems provides a good lesson. Such systems are limited because they often focus on events that have already occurred while providing less information about current conditions (Pfeffer, 1998b). With respect to safety, therefore, focusing on current safety conditions and employee attitudes and behaviors that predict subsequent safety performance would provide more relevant information for the prevention of future safety incidents than would focusing on past safety incidents. This is not to say that the number of safety incidents need not be considered. Rather, considering process-oriented measures would provide an organization with much richer information regarding safety. What is needed to enhance safety in the long term, therefore, is the measurement of varied aspects of safety, including the proximal causes of injuries such as worker attitudes and behaviors. For instance, it would be worthwhile to measure employees' commitment to the organization, job satisfaction and trust in management, and the extent to which workers take initiative with respect to safety and participate in safety matters. Furthermore, it would be worthwhile to measure the high-performance work system discussed here by considering, for instance, perceived employment security, the amount of training received, job quality, and the extent to which employees work in teams.

HOW DO HIGH-PERFORMANCE WORK SYSTEMS INFLUENCE OCCUPATIONAL SAFETY?

In the preceding section of this chapter we discuss each of the 10 high-performance work practices and their potential effects on workplace safety. The question remains as to how these practices exert positive effects on occupational safety. We believe that these practices serve to increase employees' trust in management, affective commitment to the organization, and perceptions of safety climate—all of which improve safety performance. Each of these will be discussed in turn.

When evaluating management practices, Pfeffer (1998a) opined that the single most important factor is whether the practices convey trust or destroy trust. An examination of the literature indicates that a number of studies have demonstrated that high-performance work practices promote trust in management. For instance, it has been found that trust in management mediates the relationship between transformational leadership and follower performance (Barling, Moutinho, & Kelloway, 2001; Jung & Avolio, 2000). Similarly, employment security has been found to be another factor promoting trust in management, especially in unstable times (Brockner, 1988).

The sharing of information within an organization should also serve to increase trust between management and employees. As we mentioned earlier in this chapter, Clarke's (1999) study of British Rail employees found that both management and employees underestimated the extent to which the other group valued workplace safety. Under these conditions, we should not expect workers to trust management when safety-related issues are concerned. Encouraging the sharing of information between the two parties, however, would have promoted trust between workers and management.

There is a paucity of research examining the effects of trust on workplace safety. However, a literature review by Kramer (1999) may be informative. Kramer reported that trust results in spontaneous sociability such that employees are more likely to cooperate, act in ways that extend beyond their roles, work toward common goals, and share information—all behaviors that may well promote workplace safety.

Affective commitment to the organization (Meyer & Allen, 1997) is a further mediating mechanism through which high-performance work practices impact desirable organizational outcomes, one of which is workplace safety. Numerous studies support the role of the high-performance work practices in encouraging affective commitment. For instance, a meta-analysis conducted by Mathieu and Zajac (1990) found that leadership predicts overall loyalty to the organization, whereas transformational leadership more specifically has also been associated with greater affective commitment (Barling et al., 1996; Barling et al., 2001). Similarly, both job quality (Mathieu & Zajac, 1990) and employment security (Ashford, Lee, & Bobko, 1989; Probst & Brubaker, 2001; Tsui, Pearce, Porter, & Tripoli, 1997) have been found to be related to affective commitment to the organization.

In turn, affective commitment has been found to predict work performance (Barling et al., 2001; Meyer, Paunonen, Gellatly, Goffin, & Jackson, 1989), including safety performance. Hackett, Bycio, and Hausdorf (1994) found in a sample of bus operators that affective commitment to the organization was significantly associated with the number of accidents they experienced in a year. In addition, Parker et al. (2001) found that communication quality and the ability to work autonomously, a function of job quality, were both associated with greater affective commitment to the organization and, in turn, to safer working. In the team context, organizational commitment was negatively related to the number of days lost due to a work injury (S. G. Cohen & Ledford, 1994). Most likely on account of range restrictions in the data, organizational commitment was not found to be associated with safety levels in this study.

Finally, we argue that perceived safety climate provides a further mechanism by which the high-performance work practices impact workplace safety. Safety climate is a subset of the overall organizational climate and refers to employees' perceptions of their work environment with respect to safety policies, procedures, and rewards (Griffin & Neal, 2000). Although there is little research examining the organizational factors that promote safety climate, such practices as the provision of extensive training beyond that mandated by government regulations would serve to increase employees perceptions that their organization is committed to workplace safety. Similarly, to the extent that jobs are of high quality and, for instance, the workload is appropriate (Zohar, 1980a), positive perceptions of safety climate will be encouraged. Transformational leadership is also expected to result in more favorable perceptions of the company's safety climate (Barling et al., 2001; Zohar, 2002).

Safety climate has been found to be a proximal predictor of safety behaviors. For instance, employees in a chemical processing plant with more positive perceptions of safety climate engaged in unsafe behaviors with less frequency (Hofmann & Stetzer, 1996). Furthermore, Zohar (2002) found that safety climate predicted actual injuries in a sample of production workers, as did Barling et al. (2002) in a sample of restaurant workers.

CONCLUDING THOUGHTS

We have highlighted the value of 10 high-performance work practices, each of which (a) are associated with occupational safety and (b) encourage trust in management, affective commitment to the organization, and perceived safety climate, which in turn enhance occupational safety. Nonetheless, much of the discussion in this chapter is speculative, and it now remains for future research to focus in detail on these relationships. Given the extent to which prior studies have shown the wide-ranging effects of high-performance work systems, these effects may well extend to our understanding, prediction, and control of occupational safety.

REFERENCES

Arthur, J. B. (1992). The link between business strategy and industrial relations systems in American steel minimills. *Industrial and Labor Relations Review, 45*, 488–506.

Arthur, J. B. (1994). Effects of human resource systems on manufacturing performance and turnover. *Academy of Management Journal, 37*, 670–687.

Ashford, S., Lee, C., & Bobko, P. (1989). Content, causes and consequences of job insecurity: A theory-based measure and substantive test. *Academy of Management Journal, 32*, 803–829.

Austin, J., Kessler, M. L., Riccobono, J. E., & Bailey, J. S. (1996). Using feedback and reinforcement to improve the performance and safety of a roofing crew. *Journal of Organizational Behavior Management, 16*(2), 49–75.

Barling, J., & Hutchinson, I. (2000). Commitment vs. control-based safety practices, safety reputation, and perceived safety climate. *Canadian Journal of Administrative Sciences, 17*, 76–84.

Barling, J., Kelloway, E. K., & Iverson, R. (2003). High quality work, job satisfaction and occupational injuries. *Journal of Applied Psychology, 88*, 276–283.

Barling, J., Loughlin, C., & Kelloway, E. K. (2002). Development and test of a model linking safety-specific transformational leadership and occupational safety. *Journal of Applied Psychology, 87*, 488–496.

Barling, J., Moutinho, S., & Kelloway, E. K. (2001). *Transformational leadership and group performance: The mediating role of affective commitment*. Revised manuscript submitted for publication.

Barling, J., Weber, T., & Kelloway, E. K. (1996). Effects of transformational leadership training on attitudinal and financial outcomes: A field experiment. *Journal of Applied Psychology, 81*, 827–832.

Bass, B. M. (1998). *Transformational leadership: Industry, military and educational impact.* Mahway, NJ: Erlbaum.

Baugher, J. E., & Roberts, J. T. (1999). Perceptions and worry about hazards at work: Unions, contract maintenance, and job control in the U.S. petrochemical industry. *Industrial Relations, 38*, 522–541.

Borofsky, G. L., Bielema, M., & Hoffman, J. (1993). Accidents, turnover, and use of a preemployment screening inventory. *Psychological Reports, 73*, 1067–1076.

Brockner, J. (1988). The effects work layoffs on survivors: Research, theory and practice. In B. M. Staw & L. L. Cummings (Eds.), *Research in Organizational Behavior* (Vol. 10, pp. 213–255). Greenwich, CT: JAI Press.

Butler, M. C., & Jones, A. P. (1979). Perceived leader behavior, individual characteristics, and injury occurrence in hazardous work environment. *Journal of Applied Psychology, 64*, 299–304.

Clarke, S. (1999). Perceptions of organizational safety: Implications for the development of safety culture. *Journal of Organizational Behavior, 20*, 185–198.

Cohen, A. (1977). Factors in successful occupational safety programs. *Journal of Safety Research, 9*, 168–178.

Cohen, S. G., & Ledford, G. E., Jr. (1994). The effectiveness of self-managing teams: A quasi-experiment. *Human Relations, 47*, 13–43.

Collinson, D. L. (1999). "Surviving the rigs": Safety and surveillance on North Sea oil installations. *Organization Studies, 20*, 579–600.

Conway, H., & Svenson, J. (1998, November). Occupational injury and illness rates, 1992–96: Why they fell. *International Labor Review*, 36–58.

DeJoy, D. M. (1994). Managing safety in the workplace: An attribution theory analysis and model. *Journal of Safety Research, 25*, 3–17.

Donald, I. (1995). Psychological insights into managerial responsibility for public and employee safety. In R. Bull & D. Carson (Eds.), *Handbook of psychology in legal contexts* (pp. 625–642). New York: Wiley.

Dunbar, R. L. M. (1975). Managers influence on subordinates' thinking about safety. *Academy of Management Journal, 18*, 364–369.

Eyssen, G. M., Hoffman, J. E., & Spengler, R. (1980). Managers' attitudes and the occurrence of accidents in a telephone company. *Journal of Occupational Accidents, 2*, 291–304.

Fox, D. K., Hopkins, B. L., & Anger, W. K. (1987). The long-term effects of a token economy on safety performance in open-pit mining. *Journal of Applied Behavior Analysis, 20*, 215–224.

Frone, M. R. (1998). Predictors of work injuries among employed adolescents. *Journal of Applied Psychology, 83*, 565–576.

Geller, E. S., Roberts, D. S., & Gilmore, M. R. (1996). Predicting propensity to actively care for occupational safety. *Journal of Safety Research, 27*, 1–8.

Goodman, P. S., & Garber, S. (1988). Absenteeism and accidents in a dangerous environment: Empirical analysis of underground coal mines. *Journal of Applied Psychology, 73*, 81–86.

Griffin, M. A., & Neal, A. (2000). Perceptions of safety at work: A framework for linking safety climate to safety performance, knowledge, and motivation. *Journal of Occupational Health Psychology, 5*, 347–358.

Griffiths, D. K. (1985). Safety attitudes of management. *Ergonomics, 28*, 61–67.

Grunberg, L., Moore, S., & Greenberg, E. (1996). The relationship of employee ownership and participation to workplace safety. *Economic and Industrial Democracy, 17*, 221–241.

Hackett, R. D., Bycio, P., & Hausdorf, P. A. (1994). Further assessments of Meyer and Allen's (1991) three-component model of organizational commitment. *Journal of Applied Psychology, 79*, 15–23.

Hale, A. R. (1984). Is safety training worthwhile? *Journal of Occupational Accidents, 6*, 17–33.

Haynes, R. S., Pine, R. C., & Fitch, H. G. (1982). Reducing accident rates with organizational behavior modification. *Academy of Management Journal, 25*, 407–416.

Hechanova-Alampay, R. H., & Beehr, T. A. (2002). Empowerment, span of control and safety performance in work teams after workforce reduction. *Journal of Occupational Health Psychology, 6*, 275–282.

Hemingway, M. A., & Smith, C. S. (1999). Organizational climate and occupational stressors as predictors of withdrawal behaviors and injuries in nurses. *Journal of Occupational and Organizational Psychology, 72*, 285–299.

Hofmann, D. A., Jacobs, R., & Landy, F. (1995). High reliability process industries: Individual, micro, and macro organizational influence on safety performance. *Journal of Safety Research, 26*, 131–149.

Hofmann, D. A., & Morgeson, F. P. (1999). Safety-related behavior as a social exchange: The role of perceived organizational support and leader–member exchange. *Journal of Applied Psychology, 84*, 286–296.

Hofmann, D. A., & Stetzer, A. (1996). A cross-level investigation of factors influencing unsafe behaviors and accidents. *Personnel Psychology, 49*, 307–339.

Houston, D. M., & Allt, S. K. (1997). Psychological distress and error making among junior house officers. *British Journal of Health Psychology, 2*, 141–151.

Huselid, M. A. (1995). The impact of human resource practices on turnover, productivity, and corporate financial performance. *Academy of Management Journal, 38*, 645–672.

Ichniowski, C., Shaw, K., & Prennushi, G. (1997). The effects of human resource management practices on productivity: A study of steel finishing lines. *American Economic Review, 87*, 291–313.

Jackall, R. (1988). *Moral mazes: The world of corporate managers*. New York: Oxford.

Jones, J. W. (1991). A personnel selection approach to industrial safety. In J. W. Jones (Ed.), *Preemployment honesty testing* (pp. 185–194). New York: Quorum.

Jones, J. W., & Wuebker, L. J. (1988). Accident prevention through personnel selection. *Journal of Business and Psychology, 3*, 187–198.

Jung, D. I., & Avolio, B. J. (2000). Opening the black box: An experimental investigation of the mediating effects of trust and value congruence on transformational and transactional leadership. *Journal of Organizational Behavior, 21*, 949–964.

Kaminski, M. (2001). Unintended consequences: Organizational practices and their impact on workplace safety and productivity. *Journal of Occupational Health Psychology*, 6, 127–138.

Kincaid, W. H. (1996, August). Safety in the high-turnover environment. *Occupational Health and Safety*, *65*, 22, 24–25.

Kivimäki, M., & Kalimo, R. (1993). Risk perception among nuclear power plant personnel: A survey. *Risk Analysis*, *13*, 421–424.

Kivimäki, M., Kalimo, R., & Salminen, S. (1995). Perceived nuclear risk, organizational commitment, and appraisals of management: A study of nuclear power plant personnel. *Risk Analysis*, *15*, 391–396.

Kochan, T., Smith, M., Wells, J. C., & Rebitzer, J. (1994). Human resource strategies and contingent workers: The case of safety and health in the petrochemical industry. *Human Resource Management*, *33*, 55–77.

Kramer, R. M. (1999). Trust and distrust in organizations: Emerging perspectives, enduring questions. *Annual Review of Psychology*, *50*, 569–598.

Lawler, E. E., III. (1996). *From the ground up: Six principles for building the new logic corporation*. San Francisco: Jossey-Bass.

MacDuffie, J. P. (1995). Human resource bundles and manufacturing performance: Organizational logic and flexible production systems in the world automotive industry. *Industrial and Labor Relations Review*, *48*, 197–221.

Mathieu, J. E., & Zajac, D. (1990). A review and meta-analysis of the antecedents, correlates, and consequences of organizational commitment. *Psychological Bulletin*, *108*, 171–194.

McAfee, R. B., & Winn, A. R. (1989). The use of incentives/feedback to enhance work place safety: A critique of the literature. *Journal of Safety Research*, *20*, 7–19.

McLain, D. L. (1995). Responses to health and safety risk in the work environment. *Academy of Management Journal*, *38*, 1726–1742.

Meyer, J. P., & Allen, N. J. (1997). *Commitment in the workplace: Theory, research and application*. Thousand Oaks, CA: Sage.

Meyer, J. P., Paunonen, S. V., Gellatly, I. H., Goffin, R. D., & Jackson, D. N. (1989). Organizational commitment and job performance: It's the nature of the commitment that counts. *Journal of Applied Psychology*, *74*, 152–156.

Milanovich, D. M., Driskell, J. E., Stout, R. J., & Salas, E. (1998). Status and cockpit dynamics: A review and empirical study. *Group Dynamics: Theory, Research and Practice*, *2*, 155–167.

Nelkin, D., & Brown, M. S. (1984). Workers at risk: Voices from the workplace. Chicago: University of Chicago.

O'Dea, A., & Flin, R. (2000, August). *Safety leadership in the offshore oil and gas industry*. Paper presented at the Academy of Management Annual Meeting, Toronto, Canada.

O'Reilly, C. A., & Pfeffer, J. (2000). *Hidden value*. Cambridge, MA: Harvard Business School Press.

Parker, S. K., Axtell, C., & Turner, N. (2001). Designing a safer workplace: Importance of job autonomy, communication quality, and supportive supervisors. *Journal of Occupational Health Psychology, 6*, 211–228.

Parker, S. K., & Wall, T. D. (1998). *Job and work design: Organizing work to promote well-being and effectiveness*. Thousand Oaks, CA: Sage.

Parker, S. K., Wall, T. D., & Jackson, P. R. (1997). "That's not my job": Developing flexible employee work orientations. *Academy of Management Journal, 40*, 899–929.

Patterson, M. G., West, M. A., & Wall, T. D. (2001). *Integrated manufacturing, empowerment and company performance*. Manuscript submitted for publication.

Pfeffer, J. (1998a). *The human equation: Building profits by putting people first*. Boston: Harvard Business School Press.

Pfeffer, J. (1998b, Spring). The real keys to high performance. *Leader to Leader, 22*, 23–29.

Pfeffer, J. (1998c). Seven practices of successful organizations. *California Management Review, 40*, 96–124.

Probst, T. M., & Brubaker, T. L. (2001). The effects of job insecurity on employee safety outcomes: Cross-sectional and longitudinal explorations. *Journal of Occupational Health Psychology, 6*, 139–159.

Sells, B. (1994, March–April). What asbestos taught me about managing risk. *Harvard Business Review*, 76–90.

Simard, M., & Marchand, A. (1995). A multilevel analysis of organizational factors related to the taking of safety initiative by work groups. *Safety Science, 21*, 113–129.

Simard, M., & Marchand, A. (1997). Workgroups' propensity to comply with safety rules: The influence of micro–macro organisational factors. *Ergonomics, 40*, 172–188.

Sjöberg, L., & Drottz-Sjöberg, B. M. (1991). Knowledge and risk perception among nuclear power plant employees. *Risk Analysis, 11*, 607–618.

Smith, M. J., Cohen, H. H., Cohen, A., & Cleveland, R. J. (1978). Characteristics of successful safety programs. *Journal of Safety Research, 10*, 5–15.

Tannenbaum, S. I., Mathieu, J. E., Salas, E., & Cannon-Bowers, J. A. (1991). Meeting trainees' expectations: The influence of training fulfillment on the development of commitment, self-efficacy, and motivation. *Journal of Applied Psychology, 76*, 759–769.

Tjosvold, D. (1990). Flight crew collaboration to manage safety risks. *Group and Organization Studies, 15*, 177–191.

Trist, E. L., Susman, G. I., & Brown, G. R. (1977). An experiment in autonomous working in an American underground coal mine. *Human Relations, 30*, 201–236.

Tsui, A. S., Pearce, J. L., Porter, L. W., & Tripoli, A. M. (1997). Alternative approaches to the employee–organization relationship: Does investment in employees pay off? *Academy of Management Journal, 40*, 1089–1121.

Walker, C. (1998, July). *Behavior based safety programs, or "if it's rat psychology, who is Pied Piper and who are the rats?"* Paper presented at the Association of Workers' Compensation Boards of Canada Congress, Winnipeg, Manitoba.

Walton, R. E. (1985, March–April). From control to commitment in the work place. *Harvard Business Review, 33*, 77–84.

Weick, K. E. (1987). Organizational culture as a source of high reliability. *California Management Review, 29*, 112–127.

Wheatley, M. (1997, Summer). Goodbye, command and control. *Leader to Leader, 19*, 21–28.

Whitefield, K. (2000). High-performance workplaces, training and the distribution of skills. *Industrial Relations, 39*, 1–25.

Williams, H., Turner, N., & Parker, S. K. (2000, August). *The compensatory role of transformational leadership in promoting safety behaviors*. Paper presented at the Academy of Management Annual Meeting, Toronto, Canada.

Wood, S., & Wall, T. D. (2002). Human resource management and business performance. In P. B. Warr (Ed.), *Psychology at work* (pp. 351-374). London: Penguin.

Zohar, D. (1980). Safety climate in industrial organizations: Theoretical and applied implications. *Journal of Applied Psychology, 65*, 96–102.

Zohar, D. (2002). The effects of leadership dimensions, safety climate, and assigned priorities on minor injuries in work groups. *Journal of Organizational Behavior, 23*, 75–92.

11

THE ROLE OF TRAINING IN PROMOTING WORKPLACE SAFETY AND HEALTH

MICHAEL J. COLLIGAN AND ALEXANDER COHEN

It is obvious that people learn and that they apply what they've learned to new and varied situations. This happens so frequently, and in many cases so effortlessly, that it is easy to take the process for granted. It is not surprising, then, that so many public health interventions involve an educational component, the assumption being that once people are informed about a risk and given corrective or preventive direction, they will act appropriately. Experience has demonstrated, however, that although education may be a necessary component of behavioral change, it is not sufficient to produce it. Despite the investment of millions of dollars and decades of effort in designing information campaigns to reduce obesity, promote safe sex, eliminate cigarette smoking, and encourage annual physical examinations, a significant segment of the population continues to engage in unsafe, risky behavior. It is not until we deliberately attempt to influence specific behaviors in defined populations and contexts that we become aware of the social, political, and practical complexities of the education or training process in real-world settings. With respect to promoting occupational safety and health, for example, training can be an expensive and time-consuming proposition, impacting on production schedules, personnel staffing, record keeping, facility logistics and resources, and additional administrative burdens. Despite people's intuitive acceptance that training works, the truth is that it doesn't always work, and even when it does one might ask if it has been delivered in the most efficient manner possible. It is therefore important that researchers, as well as training practitioners, understand the factors that influence training's effectiveness in reducing workplace injury and illness.

The present chapter looks at the role of training in promoting worker safety and health from three perspectives. The first perspective deals with some of the ethical and contextual issues associated with workplace safety training. The primary concern is to examine how organizational and social/psychological considerations may affect how training is accepted and acted on in a workplace setting. The second perspective involves a review of the research literature to evaluate the efficacy of training as a safety and health intervention. Here the attempt is to identify critical components of successful training interventions with an eye toward practical recommendations. Finally, the third perspective attempts to provide some guidance for future research and program development related to workplace safety and health training.

OCCUPATIONAL SAFETY AND HEALTH TRAINING: ORGANIZATIONAL AND SOCIAL/PSYCHOLOGICAL CONSIDERATIONS

In 1998, the U.S. Occupational Safety and Health Administration (OSHA) published a draft proposed rule for the implementation and management of workplace programs designed to protect employee safety and health. The draft identified employee training as one of the five essential elements of such programs. The other elements involved employer commitment, hazard analysis or surveillance, hazard control and prevention, and program evaluation. The prominence given to training is a reflection of both ethical and pragmatic considerations. On ethical grounds, it is generally accepted that workers have a right to know what risks they are assuming as a condition of their employment and what actions can be taken at the organizational and personal level to manage those risks. In a more practical vein, it is also recognized that work-related injuries and illnesses are costly in terms of both their direct and immediate impact (e.g., worker pain and suffering, company indemnity, lost-work time, decreased productivity) and their long-range consequences (e.g., worker disability, poor morale and turnover, bad public relations, escalating insurance costs). If worker training can reduce the incidence of illness and injury, then it would seem to be in the best interests of both the worker and the employer that such training be provided. The fact that in its model program OSHA recognized the need to embed the training function in a broader system safety structure—including organizational commitment, hazard surveillance, and control implementation—has important implications for the training practitioner. Although training is generally regarded as a worthwhile endeavor, it must be remembered that training programs are always delivered within a broader organizational context or culture. This "figure–ground" relationship can influence the perceptions and expectations of managers and workers about the purpose and value of training in subtle ways that are quite unrelated to the quality of the training program itself. Some examples follow.

In attempting to manage work-related hazards, occupational health professionals rely on a hierarchy of controls, ranging from engineering design as the most preferred method, through administrative controls (e.g., job rotation, workplace regulations), to the use of personal protective equipment and recommended work practices (Olishifski, 1988). The intent of the hierarchy is to "engineer out" as many hazards as feasible in order to minimize both human error and personal burden. As an example, a noise abatement program may reduce the hazard at its source through the purchase of quieter equipment or acoustical enclosures, or it may rely on a personal protective equipment approach in which workers are trained about the consequences of excessive noise and instructed on the proper use of hearing protection. The former strategy requires considerable initial costs for the company but minimal costs to the worker, and only limited training on the use and maintenance of the machinery. The latter strategy reduces company up-front costs but increases the demands on the worker, and requires a more intensive and continuing training and safety management effort. Should workers perceive that safety training is being delivered in lieu of implementing the less invasive engineering controls, they may feel that the safety burden has been disproportionately and unfairly shifted to their shoulders. Similarly, if workers are being trained in work practices that are not being followed by floor supervisors, or employees (workers and supervisors alike) are not given the equipment, resources, and support to enact recommended practices and policies, the training program quickly loses credibility. Rather than enhancing worker protection, the training offered in this context may be seen as merely cosmetic and diversionary, actually leading to increased resentment and worker reactance (Brehm, 1966).

Training can also be a sensitive issue when conducting accident investigations and making attributions of responsibility. Although it is reasonable to assume that trained workers are better prepared than nontrained workers to deal with work-related hazards, this isn't necessarily the case. The training may be inadequate, the necessary engineering controls or supervisory support may be lacking, required protective equipment may be unavailable or poorly maintained, company production pressures may result in a de-emphasis of safety procedures, and so on. Whether or not these factors are considered in an accident investigation depends on the organization's culture and management philosophy. Lacey, Husted Medvec, and Messick (1998), for example, have suggested that when the purpose of an accident investigation is to design a prevention program, people tend to make external, system-focused causal attributions. When the purpose is to assign responsibility and blame, people tend to make internal, personal attributions. Thus, workers in an autocratic organization who have been trained may be held more accountable for their safety (and more liable for their injury) than untrained workers, regardless of the actual quality of the safety environment.

The lesson to be drawn from the discussion in the preceding paragraphs is that training is not a general panacea or quick fix for all safety and health problems within an organization. Its effectiveness will be a function of the organization's overall commitment to providing a safe work environment and the employees' perception and recognition of that commitment. Although much of the research reviewed in the following section describes isolated studies looking at specific training issues or work hazards, it is important to keep in mind the broader system's safety perspective when trying to determine what makes training work.

REVIEW OF SAFETY AND HEALTH TRAINING LITERATURE

A recent 174-page technical report (Cohen & Colligan, 1998) issued by the U.S. National Institute for Occupational Safety and Health (NIOSH) offered an extensive review of the research literature dealing with the efficacy of training in promoting workplace safety and health. Only published reports involving some form of experimental design (e.g., pretraining–posttraining comparison, trained vs. untrained control group comparison) and quantitative outcome data (e.g., knowledge gain, behavioral change, injury rate change) were considered. Most attention in the report was given to 80 studies drawn largely from the period 1980 to 1996, when training interventions were used for the purposes of reducing risks to five specific types of workplace hazards (injury, chemical, biological, physical, and ergonomic). These studies offered a critical database for evaluating whether such training proved effective and what methodological issues were of importance. Tables 11.1, 11.2, and 11.3, as discussed in the following sections, are extracted from the NIOSH report and highlight major features and findings of these studies. They, along with other materials taken from the NIOSH report, are presented in this chapter as a way of characterizing the occupational safety and health training literature as a whole. Although we mention select studies in discussing specific points, the reader may wish to consult the NIOSH report for references to and descriptions of all 80 intervention studies (see References for Web address).

TRAINING EVALUATION MEASURES

Table 11.1 sorts the 80 studies according to the type of hazard being addressed by the training, the evaluation measures used to assess the training effect, and whether a successful training outcome was achieved. Particularly notable in Table 11.1 are the variety of measures that have been used to evaluate training outcomes and the overwhelming number of entries that indicate a positive effect attributed to the training. Each of these observations deserves comment.

Training evaluation measures reported in the literature ranged from simple subjective customer satisfaction measures to more global and objective measures of organizational impact. Thus, measures in Table 11.1 under the heading of Subjective/Self-Reports include *reaction* (trainee ratings of whether the training was interesting, worthwhile, or relevant); *knowledge test* (scores on a quiz, behavioral assessment, or other inquiry to determine knowledge gain); and *application* (reports from the trainees that they have changed their work practices as a result of the training or applied the learning in some other way). The heading Objective/Surrogates refers to more independent indicators, listing direct observations of behavior change or other markers (biological, environmental measures) that can be a product of the behavior changes. Positive changes from training on either of these measures serve as surrogates for improved health and safety outcomes. The last columns of Table 11.1 refer to actual measures of health and safety experience reflecting the organizational impact of the training.

Some studies used more than one evaluative measure, so the total number of entries in Table 11.1 exceeds the 80 reports. In these latter instances, efforts to show interdependence if not causal linkage among these various measures proved elusive. For example, an attempt to tie successful training outcomes such as improved work practices with reduced injury rates found that the improvement could account for only 25% of the actual injury reduction (Saari & Nasanen, 1989). In this study, worker training focused on correcting housekeeping conditions believed responsible for an excess number of injuries based on accident reports. Posttraining observations indicated reduced injury rates far greater than those that could be accounted for by increased compliance with the better housekeeping practices that were also observed. The authors speculated that the gains in housekeeping left

TABLE 11.1
Evaluation Measures Used in Training Intervention Studies and Results Reported

	Subjective/Self-Reports			Objective/ Surrogates		Organizational Results		
Hazard or agent (number of reports)	Reaction survey	Knowledge test	Applications	Behavior change	Other markers	Reduced injury	Reduced illness	Less cost days lost
Injury (21 reports)	++++	+	+++	+++++ +++++ +/o	++	++++++++ ++++ o		+++
Chemical (22 reports)	+++++++	+++++++++ +/o	++++++++ +++++	+++ +/o	++++++	o	o	
Physical (10 reports)	+	+++	++	+++++ +/o, +/o		+	+	
Ergonomic (19 reports)	++ o, o, o	+, +/o	+/o	++++++ +/o, +/o o	o+	++++	+	+++++
Biologic (8 reports)	+	+++++ +/o	++++	+	+	o	+	
Totals (80 reports)	15(+) 3(o)	19(+) 3(+/o)	22(+) 1(+/o)	25(+) 6(+/o) 1(o)	10(+) 1(o)	17(+) 3(o)	3(+) 1(o)	8(+)

Note. (+) = Positive effect on measure; (+/o) = Mixed effect; (o) = No effect increased capacity for workers to notice other potential hazards. Since the NIOSH report (Cohen & Colligan, 1998), another study (A. Cohen, 1998), employing multiple indicators (e.g., self-ratings of competency, knowledge gain, behavior change) in evaluating firefighter training in hazardous materials response, found significant correlations among the different measures but only in a limited number of cases. Further investigative study of connections between the various training outcomes seems indicated.

Training Outcomes

In Table 11.1, 119 of 138 entries show a positive change from the training or instructional effort on the measures indicated and 10 others show at least partial success. Indeed, in only nine cases do the findings display no effect, and some of these results could be traced to constraints in the transfer of a positive learning effect to a real workplace situation. In this regard, studies by Carlton (1987) and Scholey (1983), both directed to controlling ergonomic hazards, were particularly illustrative. In Carlton's study (1987), kitchen workers in training sessions held away from their job sites learned and demonstrated ways for lifting trays that imposed less strain on the low back. Subsequent observations found that these types of lifts could not be performed in their work areas because the layout and obstacles made them assume awkward postures in handling loads of trays and the work pace precluded time to follow through on the acts required for risk reduction. Scholey (1983) reported that nurses could not transfer methods learned to ease the burden imposed in patient handling and lifting tasks because some patients refused to cooperate in moving to the edge of the bed or chair to facilitate the move. Compensating for these difficulties, other studies in the ergonomics hazard category show some of the more substantial reductions in the frequency and the cost of work-related injuries as evidence of training effectiveness. However, reports by Hilyer, Brown, Sirles, and Peoples (1990); Lepore, Olson, and Tomer (1984); and McKenzie, Storment, Van Hook, and Armstrong (1985), which offer these kinds of results, also allude to ergonomic enhancements along with the training as part of an overall intervention program aimed at hazard control. Hence, it is difficult to determine how much of the reduction in the above-named indicators was due to training by itself. Clearly workplace evaluations of training present challenges in terms of controlling for other variables that can influence both the learning process and its outcome.

Training Evaluation Variables

The NIOSH report (Cohen & Colligan, 1998) analyzed key elements of the methods employed in the 80 studies and related factors in the interventions. Table 11.2 collates the study design data and shows that in almost

TABLE 11.2
Methods Used in Training Interventions

Hazard agent (number of studies)	Work setting—at-risk groups	Methodology—evaluation conditions				
		Basic design	Training target—thrust	Variables assessed	Posttraining measures	Other
Injury (21 studies)	Hospital Factory workers Warehousing Mining Paper mill Food process Maintenance Construction Shipbuilding Small tool simulation Fishing	[3] Posttest [3] Pre/Post [1] TimeSer [4] Post/Ctrl [3] Pre/Pst/ Ctrl [7] MultBsl	[18] FndtlWk Pract [1] HazRecog-Awareness [2] WkrPartic HazRecCtrl [0] WkrEmpwr HazRecCtrl	[6] TrngOnly [6] Trng & FdBk [4] Trng & GlSet/FdBk [2] Trng & Incentives [3] TrngMode Factors	[1] One/ShTm [4] Rep/ShTm [4] One/LgTm [12] Rep/LgTm	[1] Added publicity campaign to heighten interest; [1] Effect of presence of researcher. [2] Management incentives accountability.
Chemical (22 studies)	Farmers Lumbermen Asbestos workers Factory workers Hospital workers Coke oven foundrymen Miners Public employees Lead workers Waste site	[9] Posttest [7] Pre/Post [2] TimeSer [2] Post/Ctrl [1] Pre/Pst Ctrl [1] MultBsl	[8] FndtlWk Pract [4] HazRecog-Awareness [3] WkrPartic HazRecCtrl [7] WkrEmpwr HazRecCtrl	[15] TrngOnly [1] Trng & FdBk [0] Trng & GlSet/FdBk [1] Trng & Incentives [5] TrngMode Factors	[5] One/ShTm [6] Rep/ShTm [4] One/LgTm [7] Rep/LgTm	[4] Variable management support; [1] Training part of expanded awareness program. [2] Small sample (< 10).
Physical (10 studies)	Outdoor workers Radiation workers Firefighters Factory workers Maintenance Textiles	[2] Posttest [1] Pre/Post [2] TimeSer [0] Post/Ctrl [4] Pre/Pst Ctrl [1] MultBsl	[7] FndtlWk Pract [3] HazRecog-Awareness [0] WkrPartic HazRecCtrl [0] WkrEmpwr HazRecCtrl	[4] TrngOnly [2] Trng & FdBk [0] Trng & GlSet/FdBk [1] Trng & Incentives [3] TrngMode Factors	[0] One/ShTm [3] Rep/ShTm [2] One/LgTm [5] Rep/LgTm	[4] Varied management support—surveillance. [3] Other control actions to augment training.

Ergonomic (19 studies)	Health care Food service Warehousing Assembly work Firefighters Janitorial Maintenance Miners	[2] Posttest [5] Pre/Post [2] TimeSer [3] Post/Ctrl [6] Pre/Pst/ Ctrl [1] MultBsl	[16] FndtlWk Pract [3] HazRecog- Awareness [0] WkrPartic HazRecCtrl [0] WkrEmpwr HazRecCtrl	[15] TrngOnly [1] Trng & FdBk [0] Trng & GlSet/FdBk [0] Trng & Incentives [3] TrngMode Factors	[5] One/ShTm [3] Rep/ShTm [7] One/LgTm [4] Rep/LgTm	[3] Workplace constraints to training. [2] Training augmented by ergonomic enhancement. [4] Variable management support. [5] Small sample (< 10).
Biologic (8 studies)	Hospitals Health care providers	[1] Posttest [5] Pre/Post [1] TimeSer [0] Post/Ctr [1] Pre/Pst/ Ctr [0] MultBsl	[4] FndtlWk Pract [2] HazRecog- Awareness [1] WkrPartic HazRecCtr [1] WkrEmpwr HazRecCtr	[6] TrngOnly [0] Trng & FdBk [0] Trng & GlSet/FdBk [0] Trng & Incentives [2] TrngMode Factors	[3] One/ShTm [3] Rep/ShTm [0] One/LgTm [2] Rep/LgTm	[2] More accessible protective devices; [1] Rating compliance in evaluating performance. [1] Supervisor resistance.

Note. Definitions of terms in Table 11.2 are found in the Appendix at the end of this chapter and in the NIOSH report (Cohen & Colligan, 1998).

half of the reported studies, training effects were based on either posttraining measures for a given group; pretraining and posttraining differences; or comparisons before, during, and after training, again on the same group. Since many of the interventions took the form of research projects, there is no way of separating out elements of novelty and researcher effects, which could have influenced the outcomes in these subject groups apart from any training effect. Some studies used comparable untrained groups as ways to control these kinds of factors, but other influences were present during the evaluation whose effects could not be accounted for in the results reported. In this regard, management actions deserve particular mention. In one set of studies, supervisors played roles in reinforcing and sustaining the learned protective behaviors through issuing token rewards when such actions were observed (e.g., Zohar & Fussfield, 1981). In other cases, supervisors were themselves the trainees and were used to spearhead and effect the hazard control practices subject to evaluation (e.g., Maples, Jacoby, Johnson, Ter Haar, & Buckingham, 1982). As part of the training effort reported in other studies, supervisors were directed either to increase their surveillance of safe work practices (Millican, Baker, & Cook, 1981) or to consider staff compliance in performance evaluations (Lynch et al., 1990). And in still other reports, the authors suggested that management's indifference to the training objectives undermined a lasting, positive effect (C. J. Fox & Sulzer-Azaroff, 1987). These actions and other, more subtle ones probably had profound effects on the evaluations, which could not be isolated from the training results per se.

There were other design weaknesses as well. More than half of the studies measured posttraining effects less than 3 months after the instruction ended or only once after a longer interval of time. Thus, questions could be raised as to the durability of the reported effects or to possible intervening events affecting the long-term measures. Several studies did consider these issues, with the results being somewhat gratifying (Zohar, Cohen, & Azar, 1980; Hopkins, 1984). In a few cases the subject groups were too small to make generalizable conclusions, and in others workers were assigned to training conditions on a nonrandom basis, which could call into question the representativeness of the findings.

Other features of the 80 studies described in Table 11.2 were deemed more complimentary in showing the versatility of training's role in workplace hazard control. Indeed, the reported studies depict training in very diverse work settings subject to an array of different hazards. The goals of training ranged from learning basic safe-work practices and aspects of hazard recognition to developing skills in worker participation and empowerment in workplace hazard control programming. Giving emphasis to training's role in enhancing worker safety and health, the NIOSH training report (Cohen & Colligan, 1998) also included findings from surveys of disabled workers and fatality investigations that implicated gaps or lapses in training as a contributing factor to these incidents. Thus, the question for deliberation became not whether training could make a difference in reducing risks from workplace hazards and/or how much did training influence such outcomes. Rather, the question was: What are the conditions that can ensure or maximize positive training effects?

FACTORS UNDERLYING EFFECTIVE OCCUPATIONAL SAFETY AND HEALTH TRAINING

Results from the 80 training intervention studies and the other literature covered in the NIOSH report (Cohen & Colligan, 1998) became the basis for making evaluative statements on the significance of certain program characteristics in successfully meeting occupational safety and health objectives. Table 11.3 shows the kind of entries identified with various

TABLE 11.3
Summary Evidence for Distinctive Training/Extratraining Factors

Factor	References (hazard agent-setting)	Evaluative findings
Group size	Saarela, 1990 [Injury-Shipyard]; Robins et al., 1990 [Chemical-Manufacturing]	Small groups (fewer than 25), having in common similar jobs, work locations, exposures to the same hazards, offer more opportunities for effective learning experiences leading to and/or undertaking actions having positive effects in risk reduction.
Length and frequency	Parkinson et al., 1989 [Chemical-Coke Oven Plants]	Attendance at multiple training courses dealing with the recognition of workplace hazards and control actions does increase knowledge of the risks and also worker self-reports of taking more precautions to reduce apparent exposure hazards.
	Robins et al., 1990 [Chemical-Manufacturing]	Increases in trainer time per unit group of workers and use of frequent, short training sessions at the beginning of the workshift suggest more favorable outcomes in worker recognition of hazardous exposure situations, compliance with safe work practices, and prompting actions for improving other hazard control measures.

(continues)

TABLE 11.3 (Continued)
Summary Evidence for Distinctive Training/Extratraining Factors

Factor	References (hazard agent-setting)	Evaluative findings
Training mode	Saarela et al., 1989 [Injury—Shipyard]; Borland et al., 1991 [Physical—Telephone Line Repair]; Karmy & Martin, 1980 [Physical—Manufacturing Workers]	Informational campaigns involving posters, video presentations, pamphlet distribution by themselves produce some gains in assimilating fundamental safe work practices, but based on behavioral indicators, the effects are typically small and may not be durable.
	Leslie & Adams, 1973 [Injury—Punch Press Operations]; Rubinsky & Smith, 1971 [Injury—Grinding Operations]; Bosco & Wagner, 1988 [Chemical—Auto Workers]; Vaught et al., 1988 [Chemical—Miners]; Goldrick, 1989 [Biologic—Hospital]	Comparisons of written instructions/lecture versus slide/videotape presentations versus actual hands-on or interactive video techniques for learning proper work methods and fundamental safe work practices, based on knowledge tests or behavioral indicators, tend to favor the latter, more active forms of instruction.
	McQuiston et al., 1994 [Chemical—Toxic Waste Sites]; Brown & Nguyen-Scott, 1992 [Chemical—Toxic Waste Sites]; LaMontagne et al., 1992 [Chemical—Sterilizing Work]; Michaels et al., 1992 [Chemical—Public Facilities]; Weinger & Lyons, 1992 [Chemical—Farm Workers]; B. L. Cole & Brown, 1996 [Chemical—Hazardous Waste Sites]	Training methods using role play, case study of workplace safety and health problems, combined with practice in working through solutions/obstacles to improved hazard detection and control by means of individual and organizational change processes show signs of success based on reports of trainees, which note actions in overcoming shortcomings in their company hazard control program.
Transfer of training	Chhokar & Wallin, 1984 [Injury—Machine & Welding Shops]; H. H. Cohen & Jensen, 1984 [Injury—Warehousing]; C. J. Fox & Sulzer-Azaroff, 1987 [Injury—Paper Mill]; Komaki, Barwick, & Scott, 1978 [Injury—Bakery]; Ray, Purswell, & Schlegel, 1990 [Injury—Aircraft Maintenance]; Reber & Wallin, 1984 [Injury—Manufacturing]; Saarela, 1990 [Injury—Shipyard]; Saari & Nasanen, 1989 [Injury—Shipyard]	Contrasting illustrations of safe with unsafe work practices taken from actual job situations facilitates the learning of fundamental safety and housekeeping rules; however, effective carryover to the work settings in question depends on motivational-management influences found in both the training and posttraining environments.

TABLE 11.3 (Continued)
Summary Evidence for Distinctive Training/Extratraining Factors

Factor	References (hazard agent-setting)	Evaluative findings
	Carlton, 1987 [Ergonomics—Food Service]; Scholey, 1983 [Ergonomics—Hospital Wards]; St. Vincent, Tellier, & Lortie, 1989 [Ergonomics—Hospital]	Evidence of achieving well-trained safe work practices through training may fail to yield benefits in the workplace because of physical constraints or other conditions that interfere with their expression.
Motivational and promotional	Chhokar & Wallin, 1984 [Injury—Machine & Welding Shops]; H. H. Cohen & Jensen, 1984 [Injury—Warehousing]; C. J. Fox & Sulzer-Azaroff, 1987 [Injury—Paper Mill]; Komaki, Barwick, & Scott, 1978 [Injury—Bakery]; Komaki, Heinzmann, & Lawson, 1980 [Injury—Vehicle Maintenance]; Ray, Purswell, & Schlegel, 1990 [Injury—Aircraft Maintenance]; Reber & Wallin, 1984 [Injury—Manufacturing]; Saari & Nasanen, 1989 [Injury—Shipyard]; Sulzer-Azaroff et al., 1990 [Injury—Telecommunications Equipment]; University of Kansas, 1982 [Chemical—Fiberglass Product]; Maples et al., 1982 [Chemical—Chemical Manufacture]; Zohar, Cohen, & Azar, 1980 [Physical—Metal Product Manufacture]; Alavosius & Sulzer-Azaroff, 1985, 1986 [Ergonomics—Infirmary]	Setting performance goals reflecting compliance with targeted safety and health behaviors and/or providing feedback to mark progress in both the training and posttraining environment are effective methods for attaining successful training results. Of the two, feedback looms as a more potent, influential factor. The two in combination offer maximum impact.
	University of Kansas, 1982 (also Hopkins et al., 1986) [Chemical—Fiberglass Product]; D. K. Fox, Hopkins, & Anger, 1987 [Injury—Mining]; Zohar & Fussfield, 1981 [Physical—Weaving Mill]	Use of token rewards to reinforce learning of safety actions during training and when applied in the workplace found effective. Question: Does interest in awards per se cause distractions from the true intent of the training?
	Sulzer-Azaroff et al., 1990 [Injury—Telecommunications Equipment]; Lynch et al., 1990 [Biological—Hospital]	Success in OS&H training and its transfer to the workplace can also be driven by making workplace safety and health practices an element in one's performance evaluation.

(continues)

TABLE 11.3 (Continued)
Summary Evidence for Distinctive Training/Extratraining Factors

Factor	References (hazard agent-setting)	Evaluative findings
Trainer qualifications	Maples et al., 1982 [Chemical—Chemical Manufacture]; Lepore, Olson, & Tomer, 1984 [Ergonomics—Air/Space Technology]; McKenzie et al., 1985 [Ergonomics—Telecomm. Product Assembly]	Training supervisors/foremen in hazard recognition and means for control not only appears to produce positive workplace changes, but as trainers and change agents, to raise the level of worker safety performance. Best results for train-the-trainer approaches occur where other safety and health program practices are well established.
	Fiedler, Bell, Chemers, & Patrick, 1984 [Injury—Mining]; Fiedler, 1987 [Injury—Mining]; Smith, Anger, & Uslan, 1978 [Injury—Shipyard]; Sulzer-Azaroff et al., 1990 [Injury—Telecommunications Equipment]	Supervisor/foremen training for improving skills in team building, resolving personnel/conflict issues and performance management has the potential for effecting improved safety performance among the rank-and-file workforce.
	Askari & Mehring, 1992 [Biologic—Hospital]; McCarthy, Schietinger, & Fitzhugh, 1988 [Biologic—Health Care]; Saari, Bedard, Dufort, Hryniewiecki, & Theriault, 1994 [Chemical—Manufacturing]	Train-the-trainer programs targeting workplace safety and health concerns and focused on ways to promote worker actions aimed at hazard prevention and control show promise based on follow-on experiences of those who received training; however, management's acceptance of workers as trainers looms as a possible impediment.
Management role	Ray, Purswell, & Schlegel, 1990 [Injury—Aircraft Maintenance]; Saarela, 1990 [Injury—Shipyard]; Ewigman, Kivlahan, Hosokawa, & Horman, 1990 [Physical—Firefighting]; Zohar, Cohen, & Azar, 1980 [Physical—Metal Product Manufacture]; Zohar & Fussfield, 1981 [Physical—Weaving Mill]; Cole & Brown, 1996 [Chemical—Toxic Waste Sites]	The level of management support to workplace safety and health training and/or its application in the posttraining job environment greatly affects the nature and durability of its impact.

TABLE 11.3 (Continued)
Summary Evidence for Distinctive Training/Extratraining Factors

Factor	References (hazard agent-setting)	Evaluative findings
	Sulzer-Azaroff et al. [1990 Injury—Telecommunications Equipment]; Lynch et al., 1990 [Biologic—Hospital]	Initiatives giving hazard control program practices high priority and accountability measures for assuring effective efforts being undertaken at all levels of the workforce can do much to reinforce and sustain positive training outcomes.
	C. J. Fox & Sulzer-Azaroff, 1987 [Injury—Paper Mill]; Hopkins, 1984 [Chemical—Fiberglass Product Manufacture]	Indifference on the part of management can extinguish gains from training and specific efforts at enhancing safety and health practices.
Other factors	Linneman, Cannon, DeRonde, & Lamphear, 1991 [Biologic—Hospital]; Lynch et al., 1990 [Biologic—Hospital]; Seto, Ching, Chu, & Fielding, 1990 [Biologic—Hospital]; Wong et al., 1991 [Biologic—Hospital]	Making required safety related materials and personal protective items more accessible and convenient for use can facilitate the training effort and ease the burden of complying with safe work practices when at the jobsite.
	University of Kansas, 1982 [Chemical—Fiberglass Product Manufacture]; Cheng, Yang, & Wu, 1982 [Physical—Ionizing Radiation Uses]; Millican, Baker, & Cook, 1981 [Physical—Gas Diffusion Plant]; McKenzie et al., 1985 [Ergonomic—Telecommunications Products]; Parenmark, Engvall, & Malmkvist, 1988 [Ergonomic—Product Assembly Line]	Worker training may be needed to complement engineering, physical, and ergonomic solutions to control workplace hazards or to otherwise exploit their capabilities for ensuring assuring maximal health and safety protection.

factors. Those clearly linked to training—such as size of group, frequency and length of sessions, mode of instruction, and trainer qualifications—are noted, as are statements alluding to their most favorable conditions based on the literature reviewed. Also noted are factors that extend beyond the training process or reflect extratraining considerations. Those listed include aspects of transfer of training, motivation and promotional factors, management support, and some ancillary issues. The evaluative statements tied to these factors testify to their influence on positive training outcomes. That they are integral to a successful training effort is without question.

One impression of the contents of Table 11.3 is that the documentation for statements identified with the various factors is quite disproportional. Specifically, the amount of occupational safety and health literature in support of the evaluative statements on the factors of group size and length and frequency of instruction is quite small, whereas that shown for motivational and promotional factors is quite large. The extent of literature underlying the statements offered for the other factors falls in between. The supportive findings for the statements on group size and length and frequency of instruction are further limited by the fact that they were derived from a post hoc analysis in the few cited reports, not from efforts to vary parameters of each factor to ascertain their significance. In contrast, nearly all of the numerous reports underlying the statements for motivational and promotional factors employed systematic, controlled manipulation of relevant variables.

Another impression is that many of the statements agree with or are consonant with well-known concepts in the psychological literature on learning and motivation (Deese, 1952; Ruch, 1963). For example, captured in the statements in Table 11.3 are the benefits of increasing training time, repeated practice sessions, opportunities for more individualized instruction (through lower student-to-teacher ratios), and active rather than passive learning experiences focused on conditions that can promote transfer to the areas in need. Similarly, the statements on goal setting, feedback, and token rewards to help learn safe work practices and to strengthen such acts in the workplace are elements of the behavior reinforcement literature in psychology. Some extensions or illustrations of certain factors are also contained in the statements. That supervisors or foremen are key players in ensuring safe and healthful workplace conditions is well taken. Perhaps the trainer role further impresses upon the supervisor or foreman the importance of safety at work and ways to meet production goals without having to forgo hazard control measures or permit workers to take undue injury or health risks. As noted, added instruction for supervisors or foremen in both workplace safety and health as well as interpersonal relations enhances this result.

In assessing the intervention research literature, Sechrest and Figuerado (1993) made the point that more attention has to be paid to variations in the strength-of-treatment factors and how they can alter outcomes as opposed to just a treatment-versus-no-treatment (or control) approach.

Frequency and length of instruction, one of the two types of factors where the literature in occupational safety and health training proved sparse, would offer certain ways for varying the strength of a training intervention. The importance of the frequency-and-length factor cannot be stressed enough. It is the basis for defining refresher training needs as well as establishing the type of training regimen necessary to meet and sustain standards of performance in critical-skill/emergency situations. Some research on retention of job skill training over time (Sitterley, Pietan, & Metaftin,

1974) suggests that without practice, or for tasks seldom performed, high-level job skills can deteriorate much sooner than anticipated (e.g., between 1 and 4 months for piloting aircraft or firefighting); the research also suggests that different skills degrade at different rates (e.g., losses for performing complex procedural tasks are greater than those for simple or straightforward manual operations). In terms of retraining, the same studies found that methods offering dynamic, pictorial representations of the task situation (i.e., movies, videotapes) are as effective as hands-on practice, although the combination resulted in the greatest recovery. Implications of these findings for retraining issues in occupational safety and health are obvious. For example, the findings suggest that priority candidates for more frequent safety and health refresher training would be those procedures that are rarely used but are nevertheless critical when situations arise demanding appropriate action. Emergency events, which would fit this category, would justify frequent practice and drills to offset any loss in the knowledge and performance of actions to be taken. Reports by H. P. Cole et al. (1988) and Vaught, Brinch, and Kellner (1988) discuss this need in connection with evaluating miner skills in donning the self-contained self-rescuer breathing device used in cases of mine fire and explosion.

Similar needs for frequent drill may also exist for prescribed safety and health practices that are counter to natural behaviors or that add extra steps to task performance, especially when the hazard risks are not that apparent. Procedures to ensure safe performance in confined-space work and rescue actions would appear to fall in this category. For other situations, the basis for establishing training and retraining schedules is less clear. Presumably, criteria for determining such scheduling would take account of the complexity of the hazard control measures to be taught, the degree to which they are integrated into everyday work routines (and thus afford opportunities for practice), and local and industrywide injury and disease incident data for the work in question, among other considerations. The development of a decision logic for occupational safety and health training that dictates selective scheduling of training and retraining when appropriate, not sooner or later or more or less frequently than required, would seem a worthwhile effort. As part of this exercise, it would also be important to acknowledge the kinds of retraining or refresher experience that can best sustain the desired outcome.

Table 11.3 suggests conditions for the various factors that favor effective training, reinforcement of behaviors once learned, and success in their transfer to the job site. The question of some factors and conditions being more important than others is not addressed and, in actuality, may depend on situation-specific circumstances. One could argue, for example, that group size, length and frequency of instruction, and mode of instruction factors may be less important in training aimed at making workers aware of and observant of fundamental safety rules or housekeeping measures in jobs that are routine in nature, performed at fixed locations, and performed under well-defined conditions. In these circumstances, the emphasis can be less on how the training is conducted than on those factors or conditions that can motivate continued adherence to such practices. In contrast, factors in how the training is administered can be a major concern when knowledge of hazard recognition and control measures plus safe work practices becomes more formidable because of varied job operations and/or uncertain, changing workplace conditions. This would suggest that training regulations be performance or outcome based rather than overly prescriptive, allowing employers to develop a training plan that can accomplish the safety training objectives required for their job operations or work sites. At this stage, and speaking in general terms, the statements on various factors noted in Table 11.3 offer aids in structuring an effective training program based on the current literature. At the same time, and as already mentioned, the knowledge base is not strong regarding certain factors underlying the training process (e.g., length and frequency of instruction). Added efforts to address these kinds of needs and others mentioned in the course of this review seem evident.

NEEDS FOR FURTHER RESEARCH

The NIOSH literature review (Cohen & Colligan, 1998) noted gaps in available information and other limitations that suggested a need for further research into the effectiveness of training interventions in attaining occupational safety and health objectives. The following paragraphs elaborate on those gaps and limitations that were mentioned earlier in this chapter.

The training intervention studies reported in this review and showing evident success in meeting their objectives were based mainly on measures reflecting gains in safety knowledge or increased preventive actions. Relationships between these measures and actual reductions in injuries or disease, if also reported, were not clearly shown. Other criticisms were that most of these intervention studies were undertaken as site-specific research projects and therefore either were subject to experimenter novelty effects, could not account for other variables (management influence), and/or had a limited time frame for observing results. Research offering a broader base for evaluation would seem indicated where these limitations could be minimized or factored out by appropriate statistical techniques. In this regard, one approach would be to focus on the most prevalent types of injuries and diseases and the select industries or work operations where they occur. Applicable training requirements for those industries or operations would be noted along with the actual training practices being followed at different sites directed to those specific injury or disease risks. Differences in how the mandated training rules were met at the various sites selected for study and apparent linkages between the training undertaken and specific injury or disease control measures would be analyzed to assess the effect of training and identify strong or weak practices. Characterizations of management roles and concomitant risk control measures would be evaluated in viewing those sites showing the best versus the poorest work-related injury or illness records.

Related to the preceding proposal, little information exists of how industry responds or has responded to various OSHA training rules and the quality of such efforts. Some information of this nature is certain to exist in data files of NIOSH, OSHA, and the Mine Safety and Health Administration (MSHA) as a result of other programmed work undertaken by these agencies. Records from NIOSH health hazard evaluations, fatal accident investigations, and the NIOSH national occupational exposure survey databank (Pedersen & Sieber, 1988) would be good sources to tap, as would OSHA and MSHA compliance officer inspection files. Efforts to extract and assemble training-related data from these files could provide a status statement on occupational safety and health training practices. Depending on the data actually available, one could determine the extent to which the training follows the OSHA voluntary guidelines or other frameworks and any resultant experiences in attaining hazard control objectives.

A major point in the literature review in this chapter is that the success of occupational safety and health training in reducing work-related injury and disease can depend on factors lying outside the instruction process. Aside from motivational factors such as goal setting, feedback that reinforces the learning, and its transfer to the job site, there are engineering, administrative, and organizational practices that can also affect the training activity and its objectives. Survey studies (A. Cohen, 1977; Simonds & Shafai-Sahrai, 1977) comparing companies with high versus low injury rates have found that top management commitment to workplace safety and health, supervisor–worker communications on safety and other issues, and accountability stressing both safe performance and productivity can greatly influence the degree to which the lessons learned in the training program can be transferred to the shop floor. However, further in-depth studies are needed to examine which types of training practices by themselves or in combination with these other factors can yield the most positive effects. One proposal is to focus on those establishments with exemplary hazard control programs as demonstrated by their consistently low rates of injury and illness. The OSHA "STAR" companies in their Voluntary Protection Programs (OSHA, 1988) would appear to offer a sample of suitable candidates for studying. Although seemingly repeating other work aimed at defining successful program practices, this proposal would look more critically at the training activities, both in isolation and in their interactions with other factors, to define the dominant or controlling influences on the training outcomes. Similar analytical study of companies matched in size and industry to the OSHA STAR companies but demonstrating poorer safety/health performance records would be undertaken to ascertain differences in training approaches and related program issues. The end product of this work is envisioned as a set of case study reports highlighting the characteristics of effective training programs and ways to integrate them with other hazard control measures so as to produce the maximal benefit.

Other possibilities for assessing the importance of training could take the form of retrospective studies comparing differences in the level of training of workers who are injured on the job or afflicted with an occupational disease with comparable groups not so affected. The reports cited in this chapter give uncertain results as to whether appropriate training could have prevented these occurrences. One major drawback of these studies is the absence of more factual information on the extent or nature of the training actually received as it relates to the problem. Added efforts to determine whether training gaps or weaknesses can contribute to injury or disease incidents would be worthwhile. Critical to this work, however, would be the need to detail the training practices in place as well as other factors that could moderate their effects. Case-control or cohort approaches would offer ways of obtaining comparative data on affected versus nonaffected persons, given the need to separate out many nontraining factors that could be responsible for the differences.

In considering factors inherent to the training process, issues related to duration and frequency of instruction drew special attention. The occupational safety and health literature contained few references to this subject despite its significance to scheduling instruction, for both initial and refresher training purposes, so as to ensure a maximal, durable effect. The skills training literature offers some ideas about major variables that should be weighed in deciding on an appropriate schedule, and these could be the starting point for deriving a decision logic to address workplace safety and health training objectives. For this purpose, it is suggested that workshops be convened to discuss this as well as other matters relating to the effectiveness of occupational safety and health training. Invitees would include experts and practitioners conversant with job skill and occupational safety and health training plus others engaged in basic learning research, program evaluation, health education, and organizational behavior representing both the private and public sectors. This would be one way to tap the diverse knowledge applicable to answering the questions raised in this report. In addition to the topic of a decision logic for training schedules, a workshop agenda could include the following items:

1. The adequacy of the current regulatory language on training requirements.
2. Future training challenges owing to changing workplace technologies or job demands and their related hazards, worker

demographics, and emergent occupational injury or illness problems.
3. New training technologies and evaluation strategies for measuring training outcomes.
4. Desirability of merging independent training domains (skills training, occupational safety and health training, health promotion).

Outputs from the workshop would be state-of-the-art information on how training might fulfill its role in contributing to improved occupational safety and health conditions, both now and in the future.

APPENDIX: EXPLANATION OF HEADINGS/DESCRIPTOR TERMS IN TABLE 11.2

The *Hazard Agent* and *Work Setting—At-Risk* columns acknowledge the specific type of hazard addressed by the training effort and the particular situation and/or at-risk groups that were subjects for the intervention. Under the subheading *Basic Design* in the *Methodology—Evaluation Conditions* are entries defining whether the data collection for assessment involved only posttraining measures on trainees (Posttest), or comparisons between pre- and posttraining measures on a trainee group (Pre/Post), or repeated measures on a trainee group before, during, and after training, referred to as a time series design (TimeSer). In none of these instances was an untrained or control group used in the evaluation. The remaining designs did so, one form comparing posttest measures on a trained group versus a nontrained control group (Post/Ctr); another pre- and posttraining measures for a trained group compared with similar measures for a nontrained group (Pre/Post/Ctrl); and, last, a multiple baseline method (MultBsl) where repeated measures were taken on different groups before, during, and after training and upon the introduction of other factors that may influence the training outcome. Through staggering the schedule of treatments, the latter method enabled the measures on one group to serve as an added control for measuring training plus other factors affecting other groups.

The *Training Target* column in Table 11.2 uses the Office of Technology Assessment (1985) categorization for defining the actual training objectives of the study. These were learning fundamental work practices (FndtWkPract), training in hazard recognition/awareness (HazRecog), worker-directed or participative efforts in hazard recognition and control (WkrPartic), and worker empowerment training for the same purpose (WkrEmpwr). The *Variables Assessed* column describes the nature of the training conditions and/or other factors that were manipulated during training or in the follow-on evaluation. Those noted refer to training only (TrngOnly), feedback

(FdBk) with and without goal setting (GlSet), use of incentives (Trng&Incentives), and where the training plan called for specific manipulations of training content/delivery variables (TrngMode).

The *Posttraining Measures* column takes note of the frequency and time span of the data collected on the evaluative measures. The basic categories are short term versus long term (ShTm, LgTm), the latter referring to a posttraining period exceeding 3 months, and whether the data were collected one or more times (One or Rep). The last column, headed *Other*, identifies other conditions described in the studies that deserved special mention in light of their likely effect on the results. Nature of management support, workplace constraints, and aids to facilitating the training or its application in the posttraining environment are among the factors noted.

REFERENCES

Alavosius, M. P., & Sulzer-Azaroff, B. (1985). An on-the-job method to evaluate patient lifting technique. *Applied Ergonomics*, *16*, 307–311.

Alavosius, M. P., & Sulzer-Azaroff, B. (1986). The effects of performance feedback on the safety of client lifting and transfer. *Journal of Applied Behavioral Analysis*, *19*, 261–267.

Askari, E., & Mehring., J. (1992). Human immunodeficiency virus/acquired immunodeficiency syndrome training from a union perspective. *American Journal of Industrial Medicine*, *22*, 711–720.

Borland, R., Hocking, B., Godkin, G. A., Gibbs, A. F., & Hill, D. J. (1991). The impact of a skin cancer control education package for outdoor workers. *Medical Journal of Australia*, *154*, 686–688.

Bosco, J., & Wagner, J. (1988). A comparison of the effectiveness of interactive laser disc and classroom video tape for safety instruction of General Motors workers. *Educational Technology*, *28*, 15–22.

Brehm, J. (1966). *A theory of psychological reactance*. New York: Academic Press.

Brown, M. P., & Nguyen-Scott, N. (1992). Evaluating a training-for-action job health and safety program. *American Journal of Industrial Medicine*, *22*, 739–749.

Carlton, R. S. (1987). The effects of body mechanics instruction on work performance. *American Journal of Occupational Therapy*, *41*, 16–20.

Cheng, C., Yang Y., & Wu T. (1982). Radiation control through licensing and intensive training. *Health Physics*, *43*, 803–811.

Chhokar, J. S., & Wallin, J. A. (1984). A field study of the effect of feedback frequency on performance. *Journal of Applied Psychology*, *69*, 524–530.

Cohen, A. (1977). Factors in successful occupational safety programs. *Journal of Safety Research*, *9*, 168–178.

Cohen, A. (1998). A multidimensional evaluation of fire fighter training for hazardous materials response: First results from the IAFF program. *American Journal of Industrial Medicine, 34*, 331–341.

Cohen, A., & Colligan, M. J. (1998). *Assessing occupational safety and health training: A literature review* (Rèport No. 98145). Cincinnati, OH: National Institute for Occupational Safety and Health. Retrieved May 5, 2003, from http://www.cdc.gov/niosh

Cohen, H. H., & Jensen, R. C. (1984). Measuring the effectiveness of an industrial lift truck safety training program. *Journal of Safety Research, 15*, 125–135.

Cole, B. L., & Brown, M. P. (1996). Action on worksite health and safety problems: A follow-up survey of workers participating in a hazardous waste worker training program. *American Journal of Industrial Medicine, 30*, 730–743.

Cole, H. P., Mallett, L. G., Haley, J. V., Berger, P. K., Lacefield,W. E., Waslielewski, R. D., et al. (1988). A new SCSR donning procedure. In Bureau of Mines, U.S. Department of Interior (Ed.), *Research and evaluation methods for measuring non-routine mine health and safety skills* (Vol. 1, pp. 120–163). Lexington: University of Kentucky.

Deese, J. (1952). *The psychology of learning*. New York: McGraw-Hill.

Ewigman, B. G., Kivlahan, C. H., Hosokawa, M. C., & Horman, D. (1990). Efficacy of an intervention to promote use of hearing protective devices by firefighters. *Public Health Reports, 105*, 53–59.

Fiedler, F. E. (1987, July). *Structured management training in underground mining: Five years later* (Information Circular 9145). Bureau of Mines Technology Seminar, Pittsburgh, PA.

Fiedler, F. E., Bell, C. H., Jr., Chemers, M. M., & Patrick, D. (1984). Increasing mine productivity and safety through management training and organization development: A comparative study. *Basic and Applied Psychology, 5*, 1–18.

Fox, C. J., & Sulzer-Azaroff, B. (1987). Increasing completion of accident reports. *Journal of Safety Research, 18*, 65–71.

Fox, D. K., Hopkins, B. L., & Anger, W. K. (1987). The long-term effects of a token economy on safety performance in open-pit mining. *Journal of Applied Behavioral Analysis, 20*, 215–224.

Goldrick, B. A. (1989). Programmed instruction revisited: A solution to infection control in service education. *Journal of Continuing Education in Nursing, 20*, 222–226.

Hilyer, J. C., Brown, K. C., Sirles, A. T., & Peoples, L. (1990). A flexibility intervention to reduce the incidence and severity of joint injuries among municipal firefighters. *Journal of Occupational Medicine, 32*, 631–637.

Hopkins, B. L. (1984, October). *An investigation of the durability of behavioral procedures for reducing workers' exposures to a suspect carcinogen*. Special Report to the National Institute for Occupational Safety and Health, Cincinnati, OH.

Hopkins, B. L., Conard, R. J., Dangel, R. F., Fitch, H. G., Smith, M. J., & Anger, W. K. (1986). Behavioral technology for reducing occupational exposures to styrene. *Journal of Applied Behavioral Analysis, 19*, 3–11

Karmy, S. J., & Martin, A. M. (1980). Employee attitudes towards hearing protection as affected by serial audiometry. In P. W. Alberti (Ed.), *Personal hearing protection in industry* (pp. 491–501). New York: Raven Press.

Komaki, J., Barwick, K. D., & Scott, L. R. (1978). A behavioral approach to occupational safety pinpointing and reinforcing safe performance in a food manufacturing plant. *Journal of Safety Research, 63*, 434–445.

Komaki, J., Heinzmann, A. T., & Lawson, L. (1980). Effect of training and a component analysis of a behavioral safety program. *Journal of Applied Psychology, 65*, 261–270.

Lacey, R., Husted Medvec, V., & Messick, D. (1988, August). *Blame and prevention as sources of bias in causal explanations of accidents.* Presentation at the Annual Meeting of the Academy of Management, San Diego.

LaMontagne, A. D., Kelsey, K. T., Ryan, C. M., & Christiani, D. C. (1992). A participatory workplace health and safety training program for ethylene oxide. *American Journal of Industrial Medicine, 22*, 651–664.

Lepore, B. A., Olson, C. N., & Tomer, G. M. (1984). The dollars and sense of occupational back injury prevention training. *Clinical Management, 4*, 38–40.

Leslie, J. H., Jr. ,& Adams, S. K. (1973). Programmed safety through programmed learning. *Human Factors, 15*, 223–236.

Linneman, C. C., Jr., Cannon, C., DeRonde, M., & Lamphear, B. (1991). Effect of educational programs, rigid sharps containers and universal precautions on reported needle stick injuries in health care workers. *Infection Control and Hospital Epidemiology, 12*, 214–219.

Lynch, P., Cummings, M. J., Roberts, P. L., Herriott, M. J., Yates, B., & Stamm, W. E. (1990). Implementing and evaluating a system of generic infection precautions: Body substance isolation. *American Journal of Infection Control, 18*, 1–12.

Maples, T. W., Jacoby, J. A., Johnson, D. E., Ter Haar, G. L., & Buckingham, F. M. (1982). Effectiveness of employee training and motivation programs in reducing exposure to inorganic lead and lead alkyls. *American Industrial Hygiene Association Journal, 43*, 692–694.

McCarthy, P. K., Schietinger, H., & Fitzhugh, Z. A. (1988). AIDS education and training for health care providers. *Health Education Research, 3*, 97–103.

McKenzie, F., Storment, J., Van Hook, P., & Armstrong, T. J. (1985). A program for control of repetitive trauma disorders associated with hand tool operations in a telecommunications manufacturing facility. *American Industrial Hygiene Association Journal, 46*, 674–678.

McQuiston, T. H., Coleman, P., Wallerstein, N. B., Marcus, A. G., Morawetz, J. S., & Ortleib, D. W. (1994). Hazardous waste worker education: Long-term effects. *Journal of Occupational Medicine, 36*(12), 1310–1323.

Michaels, D., Zoloth, S., Bernstein, N., Kass, D., & Schrier, K. (1992). Workshops are not enough: Making right-to-know training lead to workplace change. *American Journal of Industrial Medicine, 22*, 637–649.

Millican, R., Baker, R. C., & Cook, G. T. (1981). Controlling heat stress: Administrative versus physical control. *American Industrial Hygiene Association Journal, 42*, 411–416.

Occupational Safety and Health Administration. (1988). Voluntary protection programs. *Federal Register, 53*(133), 26339–26348.

Occupational Safety and Health Administration. (1998). *Draft proposed safety and health program rule: 29CFR1900.1* (Docket No. S&H-0027). Washington, DC: U.S. Government Printing Office.

Office of Technology Assessment, U.S. Congress (1985). *Preventing illness and injury in the workplace* (OTA-H-256). Washington, D.C.: Author.

Olishifski, J. (1988). Overview of industrial hygiene. In B. A. Plog (Ed.), *Fundamentals of industrial hygiene*. Chicago: National Safety Council Press.

Parenmark, G., Engvall, B., & Malmkvist, A. K. (1988). Ergonomic on-the-job training of assembly workers. *Applied Ergonomics, 19*, 143–146.

Parkinson, D. K., Bromet, E. J., Dew, M. A., Dunn, L. O., Barkman, M., & Wright, M. (1989). Effectiveness of the United Steel Workers of America coke oven intervention program. *Journal of Occupational Medicine, 31*, 464–472.

Pedersen, D. H., & Sieber, W. K. (1988). *National occupational exposure survey: Analysis of management interview responses* (NIOSH Publication No. 89-103). Cincinnati, OH: National Institute for Occupational Safety and Health. Retrieved April 4, 2003, from http://www.cdc.gov/niosh/89-103.html

Ray, P. S., Purswell, J. L., & Schlegel, R. E. (1990). A behavioral approach to improve safety at the workplace. In B. Das (Ed.), *Advances in industrial ergonomics and safety* (pp. 983–988). New York: Taylor & Francis.

Reber, R. A., & Wallin, J. A. (1984). The effects of training, goal setting and knowledge of results on safe behavior: A component analysis. *Academy of Management Journal, 27*, 544–560.

Robins, T. G., Hugentobler, M. K., Kaminski, M., & Klitzman, S. (1990). Implementation of the federal hazard communication standard: Does training work? *Journal of Occupational Medicine, 32*, 1133–1140.

Rubinsky, S., & Smith, N. F. (1971). *Evaluation of accident simulation as a technique for teaching safety procedures in the use of small power tools* (PHS Contract 86-68-207). Providence: University of Rhode Island, Injury Control Research Laboratory.

Ruch, F. (1963). *Psychology and life* (6th ed., pp. 73–125). Chicago: Scott, Foresman.

Saarela, K. L. (1990). An intervention program utilizing small groups: A comparative study. *Journal of Safety Research, 21*, 149–156.

Saarela, K. L., Saari, J., & Alltonen, M. (1989). The effects of an informational safety campaign in the shipbuilding industry. *Journal of Occupational Accidents, 10*, 255–266.

Saari, J., Bedard, S., Dufort, V., Hryniewiecki, J., & Theriault, G. (1994). Successful training strategies to implement a workplace hazardous materials information system: An evaluation at 80 plants. *Journal of Occupational Medicine, 36*, 569–574.

Saari, J., & Nasanen, M. (1989). The effect of positive feedback on industrial housekeeping and accidents: A long-term study at a shipyard. *International Journal of Industrial Ergonomics, 4*, 201–211.

St. Vincent, M., Tellier, C., & Lortie, M. (1989). Training in handling: An evaluative study. *Ergonomics, 32*, 191–210.

Scholey, M. (1983). Back stress: The effects of training nurses to lift patients in a clinical situation. *International Journal of Nursing Studies, 20*, 1–13.

Sechrest, L., & Figueredo, A. J. (1993). Program evaluation. In L. W. Porter & M. W. Rosensweig (Eds.), *Annual review of psychology* (Vol. 44, pp. 645–674). Palo Alto, CA: Annual Reviews.

Seto, W. H., Ching, T. Y., Chu, Y. B., & Fielding, F. (1990). Reduction of the frequency of needle recapping by effective education: A need for conceptual alteration. *Infection Control and Hospital Epidemiology, 11*, 194–196.

Simonds, R. H., & Shafai-Sahrai, Y. (1977). Factors apparently affecting injury frequency in eleven matched pairs of companies. *Journal of Safety Research, 9*, 120–127.

Sitterley, T. E., Pietan, O. D., & Metaftin, W. E. (1974). *Firefighter skills study: Effectiveness of training and retraining methods*. Seattle, WA: Boeing Aerospace Company.

Smith, M. J., Anger, W. K., & Uslan, S. S. (1978). Behavior modification applied to occupational safety. *Journal of Safety Research, 10*, 87–88.

Sulzer-Azaroff, B., Loafman, B., Merante, R. J., & Hlavcek A. C. (1990). Improving occupational safety in a large industrial plant: A systematic replication. *Journal of Organizational Behavior and Management, 11*, 99–120.

University of Kansas. (1982). *Behavioral procedures for reducing worker exposures to carcinogens* (NIOSH Contract Report 210-77-0040). Cincinnati, OH: National Institute for Occupational Safety and Health.

Vaught, C., Brinch, M. J., & Kellner, H. J. (1988). *Instructional mode and its effect on initial self-contained self-rescuer donning attempts during training* (Report of Investigations 9208). Pittsburgh, PA: U.S. Bureau of Mines.

Weinger, M., & Lyons, M. (1992). Problem-solving in the fields: An action-oriented approach to farmworker education about pesticides. *American Journal of Industrial Medicine, 22*, 677–690.

Wong, E. S., Stotka, J. L., Chinchilli, V. M., Williams, D. S., Stuart, G., & Markowitz, S. M. (1991). Are universal precautions effective in reducing the number of occupational exposures among health care workers? *Journal of the American Medical Association, 265*, 1123–1128.

Zohar, D., Cohen, A., & Azar, N. (1980). Promoting increased use of ear protectors in noise through information feedback. *Human Factors, 22*, 69–79.

Zohar, D., & Fussfield, N. (1981). Modifying earplug wearing behavior by behavior modification techniques: An empirical evaluation. *Journal of Organizational Behavior and Management, 3*, 41–52.

12

LABOR UNIONS AND OCCUPATIONAL SAFETY: CONFLICT AND COOPERATION

E. KEVIN KELLOWAY

Issues of worker safety and workers' compensation have been the focus of unions' bargaining and advocacy for many years. Indeed, the attempt to secure improved working conditions and enhance safety is one of the bread-and-butter issues that characterize bargaining by North American unions (Barling, Fullagar, & Kelloway, 1992). Issues of worker safety have been the source of labor–management conflict, with labor accusing management of a lack of caring and management retaliating with accusations of a lack of cooperation (Wokutch, 1992). Moreover, health and safety issues may be the catalyst for union organizing as workers attempt to improve their working conditions or redress perceived injustice associated with unsafe working conditions through unionization (Cornish & Spink, 1994; Robinson, 1988).

Preparation of this chapter was supported by research grants from the Social Sciences and Humanities Research Council of Canada and the Nova Scotia Health Research Foundation. Correspondence may be addressed to the author at the Department of Management, Saint Mary's University Halifax, Nova Scotia, B3H 3C3 Canada, or by e-mail to kevin.kelloway@stmarys.ca.

These conflicts are typically understood from a pluralistic perspective that recognizes the inherent conflicts of interest in the workplace (Kelloway, Barling, & Harvey, 1998). From this perspective, management has a legitimate right to run the workplace efficiently and unions have a legitimate right to protect themselves from occupational hazards. The simultaneous pressures for efficiency and safety create a setting ripe for conflict, and the management of occupational safety emerges as a product of this conflict as played out in the collective bargaining process.

Despite this view, issues of health and safety are increasingly framed from a unitarist perspective characteristic of the new model of human resources management (Kelloway et al., 1998; Wells, 1993). Characteristically, the new model of human resources is aimed at creating win–win situations based on identifying areas in which labor and management share similar interests and can develop cooperative relationships (Della-Giustania & Della-Giustania, 1992; Verma, 1989; Verma & McKersie, 1987). For example, Wokutch (1992) notes:

> Occupational safety and health would seem to be an issue on which labor, management and government would want to cooperate. The costs provide an economic incentive for management to limit injuries and illnesses; workers' interests are obvious since they are the ones who get sick or injured. (p. 6)

Similarly, occupational health and safety textbooks (e.g., Montgomery & Kelloway, 2002) identify labor unions as one of the three partners (i.e., with management and government) in establishing safe workplaces. Although some authors have raised reservations about these cooperative relationships, suggesting that they may subvert traditional bargaining roles (e.g., Verma, 1989; Verma & McKerzie, 1987; Wells, 1993), labor leaders typically see little conflict between their collective bargaining and cooperative roles under the new model of human resources management (Beaumont, 1995; Pupo & White, 1994).

These contrasting views highlight the complex role that labor unions play in occupational safety, simultaneously being in conflict with and cooperating with employers to enhance workers' safety. In the remainder of this chapter I focus on these two main roles of unions relative to occupational safety. First, I review the collective bargaining and advocacy roles of unions that set the framework for how occupational safety is managed in the workplace. Second, I review the role of the union in cooperating with management in enhancing working conditions. Finally, the chapter concludes with an assessment of the effect of unionization on occupational safety.

The Collective Bargaining Role

A collective agreement is a legal contract between an employer and the bargaining agent for the employees, most typically a labor union. The traditional, and perhaps most familiar, role for labor unions in the promotion of occupational safety is played out in the collective bargaining process. In the United States, occupational safety was established as a legitimate focus for collective bargaining by the National Labor Relations Board (NLRB) in 1966. This was an important decision because, prior to 1966, employers argued that issues of health and safety dealt with the operation of the workplace and hence fell under the management rights clause of the collective agreement. That is, employers argued that they had no obligation to discuss issues of health and safety with the union. The NLRB decision struck down that position. Then, in 1978 the American Federation of Labor–Congress of Industrial Organizations (AFL-CIO) formed a department of occupational safety and health (Nelkin & Brown, 1984), and by 1987 the majority of collective agreements contained at least some form of a safety clause (Robinson, 1988).

Although unions negotiate collective agreement provisions that deal with safety, it should be noted that unions are not legally responsible for health and safety in most jurisdictions. For example, like Canadian labor law (Montgomery & Kelloway, 2002), the Occupational Health and Safety Act of 1970 in the United States specifies that the employer bears the responsibility of providing a safe workplace (Della-Giustania & Della-Giustania, 1992). The union's interest in negotiating safety provisions is not in supplanting the primary responsibility of employers. Rather, unions attempt to establish the three basic rights of workers: the right to know (about hazards in the workplace), the right to participate (in removing hazards and improving workplace safety), and the right to refuse unsafe work (Montgomery & Kelloway, 2002).

Negotiating safety provisions in a collective agreement has both manifest and latent functions. Manifestly, unions negotiate for specific provisions or workplace practices that would improve workplace safety. For example, Gray, Myers, and Myers (1998) reviewed collective agreement provisions and identified provisions dealing with five major categories of safety issues: ergonomics, union–management relations, occupational health, occupational safety and technology, and maintenance and operations (see Table 12.1 for sample safety provisions). The three most prevalent clauses pledged union cooperation on matters of health and safety

TABLE 12.1
The Eight Most Frequently Occurring Safety Provisions in Collective Agreements

	Provision for	Number of contracts	Percentage of contracts*
1.	Protective clothing	177	41
2.	Dos and don'ts	156	36
3.	Reporting health and safety needs	136	31
4.	First aid and medical facilities	104	24
5.	Refusal of hazardous work	97	22
6.	Conducting surveys	73	17
7.	Safety and health testing	76	17
8.	Assignment of employees with occupational illness or injury	67	15

* Based on private sector collective agreements expiring between August 1997 and July 1997. $N = 433$.
Note. From G. R. Gray, D. W. Myers, & P. S. Myers, "Collective Bargaining Agreements: Safety and Health Provisions," *Monthly Labor Review,* May 1998, pp. 13–35. Copyright 1998. Adapted with permission.

(64% of all agreements), established local labor–management health and safety committees (50% of all agreements), and dealt with the provision of personal protective equipment to employees (41% of all agreements).

In addition to the manifest function (i.e., the establishment of specific safety-related provisions), the negotiation of health and safety provisions has a latent function. In negotiating such provisions, issues of worker safety are brought into the grievance and arbitration system. A grievance is a complaint that one party (typically the employer since most grievances are filed by unions) has violated some provision of the collective agreement. Most typically, collective agreements will articulate a mechanism for resolving grievances that culminates in submission of the complaint to an arbitrator.

The most frequent form of health and safety clause in collective agreements is known as the general duty clause (Robinson, 1988). General duty clauses recognize management's responsibility to provide a safe workplace and may reference the applicable legislation indicating management's willingness to abide by the provisions of the legislation. Clauses pledging union cooperation in health and safety programming are also a form of general duty clause; although such clauses may read like concessions to management interests, they are frequently worded so as to bring issues of health and safety into the workplace justice system. Although apparently devoid of content, such clauses suffice to bring issues of workplace safety under the grievance system and thereby provide the union with a means of voicing and obtaining redress for complaints.

In many cases, provisions negotiated in collective bargaining may simply reiterate provisions of health and safety legislation. For example, clauses that mandate participation in joint labor–management safety committees or establish the right to refuse unsafe work merely reflect current safety law. Moreover, a collective agreement that establishes a lesser standard of safety than enshrined in legislation would not be valid in most jurisdictions (i.e., a union cannot negotiate away rights established under legislation). However, in light of strong pressures to deregulate and avoid placing a burden on employers, many jurisdictions have begun the process of watering down health and safety legislation and cutting the resources devoted to enforcing these regulations (Bain, 1997). In this environment it becomes increasingly important for the union to negotiate specific safety provisions in the collective agreement to bring those provisions under the grievance and arbitration clauses of that agreement.

Although proponents of cooperation cite health and safety as an area of mutual concern for management and labor, collective bargaining is an inherently adversarial process. There is some evidence that management and unions are driven by different sets of concerns when negotiating health and safety provisions. Unions, by virtue of their nature, tend to respond to the concerns of members with high seniority (see Barling et al., 1992), whereas management tends to respond to the concerns of young, highly mobile employees (Kahn, 1990). Similarly, unions and management may be attempting to achieve the same ends (e.g., improved safety) for vastly different reasons (e.g., reduced cost vs. individual safety; Wokutch, 1992), and these differing motives may lead to conflict on how extensive safety initiatives will be.

The adversarial nature of collective bargaining naturally leads to the question of whether issues of occupational safety are used as bargaining chips in the negotiation process. That is, do unions raise the threat of health and safety complaints as a weapon against the employer to attain other gains at the bargaining table? Northrup (1996) cites unions' use of the Occupational Health and Safety Administration (OSHA) as a form of corporate campaign. A corporate campaign is a campaign against an employer that is designed to disrupt operations, cost the firm money, and generally keep management disoriented. Northrup's suggestion is that unions may file numerous OSHA complaints against a given employer as a form of harassment.

Although Northrup (1996) does not offer any empirical evidence for his allegations, Eaton and Kriesky (1994) found that an increased number of OSHA complaints were initiated by unions after the start of a strike as part of a corporate campaign designed to reduce the use of replacement workers. In contrast, Smith (1986) found no correlation between OSHA complaints and strike activity.

The use of health and safety issues as a collective bargaining weapon was addressed directly in Hebdon and Hyatt's (1998) study of industrial relations conflict. In particular the authors investigated "the proposition that workers exploit the rights to refuse unsafe work and to make anonymous health and safety complaints to achieve industrial relations objectives beyond workplace health and safety" (p. 580). Thus, the authors suggested that unions use health and safety complaints in a manner similar to other grievances (Hebdon & Stern, 1998) as an expression of industrial relations climate. The authors reviewed data from more than 10,000 bargaining units in Ontario, Canada, but found no support for the proposition that work refusals were more likely around the time of collective bargaining. There were higher rates of refusal when there was evidence of a high degree of collective bargaining conflict (e.g., a strike or interest arbitration), although it should be noted refusal rates and strikes may share a common cause. Overall, the authors concluded that "the exercise of health and safety-related employee rights was modest and not indicative of excessive use or suggestive of abuse" (Hebdon & Hyatt, 1998, p. 590).

Although there is little evidence to support the proposition that unions use health and safety issues and health and safety regulatory agencies as political weapons in collective bargaining, there is some evidence that the presence of a union does affect the functioning of regulatory agencies. For example, injured workers who are unionized are more likely to receive workers' compensation benefits than their nonunionized counterparts (Meng & Smith, 1993). In their study of OSHA inspections, Scherer, Kaufman, and Ainina (1993) found that "overall, the inspection phase of the complaint investigation and resolution process appears to be more vigorous in firms which have a union. . . . More citations were issued per employee in union firms as well as greater time spent on-site and time for the total investigations to be completed" (p. 59). Similarly, Weil (1996) suggested that the presence of a union raised the level of regulatory activity from a variety of agencies, including OSHA, and in several studies documented large union–nonunion differences in OSHA inspections (Weil, 1991; 1992).

Conflicting Perspectives: Behavior-Based Safety

Although unions and management share interests in enhancing worker safety, it is also clear that they adopt vastly different perspectives on how to achieve this goal. Perhaps no area in safety programming is more illustrative of this point than the introduction of behavior-based safety programs.

The use of behavior modification principles to increase safety behaviors in the workplace has been called one of the most successful innovations in the last 20 years (Saari, 1994). Although this claim may be overstated, there is ample evidence to suggest that behavioral programs are effective in promoting safe working behaviors and are associated with substantial reductions in accident rates. For example, Saari and Nasanen (1989) reported reductions in accident rates of up to 80% as a result of applying behavioral principles. Following implementation of a feedback-based program in a paper mill, Fellner and Sulzer-Azaroff (1984) reported a decrease in accident rates to less than half of the preimplementation rate. These results are particularly impressive in light of the nominal costs of operating the program (estimated at $14.00/week; Fellner & Sulzer-Azaroff, 1984). Moreover, behavioral interventions have been shown to be effective in promoting safety behaviors in a wide range of industrial environments, including research laboratories (Sulzer-Azaroff, 1982); plastics manufacturing firms (Sulzer-Azaroff & DeSantamaria, 1980); metal fabrication plants (Zohar, Cohen, & Azar, 1980); paper mills (Fellner & Sulzer-Azaroff, 1984); bakeries (Komaki, Barwick, & Scott, 1978); and municipal street departments (Komaki, Heinzmann, & Lawson, 1980). In sum, there are at least 40 empirical studies that make a strong case for behavioral approaches to enhancing health and safety in the workplace (Sulzer-Azaroff, Harris, & McCann, 1994).

Despite this impressive empirical support, most labor unions remain vehemently opposed to behavior-based safety programs. Spigener and Hodson (1997) identified numerous union concerns with behavior-based safety programs. Most are summarized in an unpublished pamphlet by the United Auto Workers cited by Winn, Frederick, and Church (1999) as comprising three central issues. First, behavior-based safety programs are seen as blaming the victim because they focus on individual safety behaviors to the exclusion of product and process design even though a focus on the latter might result in a higher safety payoff. Second, environmental and ambient hazards (e.g., dust, noise, ergonomic stressors) that do not result from unsafe behaviors but are a result of the manufacturing process are ignored when companies rely on behavior-based safety programs. Finally, behavior-based programs are seen as reducing worker participation and enhancing the opportunities for management to discipline workers.

As these concerns illustrate, it is the focus on individual behavior that stands in opposition to the union perspective. On the basis of similar arguments, unions have argued against other individually based programs such as incentive pay (Barling et al., 1992). Proponents of behavior-based safety programs frequently claim that 90% of all accidents are caused by unsafe behaviors (hence the focus on eliminating such behaviors). In contrast, unionists point to empirical data that identify unsafe or illegal working conditions as the root cause of many accidents (Ashford, 1976). To some extent these claims may be a matter of interpretation. For example, the failure of a worker to wear a safety harness when working on roofs can be labeled an unsafe behavior and the worker can be targeted for intervention. However, the failure of management to provide easy access to safety harnesses and training in their use may be the root cause of the behavior. Similarly, the failure to wear personal protective equipment is a frequent focus of behavioral safety programs, yet such programs rarely consider the comfort of the safety equipment or the social isolation that comes with wearing hearing protectors or similar protective gear (Nelkin & Brown, 1984). Unionists decry safety investigations or programs that stop at the most proximal cause of the incident.

It is clear that these concerns are not insurmountable and that behavior-based safety programs have been successfully implemented in unionized environments (e.g., Spigener & Hodson, 1997; Winn et al., 1999). Moreover, similar concerns have been raised about other forms of safety programming such as safety incentive programs (Atkinson, 1999). Nonetheless, the implementation of behavior-based safety programs in a unionized environment is likely to remain a sensitive issue for the immediate future. Researchers and consultants interested in implementing such programs are well advised to note and respond to union concerns. In particular, it is important to ensure that such programs are not implemented in an attempt to deal with poor work processes, lack of training, or poor equipment design. Failure to take appropriate notice of such concerns is likely to result in both program failure and an increase in labor–management conflict in the workplace.

COOPERATION

Joint Labor–Management Health and Safety Committees

Joint health and safety committees are made up of representatives from both management and labor and are charged with the responsibility of enhancing health and safety in the workforce. The primary function of the joint health and safety committee is to provide a nonadversarial atmosphere in which labor and management can work together to create a safer and healthier workplace (Montgomery & Kelloway, 2002). Although by no means as widespread as in Europe (see Frick & Walters, 1998; Willim, 1987), such formalized systems for worker representation in occupational health and safety have become increasingly popular in North America.

Typically, joint committees are structured such that equal or better representation is required from workers who do not exercise managerial responsibilities. Joint health and safety committees have four principal functions: (a) to identify potential hazards, (b) to evaluate these potential hazards, (c) to recommend corrective action, and (d) to follow up implemented recommendations (Montgomery & Kelloway, 2002). Although the ostensible function of joint health and safety committees is to correct specific workplace hazards, the cooperation of labor and management also sends an important message about the importance of workplace safety to both management and the union in the workplace. Management commitment to health and safety has emerged as a key requirement for the improvement of workplace health and safety (Boden, Hall, Levenstein, & Punnett, 1984; Griffiths, 1985; Hofmann, Jacobs, & Landy, 1995; Ilgen, 1990; Shannon, Mayr, & Haines, 1997; Zohar, 1980). Coworker attitudes play an equally important role in shaping individual perceptions of workplace hazards and in encouraging or inhibiting self-protective behavior (Cree and Kelloway, 1997). For example, Andriessen (1978) found that the work-group safety norm was an important predictor of safety behavior. Nelkin and Brown (1984) noted that individuals were discouraged from adopting safe working behaviors through coworker harassment and alienation.

Clearly then the premise of the joint health and safety committee is that labor and management share interests in improving occupational health and safety and can work cooperatively through the committee structure. Although some have expressed concerns that such cooperative approaches to human resource issues undermine the role of collective bargaining (Beaumont, 1995), the available data seem to indicate that the existence of joint health and safety committees leads to reduced workplace injuries (Reilly, Paci, & Holl, 1995) and that the existence of such committees supplements rather than supplants the role of unions (Weil, 1999).

As Weil (1999) points out, the existence of a joint health and safety committee does not mean that the committee is effective. Indeed, like any other workplace group, committees may take some time before they become effective. Joint committees can improve health and safety through prevention, education, and training and by providing an ongoing forum for problem resolution (Weil, 1995). Unfortunately, there are data suggesting that committees are not always effective at communicating their activities to the workforce. Walters and Haines (1988) found that most workers relied on their supervisors, rather than their safety representative, for safety information. Indeed, a substantial percentage of workers could not even identify their safety representative. In both this and a subsequent study (Walters & Denton, 1990) workers' unwillingness to invoke their right to refuse unsafe work was related to workers' lack of knowledge and to the fear that management would retaliate against those who filed complaints. Thus, although the joint committee structure seems to be an effective approach to enhancing occupational safety, the effectiveness of this approach depends on a commitment that goes beyond the formation of a committee. Indeed, the effectiveness of such committees may be limited by the ability of management and the union to effect a cultural change that supports safety initiatives.

Union Effects on Occupational Safety

Given the roles that unions play in enhancing worker safety, it is surprising to note that the question of whether unionization has a positive effect on worker safety has yet to result in a satisfactory answer. Unionists would argue that unions improve health and safety by improving working conditions, enforcing safe working procedures, giving workers a say in establishing safe working practices, and generally policing the workplace. Indeed, many would argue that without the union there would be little attention paid to health and safety issues in the workplace (Nelkin & Brown, 1984). However, there are conflicting data suggesting that union members experience more injuries and accidents than do workers in nonunion workplaces.

For example, in a controversial analysis (see also Appleton & Baker, 1985; Bennett & Passmore, 1985), Appleton and Baker (1984) tested the effect of unionization on safety in bituminous coal mines. Because the United Mine Workers of America (UMWA) has an established record of being actively involved in negotiating safety provisions and promoting worker safety, Appleton and Baker (1984) expected to find that unionized mines were safer than their nonunionized counterparts. They found just the opposite—unionized mines had higher injury rates than nonunionized mines. Reardon (1996) attempted to refine these analyses by focusing on only severe injuries but found no significant effect of unionization on workplace safety.

Similarly confusing results have emerged from analyses of safety data in the construction industry. For example, Krizan and Bradford (1995) examined OSHA data on approximately 6,000 construction-related workplace deaths collected between 1985 and 1993. In each of the 9 years studied, the authors found that unionized contractors reported a significantly higher fatality rate than did nonunion contractors. In contrast, Baker and Scherer (1997) examined data from the construction industry based on 3,000 Occupational Safety and Health Administration (OSHA) safety inspections. They found that unionized firms had fewer lost-workday injuries, fewer OSHA violations, and fewer serious violations than did their nonunionized counterparts.

Thus, data exist to support the suggestion that unions have a negative impact on safety, a nonsignificant impact, and a positive impact. There are at least three possible reasons for these disparate results. First, the differing results may be a function of sampling error (Hunter, Schmidt, & Jackson, 1982). If there is a zero correlation between safety and unionization, then one would expect to find empirical results normally distributed about a mean of zero, creating the impression of conflicting findings. The plausibility of this explanation is limited by the observation that studies of the union effects are often based on extremely large samples (e.g., 3,000–5,000) in which the effects of sampling error would be minimal.

Second, differing results across studies may be at least partially attributable to the different operationalizations of the word *safety* across studies. It is possible (and indeed most likely) that unions have differential effects on fatalities, serious injuries, and minor injuries. Therefore, differing results attributable to unionization may be a function of different operationalizations of safety.

Third, features of the data used in union effects research may disguise the effect of unionization on safety. The paradigm for union effects research (e.g., the effect of unionization on wages, safety, and organizational attitudes; see Barling et al., 1992; Freeman & Medoff, 1984) comprises analyses conducted at the organizational level using archival databases. Gordon and DeNisi's (1995) analysis of the union effect on job satisfaction points to some of the problems with this paradigm. In essence, Gordon and DeNisi showed that the consistent finding that unionized workers were more dissatisfied with their job than were nonunionized workers (for a review see Barling et al., 1992) was an artifact of the union effects paradigm that relies on organizational-level data. When the authors examined union–nonunion differences within a workplace, the differences disappeared. Although most union effects research attempts to statistically control for differences attributable to industry, geographic location, type of work, and so on, other potential confounds are ignored with the use of organizational data.

For example, unions tend to organize hazardous occupations (Leigh, 1982; Worrall & Butler, 1983)—indeed the perception that the workplace is not safe may be associated with the workers' willingness to vote for union representation (Robinson, 1988). In this view, higher rates of accidents or injuries in union workplaces are a cause rather than an effect of unionization. Second, in a related point, nonunion firms may improve health and safety in an effort to stave off unionization attempts. This is a spillover effect whereby nonunionized workers may experience benefits comparable to their nonunion counterparts in response to the threat of unionization. Finally, unions may have a politicization effect in the workplace (Barling et al., 1992). In the context of safety, unions alert their members to workplace hazards and ensure that incidents are reported. This may result in increased reporting of workplace hazards and incidents as well as the perception that the workplace is more hazardous in unionized firms than in nonunionized firms.

The question of union effects on occupational health and safety is further complicated by the recognition that unions play multiple roles in the workplace. Thus, although the presence of a union may result in a heightened awareness of occupational safety, this heightened awareness may result in a higher level of reporting of safety violations or of relatively minor accidents that may go unreported in a nonunion workplace. It should also be noted that safety concerns may be expressed in other ways in unionized firms. For example, Fairris (1992) and others (e.g., Moore & Viscusi, 1990; Olson, 1981) have provided data suggesting that unions negotiate compensating wage differentials in response to hazardous working conditions.

Given the impossibility of conducting experimental research in this area, it is possible that the answer to the question of union effects on occupational safety is to be found in longitudinal research conducted within organizations. Examining a diverse array of safety indicators within a firm prior to, and subsequent to, unionization may provide a clearer picture of union effects. Similarly, a broader array of measures that go beyond injury statistics to assess safety climate (e.g., Hofmann et al., 1995; Zohar, 1980); safety compliance; and safety initiative (e.g., Barling, Kelloway, & Zacharatos, 2002; Griffin & Neal, 2000) may prove to be useful in assessing how union activities affect occupational safety.

SUMMARY AND CONCLUSION

Labor unions have a strong interest in the safety of their members and have made extensive efforts to advance the cause of worker safety. Through collective bargaining activities unions have been brought into conflict with management. Workers join unions, negotiate agreements, take industrial action, and file grievances in order to advance their demands for safety. Unions have also cooperated with management in enhancing worker safety, most notably through the operation of joint health and safety committees. The balance between conflict and cooperation is difficult to sustain. However, as King and Alexander (1996) note with regard to another safety issue (i.e., workplace violence), "As unions move forward to negotiate, administer their collective agreements, and lobby for political action, the interactions—within the union and with management—will be difficult. They will also save lives" (p. 324).

REFERENCES

Andriessen, J. H. (1978). Safe behavior and safety motivation. *Journal of Occupational Accidents, 1*, 363–376.

Appleton, W., & Baker, J. (1984). The effect of unionization on safety in bituminous deep mines. *Journal of Labor Research, 5*, 139–147.

Appleton, W., & Baker, J. (1985). Unionization and safety in bituminous deep mines: Reply. *Journal of Labor Research*, 6, 211–216.

Ashford, N. A. (1976). *Crisis in the workplace: Occupational disease and injury.* Cambridge, MA: MIT Press.

Atkinson, W. (1999). The danger of safety incentive programs. *Electrical World, 213*, 51.

Bain, P. (1997). Human resource malpractice: The deregulation of health and safety at work in the USA and Britain. *Industrial Relations Journal, 28*, 176–191.

Baker, B., & Scherer, R. E. (1997). Construction project management and safety: Do labor unions have an effect? *Project Management Journal, 28*(3), 6–11.

Barling, J., Fullagar, C., & Kelloway, E. K. (1992). *The union and its members: A psychological approach.* New York: Oxford University Press.

Barling, J., Kelloway, E. K., & Zacharatos, A. (2002). *Occupational safety*. In P. B. Warr (Ed.), *Psychology at work* (5th ed., pp. 253–275). London: Penguin.

Beaumont, P. B. (1995). *The future of employment relations*. Thousand Oaks, CA: Sage.

Bennett, J. D., & Passmore, D. L. (1984). Unions and coal mine safety: Comment. *Journal of Labor Research*, 6, 211–216.

Boden, L. I., Hall, A. A., Levenstein, C., & Punnett, L. (1984). The impact of health and safety committees. *Journal of Occupational Medicine*, 26, 829–834.

Cornish, M., & Spink, L. (1994). *Organizing Unions*. Toronto, Ontario: Second Story Press.

Cree, T., & Kelloway, E. K. (1997). Responses to occupational hazards: Exit and participation. *Journal of Occupational Health Psychology*, 2, 304–311.

Della-Giustania, D. E., & Della-Giustania, J. L. (1992). Union demands for safe and healthful workplaces. *Professional Safety*, 37, 29–32.

Eaton, A., & Kriesky, J. (1994). Collective bargaining in the paper industry: Developments since 1979. In P. Voos (Ed.), *Contemporary collective bargaining in the private sector* (pp. 25–62). Madison, WI: Industrial Relations Research Association.

Fairris, D. (1992). Compensating payments and hazardous work in union and non-union settings. *Journal of Labor Research*, 8, 205–221.

Fellner, D. J., & Sulzer-Azaroff, B. (1984). Increasing industrial safety practices and conditions through posted feedback. *Journal of Safety Research*, 15, 125–135.

Freeman, R., & Medoff, J. (1984). *What do unions do?* New York: Basic Books.

Frick, K., & Walters, D. (1998). Worker representation on health and safety in small enterprises: Lessons from a Swedish approach. *International Labor Review*, 137, 367–389.

Gordon, M. E., & DeNisi, A. S. (1995). A re-examination of the relationship between union membership and job satisfaction. *Industrial and Labor Relations Review*, 48, 222–236.

Gray, G. R., Myers, D. W., & Myers, P. S. (1998, May). Collective bargaining agreements: Safety and health provisions. *Monthly Labor Review*, 13–35.

Griffin, M. A., & Neal, A. (2000). Perception of safety at work: A framework for linking safety climate to safety performance, knowledge, and motivation. *Journal of Occupational Health Psychology*, 17, 347–358.

Griffiths, D. K. (1985). Safety attitudes of management. *Ergonomics*, 28, 61–67.

Hebdon, R., & Hyatt, D. (1998). The effects of industrial relations factors on health and safety conflict. *Industrial and Labor Relations Review*, 51, 579–593.

Hebdon, R., & Stern, R. (1998). Tradeoffs among expressions of industrial conflict: Public sector strike bans and grievance arbitrations. *Industrial and Labor Relations Review*, 51, 204–221.

Hofmann, D. A., Jacobs, R., & Landy, F. (1995). High reliability process industries: Individual, micro, and macro organizational influences on safety performance. *Journal of Safety Research, 26*, 131–149.

Hunter, J., Schmidt, F., & Jackson, S. (1982). *Meta-analysis: Cumulating results across studies*. Thousand Oaks, CA: Sage.

Ilgen, D. R. (1990). Health issues at work: Opportunities for industrial/organizational psychology. *American Psychologist, 45*, 272–283.

Kahn, S. (1990). What occupational safety tells us about political power in union firms. *RAND Journal of Economics, 21*, 481–496.

Kelloway, E. K., Barling, J., & Harvey, S. (1998). Changing employment relations: What can unions do? *Canadian Psychology, 39*, 124–132.

King, J. L., & Alexander, D. G. (1996). Unions respond to violence on the job. In G. Vandenbos & E. Q. Bulatao (Eds.), *Violence on the job: Identifying risks and identifying solutions*. Washington, DC: American Psychological Association.

Komaki, J., Barwick, K. D., & Scott, L. R. (1978). A behavioral approach to occupational safety: Pinpointing and reinforcing safe performance in a food-manufacturing plant. *Journal of Applied Psychology, 63*, 434–445.

Komaki, J., Heinzmann, A. T., & Lawson, L. (1980). Effect of training and feedback: Component analysis of a behavioral safety program. *Journal of Applied Psychology, 65*, 261–270.

Krizan, W., & Bradford, J. (1995). Is the union fatality rate higher than non-union? *Engineering News-Record, 234*(22), 10–11.

Leigh, J. P. (1982). Are unionized blue-collar jobs more hazardous than non-unionzed blue-collar jobs? *Journal of Labor Research, 3*, 349–357.

Meng, R., & Smith, D. (1993). Union impacts on the receipt of workers' compensation benefits. *Relations Industrielles, 48*, 503.

Montgomery, J., & Kelloway, E. K. (2002). *Managing occupational health and safety*. Toronto, Ontario: Nelson.

Moore, M. J., & Viscusi, W. K. (1990). *Compensation mechanisms for job risks*. Princeton, NJ: Princeton University Press.

Nelkin, D., & Brown, S. (1984). *Workers at risk: Voices from the workplace*. Chicago: University of Chicago Press.

Northrup, H. R. (1996). Corporate campaigns: The perversion of the regulatory process. *Journal of Labour Research, 17*, 345–358.

Olson, C. A. (1981). An analysis of wage differentials received by workers on dangerous jobs. *Journal of Human Resources, 16*, 167–185.

Pupo, N., & White, J. (1994). Union leaders and the economic crisis: Responses to restructuring. *Relations Industrielles, 49*, 821–845.

Reardon, J. (1996). The effect of the United Mine Workers of America on the probability of severe injury in underground coal mines. *Journal of Labor Research, 17*, 239–252.

Reilly, B., Paci, P., & Holl, P. (1995). Unions, safety committees, and workplace injuries. *British Journal of Industrial Relations*, *33*, 275–288.

Robinson, J. C. (1988). Workplace hazards and worker's desires for union representation. *Journal of Labor Research*, *9*, 237–249.

Saari, J. (1994). When does behavior modification prevent accidents? *Leadership and Organizational Development Journal*, *15*, 11–15.

Saari, J., & Nasanen, M. (1989). The effect of positive feedback on industrial housekeeping and accidents: A long-term study at a shipyard. *International Journal of Industrial Ergonomics*, *4*, 201–211.

Scherer, R. F., Kaufman, D. J., & Ainina, M. F. (1993). Resolution of complaints by OSHA in union and non-union manufacturing organizations. *Journal of Applied Business Research*, *9*, 55–61.

Shannon, H., Mayr, J., & Haines, T. (1997). Overview of the relationship between organizational and workplace factors and injury rates. *Safety Science*, *26*, 201–217.

Smith, R. S. (1986). Greasing the squeaky wheel: The relative productivity of OSHA complaint inspections. *Industrial and Labor Relations Review*, *40*, 35–47.

Spigener, J. B., & Hodson, S. J. (1997). Are labor unions in danger of losing their leadership position in safety? *Professional Safety*, *42*, 37–39.

Sulzer-Azaroff, B. (1982). Behavioral approaches to occupational health and safety. In L. W. Frederiksen (Ed.), *Handbook of organizational behavior management* (pp. 505–537). New York: Wiley.

Sulzer-Azaroff, B., & DeSantamaria, M. C. (1980). Industrial safety hazard reduction through performance feedback. *Journal of Applied Behavioral Analysis*, *13*, 287–295.

Sulzer-Azaroff, B., Harris, T. C., & McCann, K. B. (1994). Beyond training: Organizational performance management techniques. In M. J. Colligan (Ed.), *Occupational safety and health training* (pp. 321–340). Philadelphia: Hanley & Befus.

Verma, A. (1989). Joint participation programs: Self-help or suicide for labor? *Industrial Relations*, *28*, 401–410.

Verma, A., & McKersie, R. (1987). Employee involvement: The implication of non-involvement by unions. *Industrial and Labor Relations Review*, *40*, 556–568.

Walters, V., & Denton, M. (1990). Workers' knowledge of their legal rights and resistance to hazardous work. *Relations Industrielles*, *45*, 20–36.

Walters, V., & Haines, T. (1988). Workers' use and knowledge of the internal responsibility system: Limits to participation in occupational health and safety. *Canadian Public Policy: Analyses de Politiques*, *14*, 411–423.

Weil, D. (1991). Enforcing OSHA: The role of labor unions. *Industrial Relations*, *30*, 20–36.

Weil, D. (1992). Building safety: The role of construction unions in the enforcement of OSHA. *Journal of Labor Research, 13*, 121–132.

Weil, D. (1995). Mandating safety and health committees: Lessons from the United States. *Proceedings of the 47th annual meeting of the Industrial Relations Research Association*, 273–281.

Weil, D. (1996). Regulating the workplace: The vexing problem of implementation. In D. Lewin, B. Kaufman, & D. Sockell (Eds.), *Advances in industrial and labor relations* (Vol. 7, pp. 247–286). Greenwich, CT: JAI Press.

Weil, D. (1999). Are mandated health and safety committees substitutes for or supplements to labor unions? *Industrial and Labor Relations Review, 52*, 339–361.

Wells, D. (1993). Are strong unions compatible with the new model of human resource management? *Relations Industrielles*, *48*, 56–85.

Willim, H. (1987). Trade unions and occupational safety in the German Democratic Republic. *International Labour Review, 126*, 329–336.

Winn, G. L., Frederick, L. J., & Church, G. M. (1999). Unions and behavior-based safety: Always the odd couple? *Professional Safety*, *44*, 32–34.

Wokutch, R. E. (1992). *Worker protection Japanese style: Occupational safety and health in the auto industry*. Ithaca, NY: ILR Press.

Worrall, J. D., & Butler, R. J. (1983). Health conditions and job hazards: Union and nonunion jobs. *Journal of Labor Research*, *4*, 340–346.

Zohar, D. (1980). Safety climate in industrial organizations: Theoretical and applied implications. *Journal of Applied Psychology*, *65*, 96–102.

Zohar, D., Cohen, A., & Azar, N. (1980). Promoting the increased use ear protectors in noise through information feedback. *Human Factors*, *22*, 69–79.

13

RETURNING TO WORK AFTER OCCUPATIONAL INJURY

NIKLAS KRAUSE AND THOMAS LUND

Facilitating return to work (RTW) after occupational injury is increasingly recognized as an integral part of injury prevention and workplace safety programs. To help workers with disabling injuries return to work, it is important to understand the factors that either impede or facilitate employees' reintegration into the active labor force. This chapter describes the relevant empirical evidence.[1] During the past decade research inquiring into the determinants of disability and RTW after occupational and nonoccupational injuries and illnesses is gaining interest among health and safety researchers in different fields, including epidemiology, medicine, psychology, and the

[1]Although the focus of this book is on traumatic injuries, most of the available injury statistics and the majority of the relevant literature do not offer a valid and reliable method to differentiate traumatic injuries from nontraumatic, cumulative trauma, repetitive strain, or other types of occupational injuries or illnesses. In fact, the vast majority of compensated work injuries, including injuries of the lower back and the upper extremities, has been linked to chronic repeat exposures (Bernard, 1997). The concept of precipitating events and the medicolegal notion of aggravation of a preexisting condition further complicate the classification of occupational injuries. For these reasons this chapter often use the term *injuries and illnesses* to appropriately reflect the basis of the available evidence.

social sciences. Driving forces behind this intensified interest in disability research are demographic shifts in most developed countries resulting in an aging workforce (Ilmarinen & Tuomi, 1992); anticipated labor shortages when the baby boomers retire; increased labor force participation of women with caretaking responsibilities at home (Biddle & Blanciforti, 1999); the epidemic rise in work-related soft-tissue injuries of the musculoskeletal system during past decades; and the enormous societal costs associated with these conditions, especially when they become chronically disabling (Courtney & Webster, 1999; Hashemi, Webster, & Clancy, 1998; Leigh, Markowitz, Fahs, Shin, & Landrigan, 1997).

The total of direct and indirect costs for occupational injuries and illnesses in the United States was estimated to be $149 billion in 1992, about 5 times as high as all costs associated with AIDS ($30 billion) and comparable to the total costs for all circulatory diseases ($164 billion) and cancer ($170 billion; Leigh et al., 1997). Expenses for indemnity insurance administration were $8.9 billion, and indirect costs associated with work disability were $96.2 billion; the latter included $68 billion in wage losses, $14 billion in fringe benefits, $8 billion in home production losses, and $5.2 billion for workplace training (Leigh et al., 1997). Thirty-four percent of the total costs, or $49 billion in 1992 dollars, is attributable to work-related back pain alone (Leigh et al., 1997). Low-back pain and, more recently, upper-extremity disorders have become the leading diagnoses of disabling work-related injuries in the United States and Canada (Kraut, 1994; U.S. Department of Labor, 1997). For these conditions indemnity benefits constitute the majority of direct costs, and long-term work disability is responsible for most of these costs. Seven percent of claims, with a length of disability greater than 1 year, account for 75 percent of the cost and 84 percent of total disability days due to low-back pain (Hashemi, Webster, Clancy, & Volinn, 1997) and 60 percent of the cost and 75 percent of total disability days due to upper-extremity disorders (Hashemi, Webster, Clancy, & Courtney, 1998)

Clearly, understanding the reasons for prolonged work disability due to work-related injury, and especially musculoskeletal conditions, is an important task for society and occupational health and safety researchers. Primary prevention of injuries needs to be complemented by secondary and tertiary prevention of work disability. At a historical time when widespread neoconservative politics shift more responsibility for injury loss control and disability prevention from state agencies to individual employers (Estes, 2001, p. 102), the opportunity and the necessity for the development, implementation, and evaluation of employer-based RTW programs has arisen. The main goal of these workplace interventions is to facilitate safe and timely RTW after a disabling injury or illness. Secondary goals include not only increased productivity but also quality-of-life and economic security for injured workers and their families.

Work disability and RTW are not uniquely biomedical outcomes, but are processes influenced by a variety of social, psychological, and economic factors not necessarily specific to the underlying or precipitating injury or illness (Krause, Frank, et al., 2001; Krause & Ragland, 1994; Lawrence & Jette, 1996; Pope & Tarlov, 1991; Verbrugge & Jette, 1994). Putative determinants of RTW include characteristics of the injured worker; the quality of medical care and vocational rehabilitation; physical and psychosocial job characteristics; workplace accommodation; employer factors; disability insurance factors (especially workers' compensation); and societal factors, including the legal framework, discrimination, local characteristics of the labor market, and macroeconomic conditions. Only a few systematic reviews covering some of these domains are available; a comprehensive review should be an integral part of future research activities (Krause, Frank, et al., 2001).

This chapter focuses on workplace factors that are under direct control of the employer and the employees or their representatives and may influence RTW outcomes. After a short overview of different RTW outcome measures, we present two lines of empirical research. First, we present the scientific evidence for an association between workplace factors and duration of disability or RTW. Second, we evaluate workplace interventions aimed to facilitate RTW after a work disability occurred. Finally, we discuss directions for future research on RTW. This chapter does not review the role of behavioral factors and mental health of the individual employee. Of course, appropriate clinical practices need to consider individual psychological factors because they may influence the employee's capacity to successfully deal with the process of rehabilitation and RTW. However, the collection and use of the necessary in-depth information about the individual worker's private life in the context of an employment contractual relationship faces many thorny ethical issues that in fact limit the usefulness of such information in a workplace setting. The reader should also be aware of the fact that this is not a comprehensive and systematic review of the literature on RTW but rather a presentation of an illustrative selection of the most relevant empirical scientific literature on workplace predictors of RTW. Some studies using other outcomes such as absenteeism, job retention, employer change, disability retirement, and unemployment are also cited here because predictors of these outcomes are conceptually related to predictors of RTW outcomes.

DEFINITION AND MEASUREMENT OF RTW OUTCOMES

The term *return to work* or *return-to-work outcomes* refers to a variety of related concepts and definitions of vocational outcomes after disabling injury or illness and is used to describe the duration or extent of an inability to work due to impaired health or functional limitations. Duration of work disability can be defined (a) cumulatively, as the duration of all days lost from work beginning with the date of injury; (b) categorically (e.g., return to work ever [yes/no]; working at time *x* [yes/no]); or (c) continuously, as time to RTW (e.g., calendar time from date of injury to date of first RTW or to sustained RTW, i.e., the end of the last missed workday after a series of disability episodes). RTW may be qualified as return to the preinjury employer or the preinjury job, implying a comparison with the preinjury situation. Measurements may be based on actual RTW, ability to reutrn to work, time receiving workers' compensation wage replacement benefits, earnings data, the presence of a job offer, sick leave that is not paid for by workers' compensation but is sometimes a result of occupational injury or illness, or various sequential combinations of different RTW outcomes ("RTW patterns") with or without gaps in disability. Several types of job separation may be used as indirect measures of RTW outcomes, including involuntary termination, unemployment, or retirement. Operationalizations of all these definitions vary considerably. The need for clear definitions of RTW outcomes and the pros and cons of some commonly used outcomes are discussed in detail elsewhere (Dasinger, Krause, Deegan, Brand, & Rudolph, 1999; Krause Dasinger, et al., 1999; Krause, Frank, et al., 2001).

Regardless of the outcome measure used in any study, it is appropriate to conceptualize outcomes and risk factor effects along a continuum of stages or phases in the processes of disablement and RTW (Krause, Frank, et al., 2001). As examples cited throughout this chapter demonstrate, the size and direction of the effects of various risk factors and interventions may depend on the disability phase under study. In the context of occupational health, disability phases are typically defined by duration of work disability since the date of injury, following a classification originally suggested for disabling low-back injuries (Krause & Ragland, 1994). Three main disability phases are commonly distinguished: an acute phase (1–30 days of work disability), a subacute phase (31–90 days of work disability), and a chronic disability phase (more than 90 days of work disability). The following sections describing work-related predictors of RTW will note the phase specificity of any observed effects whenever available.

JOB CHARACTERISTICS CONSTITUTING BARRIERS TO RTW

Injured employees may only partially or gradually recover from their injury or illness. The decisions of whether and when to return to work therefore depend in part on the actual job demands employees face upon their return. Several studies have identified physical and organizational job characteristics that may constitute significant barriers for RTW. The following physical job characteristics have been associated with duration of work disability in the literature: heavy physical labor (Andersson, Svensson, & Oden, 1983; Danchin, David, Robert, & Bourassa, 1982; Dasinger, Krause, Deegan, Brand, & Rudolph, 2000; Høgelund, 2000; Krause, Dasinger, Deegan, Brand, & Rudolph, 2001; Krause, Lynch, et al., 1997; Lanier & Stockton, 1988; MacKenzie et al., 1998; Ronnevik, 1988); repetitive or continuous strain, musculoskeletal strain, uncomfortable working position, and crouching (Krause, Lynch, et al., 1997); bending, twisting, or working in fixed positions (Bergquist-Ullman & Larsson, 1977); and construction work (Cheadle et al., 1994; Hogg-Johnson, Frank, & Rael, 1994; W. G. Johnson & Ondrich, 1990; McIntosh, Frank, Hogg-Johnson, Bombardier, & Hall, 2000; Oleinick, Gluck, & Guire, 1996). Interactions were observed between physical job demands and physical limitations of the employee (Yelin, 1986; Yelin, Henke, & Epstein, 1986), and between physical job demands and place of residency (Maeland & Havik, 1986). These studies provide broad evidence that high physical job demands of various forms constitute an important barrier to RTW. Only one study reviewed here (Infante-Rivard & Lortie, 1996) found no effect of heavy physical labor on RTW.

Regarding psychosocial job characteristics, the following factors have been associated with prolonged work disability: low worker control over the job (Krause, Dasinger, et al., 2001; Marklund, 1995; Yelin, 1986) and, especially, over the work and rest schedule (Infante-Rivard & Lortie, 1996; Krause, Dasinger, et al., 2001; Kristensen, 1991); long work hours (Krause, Lynch, et al., 1997); high psychological job demands (Krause, Dasinger, et al., 2001; Krause, Lynch, et al., 1997; Marklund, 1995); monotonous work (Kristensen, 1991); low skill discretion (Lund, Iversen & Poulsen, 2001); and high job stress or job strain (Krause, Dasinger, et al., 2001; Maeland & Havik, 1986; Marklund, 1995; Theorell, Harms-Ringdahl, Ahlberg-Hulten, & Westin, 1991; Yelin, 1986). Inconsistent findings are reported for social support at work; some studies linked low supervisor support (Krause, Dasinger, et al., 2001; Krause, Lynch, et al., 1997) or low coworker support (Bergquist-Ullman & Larsson, 1977; van der Weide, Verbeck, Sallé, & van Dijk, 1999) to prolonged disability, whereas other studies reported no effect (e.g., Marklund, 1995). In a retrospective cohort study of 434 workers' compensation low-back-pain claimants with 1–4 years of follow-up, low supervisor support reduced RTW rates by up to 21 percent even after adjustment for injury severity, physical workload, and other confounding factors, but no significant effect was seen for coworker support (Krause, Dasinger, et al., 2001). Mixed results regarding coworker support may be due to various forms of such support: Coworkers may provide support by cooperating with the injured employee in his or her modified work program, or they may also support the injured employee in his or her pain behavior by suggesting that the employee stay off work until he or she is 100 percent recovered. Variation in the amount of contact with coworkers as a function of the type of work may also contribute to null findings regarding coworker support. Job dissatisfaction was positively associated with work disability in some studies (Bergquist-Ullman & Larsson, 1977; Krause, Lynch, et al., 1997), but not in others (Krause, Dasinger, et al., 2001; MacKenzie et al., 1998). Low job seniority is consistently associated with longer duration of work disability even after controlling for age (Dasinger et al., 2000; W. G. Johnson & Ondrich, 1990; Krause, Dasinger, et al., 2001; Polatin et al., 1989; Tate, 1992).

Only a few studies investigated the phase specificity of these risk factors, and all of them for duration of disability after low-back pain (Dasinger et al., 2000; Krause, Dasinger, et al., 2001; MacKenzie et al., 1998). Both high physical and psychological job demands appear as independent barriers to RTW during the acute and subacute/chronic disability phases, whereas supervisor support, low job control, and low control over the work and rest schedule seem to be especially strong predictors during the subacute/chronic disability phase (Krause, Dasinger, et al., 2001).

COMPANY ORGANIZATIONAL STRUCTURES AS BARRIERS TO RTW

Only a few and relatively recent studies have identified global measures of organizational-level characteristics. Some studies focus on dimensions related to the policy level of the company, such as safety policy or company culture, factors potentially subject to employer change. Others focus on organizational characteristics of a less changeable and more descriptive nature, such as company size and firm ownership. Regarding the latter, disability seems to be prolonged if the company is private versus public (Cheadle et al., 1994; Galizzi & Boden, 1996; Infante-Rivard & Lortie, 1996). These studies control for a number of important potential confounders, the latter two also for type of industry and occupation. The effect of firm ownership is probably not a direct effect but rather mediated by a different set of job types with different specific exposures.

In a study of time to first RTW among 188,965 employees (Galizzi & Boden, 1996), an employer size smaller than 50 employees versus more than 1,000 is related to length of disability period in a disability phase-specific way. For employees with periods of disability less than 30 days, time to first RTW is shorter if the employer is small, whereas the opposite is the case if the disability period exceeds 30 days: Small employer size lengthens time to first RTW. The positive effects on RTW of large employer size for long-term disabled workers may be related to the better opportunities of the larger organization to find alternative work for a disabled employee.

Inconsistent results have been reported regarding the effect of self-employment on RTW. A population-based prospective study of working

Finnish men showed that self-employed farmers and entrepreneurs were at a threefold to almost fivefold higher risk of permanently leaving the workforce due to disability retirement (Krause, Lynch, et al., 1997). A cross-sectional study of 180 patients with rheumatoid arthritis showed that work disability was about twice as common among people working for somebody else compared to self-employed (Yelin, Meenan, Nevitt, & Epstein, 1980). While the latter finding was partially explained by higher control over the pace of work and flexibility in the activities of work among self-employed patients, the former finding among Finnish men remains unexplained.

Other studies have addressed measures related to company policies, practices, and culture (Amick et al., 2000; Habeck, Leahy, Hunt, & Chan, 1991). Amick and colleagues studied RTW among 197 workers treated with carpal tunnel surgery. Disability was shortened in companies comprising a workplace culture emphasizing an interpersonal and value-focused environment. Such a culture was typically expressed in the provision of worker training programs in safe job practices; company policies aiming at reducing the biomechanical workload (heavy lifting and repetitive movements); and policies and practices stressing early intervention, communication, and coordination in disability case management along with a proactive RTW policy (education and accommodation of employees returning to work after disability). The combined effects of these practices were independent of employee age, gender, and symptom severity. The study by Habeck and colleagues presented unadjusted effects of organizational-level variables related to the scales suggested by Amick and colleagues: safety monitoring and training, employee assistance programs, and RTW programs. These factors were all related to whether the firms under study had a high or a low rate of closed disability compensation claims. The studies provide some evidence for an association between organizational policies and practices and RTW. However, none of the studies addressing organizational policies and practices simultaneously addressed the effects of work-site exposures occurring on the individual level, which may be altered by these policies. In order to estimate simultaneously the effects of organizational-level and individual-task-level exposures, future studies should use expertise from multiple disciplines to collect valid data at both levels. Analytic methods to approach such complex relationships include, for example, path analyses and structural equation modeling for multistep processes as well as hierarchical regression modeling for multilevel analyses (Diez-Roux, 1998).

EMPLOYER-BASED RTW PROGRAMS

Employer-based RTW programs for injured employees may include active case management, provision of medical treatment and rehabilitation services, provision of transitional work opportunities, modified work, ergonomic job redesign, changes in attitude or corporate culture, and financial incentives for employees or managers—all with or without involvement of third parties. Currently, no representative data exist to describe the distribution and structure of such programs in the work environment. Knowledge about existing programs is mostly anecdotal and company specific, or based on self-reports from employer or employee surveys of limited scope. Only a few RTW programs have been described in the scientific literature, and even fewer have been properly evaluated using scientific methods. The following review of RTW programs is limited to those that have been reported in the scientific literature and fulfill at least 2.5 of 5 criteria of methodological quality. In congruence with what we know about the job-related determinants of RTW, we included in this review only those programs that, in at least one of their program elements, address and modify some features of the worker's actual job, work tasks, equipment, work station, work schedule, or the mode of interaction with coworkers and supervisors. RTW programs limited to economic incentives or medical treatment alone are not reviewed in this chapter. The quality of each study was rated with a score between 0 and 5 based on five methodological criteria covering possible selection bias, measurement of exposure and outcome variables, the temporal relationship between exposure and outcome, control for confounding factors, and design and statistical analysis. This list of studies partly stems from an earlier systematic review of the literature (Krause, Dasinger, & Wiegand, 1997; Krause, Dasinger, & Neuhauser, 1998) and was updated for this chapter in April 2001 using the same review process.

It is difficult to synthesize the information on the effectiveness of modified work from the available literature for several reasons. First, heterogeneity in terms of design, methodological quality, type of injury, and type of modified work precludes the estimation of an overall success rate via a formalized meta-analysis. Second, modified work is usually combined with other concurrent RTW policies possibly interacting with the effect of modified work. Only a few studies (Baldwin, Johnson, & Butler, 1996; Butler, Johnson, & Baldwin, 1995; Gice & Tompkins, 1989; R. Johnson, 1987; Schmidt, Oort-Marburger, & Meijman, 1995) allow for separate analyses of the effects of modified work programs. Third, the main goal of most studies was not to evaluate modified work itself, but rather to evaluate a wide range of often multidisciplinary programs combining quite different interventions of which modified work was but one element. Finally, although the magnitude of association could be used in conjunction with methodological quality to evaluate the effectiveness of modified work programs, outcome definitions and measures of association varied widely among the reviewed studies. This prohibits a direct comparison of effect sizes. Nearly all of the studies we discuss claimed a positive association between modified work and

RTW and/or cost containment. However, we base our assessment of effectiveness and efficiency only on the higher-quality studies (quality ranking of 4 or 5).

The evaluation of these studies strongly suggests effectiveness of employer-based modified work programs. The best-quality studies (ranking 4 or higher) reported at least a doubling of RTW rates and/or the number of days worked when modified work programs were offered to injured workers (Baldwin et al., 1996; Bernacki, Guidera, Schaefer, & Tsai, 2000; Bloch & Prins, 2001; Butler et al., 1995; Høgelund, 2000; Loisel et al., 1994; Loisel, Durand, Gosselin, Simard, & Turcotte, 1996). Bloch and Prins reported on six prospective studies of highest methodological quality (ranking 5) with 2-year follow-up conducted as part of a cross-national study in Denmark, Germany, Israel, the Netherlands, Sweden, and the United States. They summarized the main predictors of RTW as follows:

> Viewing all national cohorts together, four factors appear to be especially important. Higher perceived work ability and lower pain intensity at the outset were important predictors of RTW at both one and two years, while advancing age and greater physical job demands operated against work resumption. Among interventions, work place accommodations appeared to be the most successful intervention across countries. (pp. 250–251)

Bloch and Prins estimated the improvement factor of RTW rates due to workplace accommodations to be between 2.47 (Netherlands) and 21.77 (Denmark) at 1-year follow-up, and between 6.71 (Denmark) and 12.56 (Netherlands) at 2-year follow-up (Bloch & Prins, 2001, pp. 266, 271). For some countries effects were not reported because they were not statistically significant or were based on data with more than 15 percent missing values. Medical interventions and a host of other factors evaluated simultaneously did not make much of a difference in any of these studies. It follows from these studies that modified work programs are able to considerably improve RTW rates by a factor between 2 and 22. Clearly, this result indicates a very significant potential for disability prevention through adaptation of the working environment to the needs of the injured employee. Future research needs to identify the essential program elements and the cost-effectiveness of these programs. The following sections try to give some preliminary answers based on the scientific evidence that is currently available.

Which RTW Outcomes Are Influenced by Employer-Based Interventions?

Benefits of employer-based RTW programs have been observed for various outcome measures, ranging from first RTW to sustained RTW during multiple years of follow-up. Two exemplary studies illustrate this. Crook,

Moldofsky, and Shannon (1998), over a 21-month period, repeatedly interviewed a cohort of 148 randomly selected workers' compensation claimants with 90–97 days of continuous work disability. This high-quality prospective study found a 1.9 times increased rate of self-reported first RTW when modified work was available, after adjustment for age, sex, and functional disability. This finding supports the conclusion that claimants who have entered the chronic disability phase can significantly improve their chances of first RTW if modified work is made available to them. Unfortunately, this study did not investigate whether claimants remained at work after their first RTW. Other studies have shown that first RTW is often followed by repeated episodes of long-term disability, and several authors (Baldwin et al., 1996; Butler et al., 1995; Krause, Dasinger, Deegan, Brand, & Rudolph, 1999; Krause, Frank, et al., 2001) have suggested that it is necessary to use other RTW outcome measures to evaluate the effectiveness of disability prevention programs.

Baldwin, Butler, and Johnson (Baldwin et al., 1996; Butler et al., 1995) published analyses of data from participants in the Ontario Survey of Workers with Permanent Impairments. This study included 1,850 workers who were interviewed 3 to 15 years after their injury. From the data collected, four mutually exclusive RTW patterns were identified: (1) successful RTW after a single absence, (2) unsuccessful RTW after a single absence, (3) successful RTW after multiple absences, and (4) unsuccessful RTW after multiple absences. RTW was defined as successful if the claimant was working at the time of the follow-up interview. The probability of each of these outcomes was estimated as a multivariate function of age, sex, education, type of injury, union membership, postinjury wage rate, workers' compensation rate, unemployment rate, and work accommodations. Different types of work accommodations were independently and simultaneously analyzed, including provision of reduced work hours, flexible work schedules, special training, modification of equipment and tools, and assignment to light duty. Using the employment pattern of unsuccessful RTW with multiple absences as the reference outcome, the following differences were observed: Workers with light duty assignment were 2 times as likely to have a successful return after a single absence ($p < .01$) and 1.6 times as likely to experience a successful return after multiple absences ($p < .05$). Workers with reduced hours were 1.9 times as likely to have a successful return after a single absence ($p < .10$) and 1.6 times as likely to have a successful return after multiple absences (*ns*). Workers provided with modified equipment were twice as likely to have a successful return after a single absence (*ns*) but somewhat less likely to have a successful return after multiple absences (*ns*). The study was based on workers' postinjury employment history in a wide variety of occupations. Misclassification of employment histories due to recall problems was a possible weakness of this study that was not addressed by the authors. The study's methodological strengths included control of a broad range of confounding factors, the ability to assess the independent effects of different work modifications, and the ability to measure sustained RTW over a long period after the date of injury. The study suggests that light duty assignment, reduced hours, and equipment modification are all significant and independent predictors of successful RTW across a variety of industries and jobs.

Are RTW Programs Cost-Efficient?

It is premature to draw conclusions about the efficiency of modified work programs on the basis of the review of higher-quality studies. Only four studies reported any cost data (Fitzler & Berger, 1982, 1983; Wiesel, Boden, & Feffer, 1994; Yassi, Tate, et al., 1995). Only one of these studies (Yassi, Tate, et al., 1995) took program costs into account, and none of the studies provided a complete cost–benefit analysis taking administrative and other indirect costs into account. The available data on direct cost savings range between 8 percent, taking program costs into account (Yassi, Tate, et al., 1995), and 90 percent, not taking program costs into account (Fitzler & Berger, 1983). A cost–benefit analysis limited to direct costs is most likely to underestimate total savings because indirect injury-related costs have been estimated to be about 4 times higher than direct costs (Snook & Webster, 1987). Tertiary prevention through modified work programs, including permanent organizational or ergonomic improvements, sometimes also has a primary prevention effect in reducing the risk of reinjury and in benefiting coworkers sharing the same workplace. In fact, four of the reviewed studies reported a decrease in injury incidence after implementation of modified work (Gice & Tompkins, 1989; Ryden, Molgaard, & Bobbitt, 1988; Wiesel et al., 1994; Yassi, Khokhar, et al., 1995). Taking all these effects into account will increase current estimates of the efficiency of modified work programs.

Which Employer-Based Interventions Work Best?

It is difficult to determine which are the essential program features that are responsible for this overall effectiveness of employer-based RTW programs. Early reports of the success of employer-based interventions were based on comprehensive programs combining, for example, a shift in workplace culture with encouragement of early injury reporting, provision of light duty, and on-site physical therapy among other elements (Fitzler & Berger, 1982, 1983). At the time, no attempts were made to evaluate the specific contribution of any program element. Two decades after Fitzler and Berger's (1982, 1983) study, it is still unclear which components of RTW programs are most effective. It is conceivable that the answer will differ depending on the type and etiology of the injury. Most studies reviewed dealt with occupational disability due to low-back pain; three were concerned with several musculoskeletal conditions (Crook et al., 1998; Schmidt et al., 1995; Wiesel et al., 1994); two were concerned with head or brain injury (R. Johnson, 1987; West, 1995); and three included all types of injuries (Baldwin et al., 1996; Bernacki et al., 2000; Butler et al., 1995; Gice & Tompkins, 1989). A specific form of modified work—so-called supported employment, characterized by the provision of an employment specialist or job coach at work—is appropriate and effective for workers with mental illnesses (see also recent reviews by Crowther, Marshall, Bond, & Bourassa, 2001, and West, 1995). Overall, studies of high methodological quality and allowing for the independent assessment of ergonomic factors suggest that programs including an element of ergonomic workplace or equipment modifications can double RTW rates and cut the number of lost workdays in half (Krause et al., 1998).

Medical Case Management and Ergonomic Work Modifications

Large employers with in-house medical departments may exert control over some aspects of the RTW process by selecting the physician and other health care staff. In some jurisdictions (e.g., California) employers also have control over the choice of physicians providing occupational health services to workers' compensation claimants outside of the company. To our knowledge a possible association between employer control over physician choice and RTW outcomes has not yet been rigorously scientifically evaluated.

There has been little research addressing the role of the treating physician in facilitating RTW, especially while taking injury severity and workplace barriers to RTW (over which the physician may have no control) into account. A recent cross-sectional study showed that the patient's perception of the degree of physician understanding of the patient's work, and of physician communication about the nature of the job, work restrictions, and possible measures to prevent reinjury, reduces the time off work during the acute disability phase, even after controlling for injury severity and physical and psychological job demands (Dasinger, Krause, Thompson, Brand, & Rudolph, 2001). Additional, prospective studies, preferably with direct observation of the patient–provider interaction, are needed to assess the possible impact of physicians beyond their prescription of medical diagnostic and therapeutic modalities.

An often-cited study by Hall, McIntosh, Melles, Holowachuk, and Wai (1994) at first glance could be considered evidence against the effectiveness of modified work, as the authors themselves suggested. However, a closer look reveals that the data in fact do not support the authors' claims.

This study tested whether an emphasis on RTW without restrictions (i.e., no work modification) versus RTW with restrictions would facilitate reemployment for workers' compensation patients with an acute back injury. An initial control group of 669 patients participated in a rehabilitation program without any particular emphasis on return to unrestricted work. Members of a second group of 769 patients were recommended to return to restricted work only if a medically valid reason for imposing restrictions could be demonstrated. Pain in the absence of any objective findings did not constitute a reason for return to modified duty. The authors reported a twofold increase of RTW for the intervention group based on absolute numbers even though the RTW rates actually did not differ between the intervention and control groups. The overall RTW rates in the two groups were 81% and 82%, respectively (calculated from tables presented in the article). In contrast to the authors' interpretation of their results, this study demonstrates that limiting the recommendation of RTW with restrictions had no positive effect on the overall RTW rate. A recent study by Dasinger and colleagues (Dasinger et al., 2001) showed that patient-perceived doctor communication about work issues improved RTW rates during the first month of disability, but not thereafter. The study also showed that the impact of doctor–patient communication on RTW was partially confounded by job demands, suggesting that the evaluation of medical care interventions needs to control for physical and organizational job characteristics.

The contribution of clinical case management and specific medical interventions to the success of the multidisciplinary interventions appears uncertain, mainly because such program elements typically concur with an increased utilization of modified work (e.g., Wiesel et al., 1994). The only study able to directly compare ergonomic and medical interventions is the randomized controlled trial from Sherbrooke, Canada (Loisel et al., 1997). Interestingly, this study shows no independent effect of the clinical intervention on RTW but a significant effect of an occupational intervention embracing a participatory ergonomic approach (Loisel et al., 1997). This randomized control trial and a recent prospective study from Baltimore (Bernacki et al., 2000) both provide strong evidence for the importance of on-site ergonomic workplace evaluations by an ergonomist or industrial hygienist in the RTW process. Both studies are of high methodological quality and are described in more detail in the following paragraphs.

Loisel and coworkers are the first investigators to use a controlled randomized design to study the component effects of a multidisciplinary RTW program (Loisel et al., 1997; Loisel et al., 1994; Loisel et al., 1996). Participants were 32 employers with 105 employees experiencing chronic back injuries randomly assigned to four different intervention groups. One group of employees received a participatory on-site ergonomic evaluation and recommendations for workplace modification, which were implemented in about 50% of the cases in this group. The second group participated in a clinical functional restoration program and therapeutic RTW (a variant of graded work exposure). The third group received all interventions, and a fourth group, the control group, was assigned to usual care.

Measured over a 12-month period, differences in the median number of days absent from regular duty showed significant differences across the four groups (Log Rank Test, $p = .042$). Participants receiving the full intervention or the ergonomic intervention had median absences from work that were about half as long as those seen for the usual care and the clinic intervention only group (60 and 67 days vs. 121 and 131 days, respectively). When the three intervention groups were compared to the usual care group in a Cox regression analysis controlled for age, sex, comorbidity, and body mass index, only the complete intervention was statistically significant ($p = .014$). Subjects in this group had on average 2.41 times fewer lost days from regular duty work than subjects in the usual care group. When the effects of having an ergonomic intervention and having a clinical intervention were examined separately, only the ergonomic intervention was significant. Workers receiving the ergonomic intervention had on average 1.91 times fewer lost workdays than those without the ergonomic intervention ($p = .009$).

The results show that in the multidisciplinary program tested, the participatory ergonomic intervention was the most important in facilitating an earlier RTW. The potential effect of ergonomic modifications may have been underestimated, because a follow-up interview of injured workers, employer representatives, and union representatives 6 months after the ergonomic intervention showed that only about half of the proposed ergonomic solutions had been implemented at that time. The clinical intervention, which consisted of work hardening, behavioral-cognitive interventions, and therapeutic RTW, did not show an impact on its own. Whether this finding could be the result of a timing effect (the ergonomic intervention was administered during weeks 6 to 10, but workers did not receive the work hardening component of the clinical intervention until weeks 13 to 26 after injury) was not addressed by the authors.

The design of this study does not allow for a complete evaluation of the impact of graded work exposure on RTW. The authors defined RTW as return to full duties. Thus, injured workers in any of the four experimental groups could have been assigned to modified or light duty before recruitment into the study. The effect of randomization on this potentially confounding variable was not checked. However, the controlled randomized experimental design, the control of other confounding variables (age, sex, comorbidity, body mass index), and the comparability of the four intervention groups in terms of duration of work absence, functional disability, pain, social support, and medical scores make this study of high methodological quality. The study does not show a significant effect of graded work exposure on RTW, but the study suggests that ergonomic work modifications significantly reduce the time off work after a back injury.

Bernacki and colleagues (Bernacki et al., 2000) described an early RTW program introduced in 1992 at a large urban medical center in Baltimore, Maryland, aimed to control the incidence and costs of work-related illness and injury. The main program elements were employer and supervisor training, job redesign based on job analyses, work accommodation, and case management conferences. An industrial hygienist with training in ergonomics was designated to facilitate the placement of employees with work restrictions. The average rate of lost workdays per 100 employees was 26.3 in 4 preprogram years and 12.0 during the 7 years after introduction of the RTW program, yielding an improvement factor of 2.2. The number of lost-workday cases showed a similar drop (rate ratio 1.98). The average number of non-lost-workday cases remained stable (5.6 and 6.0 per 1,000 employees, respectively), indicating that improved RTW rates were not due to a change in reporting patterns. The number of restricted duty days increased by a factor of 20, from 0.63 to 13.4 days per 100 employees. The authors reported that involvement of the industrial hygienist increased the accommodation rate for injured employees by 54 percent. This study, similar to Loisel et al.'s (1997) study, suggests that the involvement of an ergonomic job analysis and work accommodations play a central role in improving RTW outcomes.

In conclusion, the limited available evidence indicates that ergonomic interventions seem to be superior to clinical or behavioral interventions in their ability to facilitate RTW. The involvement of ergonomically trained professionals in on-site job analysis, work accommodation, and case conferences is a common feature of the successful programs described by Loisel et al. (1997) and by Bernacki et al. (2000). Interestingly, the positive results of the Sherbrooke model were achieved despite the fact that only about half of the ergonomist's recommendations had been implemented at follow-up (Loisel et al., 1996). The full potential of ergonomic interventions needs to be assessed in future studies with longer follow-up periods to allow enough time for the complete implementation of the ergonomic intervention.

Job Security, Disability Insurance, and Other Benefit Programs

It is conceivable that employers may influence RTW decisions of their injured workers by the provision of job security, disability insurance, child care facilities, health care plans, or other benefits. Little research has been done to assess the effect of such programs. The extant literature instead focuses on the level of wage replacement benefits from state-mandated workers' compensation, unemployment, or social disability insurance. The interpretation of this literature remains controversial, and it is beyond the scope of this chapter to delineate the long-standing debate of social scientists about this topic. However, newer research suggests that interaction effects may in part explain the inconclusiveness of previous research (Bloch & Prins, 2001). In a cross-national series of prospective studies, the Work Incapacity and Reintegration Project (Bloch & Prins, 2001) compared the effectiveness of different RTW interventions used by social security systems and health care providers in six countries: Denmark, Germany, Israel, the Netherlands, Sweden, and the United States. Homogeneous cohorts of employees who were work-disabled for at least 3 months because of low-back disorders were followed over a 2-year period with data collection at baseline and at 1- and 2-year follow-ups. The association between wage replacement benefits and duration of disability was dependent on the degree of job security. The combination of extensive benefits with strong job protection predicted early RTW, whereas weak job protection combined with extensive benefits did not improve RTW. Shorter disability periods were seen for low levels of job protection combined with low levels of benefits, but mainly for new employees.

Another population-based prospective study of Finnish men has shown that disability retirement (another form of failure to return to work) is strongly influenced by the number of family members who are working and who are unemployed (Krause, Lynch, et al., 1997). Future research needs to examine how injured workers weigh the pros and cons of RTW at different levels of job security, wage replacement rates, and a range of economic and other needs of the family unit. (For a conceptual framework to study this decision-making process, see Franche & Krause, 2002, in press.) An employer may exert an influence on this decision-making process by providing benefits that assist the injured worker to balance the needs for recovery, economic security, and maintaining active occupational and recreational roles. A recent study of RTW rates after disabling low-back pain in the populations of Norway, Great Britain, and the United States found the highest RTW rates in the country with the most generous sickness benefits (Hagen & Thune, 1998), another example of a positive effect of benefit programs on RTW.

Improving the Quality of Employer–Employee Interactions

Most of the intervention studies we reviewed described and evaluated the structure (i.e., the components) of RTW programs, but less attention was given to the process (i.e., the manner in which these programs were administered). More recently, the important role of workplace interactional factors in the process of RTW has repeatedly been demonstrated in empirical studies (Krause, Dasinger, et al., 2001; Krause, Lynch, et al., 1997; van der Weide et al., 1999). To study these interactions systematically, Franche and Krause propose a new conceptual framework combining the Readiness for Change Model (Prochaska, DiClemente, & Norcross, 1992) and the Phase Model of Disability (Krause & Ragland, 1994) for a systematic examination of the impact of employee–employer interactions on the employee's decision to return to work after the occurrence of an injury (Franche & Krause, 2002, in press).

There is plentiful anecdotal evidence but also already some indirect empirical evidence for the importance of good communication among employer, employee, and other agents in the RTW process. As mentioned above, coworkers' (van der Weide et al., 1999) and supervisors' interactions with injured and ill workers need to be recognized as influential predictors of RTW (Krause, Frank, et al., 2001; Krause, Lynch, et al., 1997). For example, in a study of 120 Dutch workers with work absences of 10 days or more due to low-back pain, problematic relations with colleagues was one of four predictive factors of time off work (van der Weide et al., 1999). Another factor that has received increased attention is legitimacy (Tarasuk & Eakin, 1994, 1995). *Legitimacy* refers to the degree to which an injured employee feels believed by others regarding the authenticity of his or her injury and its symptoms. It is of particular relevance to work absences involving "invisible" injuries or illnesses. In the Canadian Early Claimant Cohort study, legitimacy was a significant predictor of duration on benefits (Smith, Tarasuk, Ferrier, & Shannon, 1996; Smith, Tarasuk, Shannon, & Ferrier, 1998). The mechanisms underlying the association between legitimacy and RTW outcomes remain speculative. If an employee feels that workplace staff questions the legitimacy of his or her symptoms, the worker may develop negative feelings toward the workplace, which will certainly weigh against returning to work in his or her decisional balance. Alternatively, the perceived expression of disbelief of one's symptoms and complaints may bring an employee to invest energy in "proving" that his or her injury and pain are real by not returning to work.

As described earlier, workplace culture (referring to a general interpersonal and value-focused atmosphere) is also associated with RTW outcomes (Amick et al., 2000; Hunt & Habeck, 1993). In one prospective study of 198 employees with carpal tunnel syndrome, an increased level of people-oriented culture was associated with higher RTW rates 6 months after being identified in community medical practices, when age, gender, and baseline carpal tunnel syndrome symptom severity were controlled for (Amick et al., 2000). People-oriented culture was a factorially derived dimension defined as "the extent to which the company involves employees in meaningful decision-making, where there is trust between management and employees, and openness to share information in a cooperative work environment" (Amick et al., 2000, p. 30).

The goal adopted in the area of RTW is generally to achieve an early and safe RTW. Risk factors contributing to a premature and unsafe RTW are seldom considered in the extant literature. It is important to consider that fear of losing one's job and financial strain will weigh in the employee's decision balance and can contribute to the decision of going back to work too soon, increasing risk of reinjury and ill health (Pransky et al., 2000).

A collaborative and respectful approach from workplace parties will clearly lead to a climate of trust much more conducive to reducing one's anxieties about returning to work and to shifting the decisional balance toward being ready for an RTW attempt. Attention to coworkers' safety, workload, and understanding of modified work will also set the stage for a modified work accommodation conducive to increased self-efficacy and successful coping with setbacks in a RTW process often characterized by repeated disability episodes months or years after the initial injury.

CHALLENGES FOR FUTURE RESEARCH

Recently, the National Institute for Occupational Safety and Health (NIOSH) invited researchers from the United States and Canada to evaluate the current state of knowledge regarding the measurement and determinants of RTW outcomes after disabling injury and illness and to identify critical data and research needs. The researchers identified five key challenges for future research and developed some general recommendations, summarized in the following sections. (For more details and a full biography, see Krause, Frank, et al., 2001; Krause, Frank, et al., 1999.)

Definition, Choice, and Measurement of RTW Outcomes

The first challenge is the definition, choice, and measurement of RTW outcomes. It needs to be acknowledged that each stakeholder in the RTW process has his or her own concept of the most appropriate outcome measure that captures the key benefit of successful RTW (Melles, McIntosh, & Hall, 1995). In some cases the RTW outcome is directly and routinely measured (e.g., workers' compensation costs, end of disability benefit payments). In other instances, the outcome is only indirectly related to RTW and is not routinely measured (e.g., improvement in pain, function, and quality of life). Limiting a research agenda to traditional RTW outcomes (such as duration of temporary disability benefits) would reflect a limited perspective bound to underestimate duration of work disability and total burden (Dasinger et al., 1999; Galizzi & Boden, 1996; Jette & Jette, 1996). More inclusive measures of the direct impact of work disability, including all indemnity benefit payments, and legal and medical costs, still represent only a fraction of the total actual economic burden of occupational injury and illness. Indirect costs borne by the employer, including productivity losses and employee substitution costs, have been estimated to be at least 2–4 times direct workers' compensation costs (Snook, 1988; Snook & Webster, 1987). Including wage-loss data helps capture even more of the economic losses suffered by the injured worker (Peterson, Reville, Kaganoff-Stern, & Barth, 1997; Reville, 1999). It is of course impossible to adequately describe in financial terms the burden of pain and suffering and reduced quality of life, and their full impact on families, coworkers, communities, and society at large. A recent appraisal estimated that all workers' compensation costs (about 60 billion dollars in 1992) represent only about 40% of the total costs associated with occupational injury and illness in the United States (Leigh et al., 1997). Although RTW outcomes cannot completely reflect the impact of occupational injury and illness in terms of work-related disability, they are important in their own right and a necessary component of any more complete assessment.

Future research should include both self-reported and administrative data on disability and RTW, in order to ensure a comprehensive assessment of work-related disability and to provide the means to assess the magnitude of reporting biases from any one data source. If workers' compensation data are used, several outcome measures may need to be reported (Krause, Dasinger, et al., 1999). The outcomes "time to the end of first temporary disability episode" and "time to first RTW" have very limited value and should always be complemented by measures more inclusive of recurrences (Butler et al., 1995; Dasinger et al., 1999; Galizzi & Boden, 1996; Krause, Dasinger, et al., 1999). Even cumulative measures of time on temporary disability are insufficient for capturing the important effects of long-term disability beyond 1 year. In jurisdictions providing permanent disability benefits and other wage replacement benefits, researchers should also provide the total number of effectively compensated days as one of their outcome measures. This measure is easily calculated from claims data and includes indemnity benefit payments during temporary disability, vocational rehabilitation, and permanent disability (Dasinger et al., 1999; Krause, Dasinger, et al., 1999; Oleinick et al., 1993). We further recommend that future studies include at least one functional or quality-of-life component, such as pain or functional limitations, and that intervention studies be designed a priori to include a comprehensive cost-effectiveness and cost–benefit analysis. Obtaining both economic and functional outcome information addresses the immediate concerns of employers regarding the costs of return-to-work programs, and of employees and their physicians regarding the risk of recurrence or exacerbation of the health condition if work is resumed too early.

Multifactorial Nature of Disability and RTW and Need for an Overarching Conceptual Framework

Another challenge for future research is the multifactorial nature of disability and RTW. In a recent review article, more than 100 putative determinants of work disability have been identified covering reports in the medical, epidemiological, psychological, and economic literature (Krause,

Frank, et al., 2001). The authors diagnose the lack of an overarching conceptual framework for a field that would encompass multiple health conditions and multiple scientific disciplines. Although there are some excellent general theories of disablement (Nagi, 1965; Stone, 1984; Verbrugge & Jette, 1994) that emphasize the importance of the social environment, these make only passing reference to the workplace specifically or to the various stakeholders in the RTW process. The development of an explicit conceptual framework for the RTW process, requiring the understanding and integration of rather different theories from the various relevant disciplines, would help unify the many streams of research in this field. Further deliberation on this topic is beyond the scope of this chapter. However, some preliminary suggestions for a more comprehensive conceptualization of occupational health, disability, and RTW have been made in the recent literature and may serve as a starting point for future developments (Franche & Krause, 2002a; Jette & Jette, 1996; Krause & Ragland, 1994).

Which Risk Factors and Interventions Should Be Examined in Future Research?

Given the multifactorial nature of the RTW process, only a few general and preliminary recommendations can be given. The following criteria may help researchers set priorities in different areas of interest.

Amenability to Change

Krause and colleagues (Krause, Frank, et al., 2001; Krause, Frank, et al., 1999) recommend directing interdisciplinary research efforts to those putative or known risk factors that are amenable to change. It is necessary to differentiate between two types of predictors of RTW: causal risk factors and risk markers (alias risk indicators). Risk markers serve to identify individuals, groups, or places at risk, and may constitute factors that cannot be changed (e.g., gender), whereas risk factors proper are amenable to change and provide the basis for interventions. Factors amenable to change include worker control over amount of work (physical workload) and organization of work tasks (work tempo and scheduling of rest breaks), the availability and type of modified work, help from coworkers, employer accommodations, and local compliance with evidence-based medical care. If there is already a fair body of knowledge on the risk factor (e.g., lack of worker control over the work and rest schedule), interventions that address the risk factor now need to be developed and then evaluated. If the effectiveness of an intervention is well established, as for example for modified work programs (Krause et al., 1998), but implementation is not as widespread as desirable, research should identify the barriers to implementation and plan interventions in order to increase uptake. If the costs of an intervention constitute a potential barrier, a cost-effectiveness or cost–benefit analysis is needed. If potential beneficiaries of a researched intervention are unaware of it (Hennessey & Muller, 1995), a needs assessment for the dissemination of such results and the development and evaluation of an appropriate research transfer/educational campaign would be the appropriate agenda.

Generalizability

The generalizability or importance of factors across disability phases, health conditions, and settings or jurisdictions is another useful criterion to identify priority topics for future research. Resources committed to the understanding and modification of factors that are involved in more than one phase of the illness–injury–disability–RTW process are likely to generate a greater return on investment. Candidate risk factors of this sort include supportive corporate climate, participatory ergonomics, control over workload, and ability to take unscheduled breaks. Increased control over job tasks and work schedule can be hypothesized to facilitate RTW in almost any setting, for almost any health condition (Karasek & Theorell, 1990; Krause, Dasinger, et al., 2001; Kristensen, 1991; Syme, 1990, 1991).

Qualitative Studies and Instrument Development

Candidate risk factors and interventions should be selected not only on the basis of the existing literature but also de novo from exploratory qualitative research. For example, focus groups of stakeholders may unearth new candidate factors. One focus of qualitative studies should be the identification of new secondary risk factors that emerge during the progressive stages of disability and become barriers to RTW during the subacute and chronic phases of disability. Further, when it is widely accepted that a construct is probably causal but is not well measured, better measures need to be developed and pretested in the field. Examples include measures of workplace health and safety culture from organizational psychology, and measures for the quality of communication and administration of RTW issues between stakeholders.

Participatory Research

Another recommended priority is to target risk factors and research to the user and setting. Funders and researchers may benefit by organizing risk factor knowledge and related research agendas by locus of action (i.e., among the various stakeholders or agents, e.g., employee, employer, health care provider, claims adjuster, disability manager) or intervention setting (e.g., workplace, health care system, insurance system, regulatory or political bodies; Frank et al., 1998; Krause & Ragland, 1994). This will facilitate the alignment of research questions with the practical decision-making needs of various stakeholders. Stakeholders can be expected to be more motivated to fund and use research on factors that they themselves deal with and that they can change or influence. The involvement of stakeholders in the planning and execution of research will also promote transfer of research results into practice (e.g., via the provision of more economic incentives to employers offering modified work).

Study Design and Analytic Strategies

The choice of appropriate design and analytic approaches is another challenge, even for quantitative studies. There is an upward trend in the methodological quality of published studies. However, it should be emphasized that the randomized controlled trial does not necessarily constitute the ideal design for the evaluation of RTW programs. It is often not feasible to randomize individual subjects to interventions that can be efficiently delivered only to entire populations (e.g., modified work programs in a workplace or jurisdictional-level workers' compensation policies). This is well recognized in other fields of public health, where the value of quasi-experimental designs is thoroughly established (e.g., early municipal water fluoridation trials and more recent community intervention trials to prevent coronary heart disease). It is laudable that some investigators have performed randomization of entire companies; in a landmark study, Loisel et al. (1997) accomplished this for 32 firms across four treatment arms. However, even with this extraordinary effort, the numbers of firms in each arm were insufficient to achieve confounder balance for key variables, especially those at the firm level. Since firm-level variables are widely thought to be critical in RTW (Frank et al., 1998), the advantages of cluster randomization at the firm level are therefore dubious in this context. Quasi-experimental designs, such as staggered time-series designs, in which various workplaces receive the intervention at intervals with pre- and postmeasures in each, can go a long way toward controlling for a full range of confounders and are much more likely to be acceptable to participating companies, local insurers, or regulatory bodies than are external concurrent control groups. In all such designs, it is desirable to have one group of subjects maintain control status for as long as possible in order to control for secular trends in either risk factors or outcomes. A novel suggestion is the use of job-matched controls, which have been successfully used in an etiological study of low-back pain at work (Kerr et al., 2001) and in a recent wage-loss study of permanently disabled workers (Peterson et al., 1997; Reville, 1999).

Another recommendation for further research is to take the disability phase specificity of risk factors and interventions into account during both the design and analysis stages of a study. *Phase specificity* refers to the fact that the impact of risk factors (or interventions) may vary across different phases of the disablement and RTW process (Dasinger et al., 2000; Krause, Dasinger, et al., 2001; Krause & Ragland, 1994). Research designs and analytic strategies not accounting for this phenomenon may be unable to detect risk factors that are predominantly associated with only certain stages of the disabling process. In fact, the failure to stratify analyses according to work disability phase may lead to the masking of important risk factor effects (Oleinick et al., 1996). Similarly, intervention researchers should recognize that suboptimal timing of interventions may be responsible for disappointing results (Frank et al., 1998; Sinclair, Hogg-Johnson, Mondloch, & Shields, 1997).

Disability phases can be defined by duration of work disability along a temporal continuum since the date of injury (Krause & Ragland, 1994). Although individual studies may choose different cut-points between phases, the differentiation of an acute phase (up to 30 days of work disability), a subacute phase (30–90 days), and chronic disability (more than 90 days) is becoming widely used. The concept of phase specificity was originally developed for the study of occupational disability due to low-back pain (Frank et al., 1996; Krause, 1993; Krause & Ragland, 1994; Spitzer et al., 1987; Von Korff, 1994); however, its usefulness has since been empirically demonstrated not only for spinal disorders (Dasinger et al., 2000; Frank et al., 1998; Frank et al., 1996; Frymoyer, Rosen, Clements, & Pope, 1985; Hogg-Johnson, Frank, & Rael, 1994; Krause, Dasinger, et al., 2001; McIntosh et al., 2000; van der Weide et al., 1999; Williams, Feuerstein, Durbin, & Pezzullo, 1998) but also for other injuries and illnesses (Galizzi & Boden, 1996; Oleinick et al., 1996; Rael, 1992).

Finally, the integration of cross-disciplinary qualitative and quantitative methods constitutes a major challenge for future RTW research because such interdisciplinary efforts do not tend to arise naturally from university-based research. Future RTW research funding needs to explicitly reward interdisciplinary teams of researchers and provide incentives for the submission of proposals by such mixed-methods consortia in the request for proposal (RFP) process.

SUMMARY

The scientific evidence for an association between workplace factors and RTW or duration of disability has been established in two ways. First, in a review of the scientific literature strong associations were found between job- and employer-related factors and RTW rates and duration of work disability. Second, the evaluation of employer-based RTW programs showed that interventions that include some form of modified work improve RTW rates consistently by a factor of about 2 and cut lost workdays in half. Therefore, one can conclude that RTW programs including modified work are effective. More comprehensive cost-effectiveness analyses are needed to provide better data on the cost-effectiveness of such interventions; however, the limited evidence suggests that they are feasible. Whereas current research has focused on the effective structural elements of RTW programs, future research needs to also address which modes of interactions between employer and employee and other stakeholders are most conducive for positive RTW outcomes. The investigation of such process variables will require interdisciplinary research using both qualitative and quantitative methods and a conceptual model integrating several theories from different research disciplines.

REFERENCES

Amick, B., Habeck, R. V., Hunt, A., Fossel, A. H., Chapin, A., Keller, R. B., et al. (2000). Measuring the impact of organizational behaviors on work disability prevention and management. *Journal of Occupational Rehabilitation, 10*, 21–38.

Andersson, G. B., Svensson, H. O., & Oden, A. (1983). The intensity of work recovery in low back pain. *Spine, 8*, 880–884.

Baldwin, M. L., Johnson, W. G., & Butler, R. J. (1996). The error of using returns-to-work to measure the outcomes of health care. *American Journal of Industrial Medicine, 29*, 632–641.

Bergquist-Ullman, M., & Larsson, U. (1977). Acute low back pain in industry: A controlled prospective study with special reference to therapy and confounding factors. *Acta Orthopaedica Scandinavia, 170*(Suppl.), 9–103.

Bernacki, E. J., Guidera, J. A., Schaefer, J. A., & Tsai, S. (2000). A facilitated early RTW program at a large urban medical center. *Journal of Occupational and Environmental Medicine, 42*, 1172–1177.

Bernard B. P. (Ed.). (1997). *Musculoskeletal disorders and workplace factors* (DHHS [NIOSH] Publication No. 97-141). Cincinnati, OH: U.S. Department of Health and Human Services, National Institute for Occupational Safety and Health.

Biddle, E. A., & Blanciforti, L. A. (1999). Impact of a changing U.S. workforce on the occupational injury and illness experience. *American Journal of Industrial Medicine, 1*(Suppl.), 7–10.

Bloch, F. S., & Prins, R. (Eds.). (2001). *Who returns to work and why? A six-country study on work incapacity and reintegration*. New Brunswick, NJ: Transaction Publishers.

Butler, R. J., Johnson, W. G., & Baldwin, M. L. (1995). Managing work disability: Why first RTW is not a measure of success. *Industrial and Labor Relations Review, 48*, 452–469.

Cheadle, A., Franklin, G., Wolfhagen, C., Savarino, J., Liu, P. Y., Salley, C., et al. (1994). Factors influencing the duration of work-related disability: A population-based study of Washington State workers' compensation. *American Journal of Public Health, 84*, 190–196.

Courtney, T. K., & Webster, B. S. (1999). Disabling occupational morbidity in the United States. An alternative way of seeing the Bureau of Labor Statistics' data. *Journal of Occupational and Environmental Medicine, 41*, 60–69.

Crook, J., Moldofsky, H., & Shannon, H. (1998). Determinants of disability after a work related musculoskeletal injury. *Journal of Rheumatology, 25*, 1570–1577.

Crowther, R. E., Marshall, M., Bond, G. R., & Huxley, P. (2001). Helping people with severe mental illness to obtain work: Systematic review. *British Medical Journal, 322*, 204–208.

Danchin, N., David, P., Robert, P., & Bourassa, M. G. (1982). Employment following aortocoronary bypass surgery in young patients. *Cardiology, 69*, 52–59.

Dasinger, L. K., Krause, N., Deegan, L. J., Brand, R. J., & Rudolph, L. (1999). Duration of work disability after low back injury: A comparison of administrative and self-reported outcomes. *American Journal of Industrial Medicine, 35*, 619–631.

Dasinger, L. K., Krause, N., Deegan, L. J., Brand, J. B., & Rudolph, L. (2000). Physical workplace factors and RTW after compensated low back injury: A disability phase-specific analysis. *Journal of Occupational and Environmental Medicine, 42*, 323–333.

Dasinger, L. K., Krause, N., Thompson, P. J., Brand, R. J., & Rudolph, L. (2001). Doctor proactive communications, RTW recommendation, and duration of disability after a compensated low back injury. *Journal of Occupational and Environmental Health, 43*, 515–525.

Diez-Roux, A. (1998). Bringing context back into epidemiology: Variables and fallacies in multilevel analyses. *American Journal of Public Health, 88*, 216–222.

Estes, C. L. (2001). *Social policy and aging: A critical perspective*. Newbury Park, CA: Sage.

Fitzler, S. L., & Berger, R. A. (1982). Attitudinal change: The Chelsea back program. *Occupational Health and Safety, 51*, 24–26.

Fitzler, S. L., & Berger, R. A. (1983). Chelsea back program: One year later. *Occupational Health and Safety, 52*, 52–54.

Franche, R. L., & Krause, N. (2002). Readiness for return to work following injury or illness: Conceptualizing the interpersonal impact of healthcare, workplace, and insurance factors. *Journal of Occupational Rehabilitation, 12*, 233–256.

Franche, R. L., & Krause, N. (in press). Supporting the employee's RTW: Critical factors in recovery and return-to-work. In J. W. Frank & T. S. Sullivan (Eds.), *Reducing work-related disability: A reader*. London: Taylor & Frances.

Frank, J. W., Brooker, A. S., DeMaio, S. E., Kerr, M. S., Maetzel, A., Shannon, H., et al. (1996). Disability resulting from occupational low back pain: Part 2. What do we know about secondary prevention? A review of the scientific evidence on prevention after disability begins. *Spine, 21*, 2918–2929.

Frank, J. W., Sinclair, S., Hogg-Johnson, S., Shannon, H., Bombardier, C., Beaton, D., et al. (1998). Preventing disability from work-related low-back pain. New evidence gives new hope—if we can just get all the players onside. *Canadian Medical Association Journal, 158*, 1625–1631.

Frymoyer, J. W., Rosen, J. C., Clements, J., & Pope, M. H. (1985). Psychologic factors in low-back-pain disability. *Clinical Orthopaedics and Related Research, 195*, 178–184.

Galizzi, M., & Boden, L. I. (1996). *What are the most important factors shaping RTW? Evidence from Wisconsin*. Cambridge, MA: Workers' Compensation Research Institute.

Gice, J. H., & Tompkins, K. (1989). RTW program in a hospital setting. *Journal of Business and Psychology, 4*, 237–243.

Habeck, R. V., Leahy, M. J., Hunt, H. A., & Chan, F. (1991). Employer factors related to workers' compensation claims and disability management. *Rehabilitation Counseling Bulletin, 34*, 210–226.

Hagen, K. B., & Thune O. (1998). Work incapacity from low back pain in the general population. *Spine, 23*, 2091–2095.

Hall, H., McIntosh, G., Melles, T., Holowachuk, B., & Wai, E. (1994). Effect of discharge recommendations on outcome. *Spine, 19*, 2033–2037.

Hashemi, L., Webster, B. S., & Clancy, E. A. (1998). Trends in disability duration and cost of workers' compensation low back pain claims (1988–1996). *Journal of Occupational and Environmental Medicine, 40*, 1110–1119.

Hashemi, L., Webster, B. S., Clancy, E. A., & Courtney, T. K. (1998). Length of disability and cost of work-related musculoskeletal disorders of the upper extremity. *Journal of Occupational and Environmental Medicine, 40*, 261–269.

Hashemi, L., Webster, B. S., Clancy, E. A., & Volinn, E. (1997). Length of disability and cost of workers' compensation low back pain claims. *Journal of Occupational and Environmental Medicine, 39*, 937–945.

Hennessey, J. C., & Muller, L. S. (1995). The effect of vocational rehabilitation and work incentives on helping the disabled-worker beneficiary back to work. *Social Security Bulletin, 58*, 15–28.

Høgelund, J. (2000). *Bringing the sick back to work: Labor market reintegration of the long-term sicklisted in the Netherlands and Denmark*. Copenhagen, Denmark: Roskilde University, Danish National Institute of Social Research.

Hogg-Johnson, S., Frank, J. W., & Rael, E. (1994). *Prognostic risk factor models for low back pain: Why they have failed and a new hypothesis* (Working Paper No. 19). Toronto, Ontario: Institute for Work and Health.

Hunt, H. A., & Habeck, R. V. (1993). *The Michigan disability prevention study*. Kalamazoo, MI: W. E. Upjohn Institute for Employment Research.

Ilmarinen, J., & Tuomi, K. (1992). Work ability index for aging workers. *Proceedings of the International Scientific Symposium on Aging and Work, 4*, 142–151.

Infante-Rivard, C., & Lortie, M. (1996). Prognostic factors for RTW after a first compensated episode of back pain. *Occupational and Environmental Medicine, 53*, 488–494.

Jette, D. U., & Jette, A. M. (1996). Health status assessment in the occupational health setting. *The Orthopedic Clinics of North America, 27*, 891–902.

Johnson, R. (1987). RTW after severe head injury. *International Disability Studies, 9*, 49–54.

Johnson, W. G., & Ondrich, J. (1990). The duration of post-injury absences from work. *Review of Economics and Statistics, 72*, 578–586.

Karasek, R., & Theorell, T. (1990). *Healthy work: Stress, productivity, and the reconstruction of working life*. New York: Basic Books.

Kerr, M., Frank, J., Shannon, H., Norman, R., Wells, R., Neumann, W., et al. (2001). Biomechanical and psychosocial risk factors for low back pain at work. *American Journal of Public Health, 91*, 1069–1075.

Krause, N. (1993). *Work disability and low back pain: A new classification model for research and intervention*. Berlin, Germany: Wissenschaftszentrum Berlin für Sozialforschung.

Krause, N., Dasinger, L. K., Deegan, L. J., Brand, R. J., & Rudolph, L. (1999). Alternative approaches for measuring duration of work disability after low back injury based on administrative workers' compensation data. *American Journal of Industrial Medicine, 35*, 604–618.

Krause, N., Dasinger, L. K., Deegan, L. J., Brand, R. J., & Rudolph, L. (2001). Psychosocial job factors and RTW after low back injury: A disability phase-specific analysis. *American Journal of Industrial Medicine, 40*, 374–392.

Krause, N., Dasinger, L. K., & Neuhauser, F. (1998). Modified work and RTW: A review of the literature. *Journal of Occupational Rehabilitation, 8*, 113–139.

Krause, N., Dasinger, L., & Wiegand, A. (1997). *Does modified work facilitate RTW for temporarily or permanently disabled workers? Review of the literature and annotated bibliography*. Report prepared for the Industrial Medical Council of the State of California and the California Commission on Health and Safety and Workers' Compensation. Berkeley: University of California.

Krause, N., Frank, J. W., Sullivan, T., Dasinger, L. K., & Sinclair, S. J. (2001). Determinants of duration of disability and RTW after work-related injury and illness: Challenges for future research. *American Journal of Industrial Medicine, 40*, 464–484.

Krause, N., Frank, J. W., Sullivan, T., Dasinger, L. K., Sinclair, S. J., & Rudolph, L. (1999). Determinants of RTW and duration of disability after work related injury and illness: Developing a research agenda. In *Functional, Economic, and Social Outcomes of Occupational Injuries and Illnesses: Integrating Social, Economic, and Health Services Research* (pp. 189–230). Denver, CO: U.S. Department of Health and Human Services, Centers for Disease Control, National Institute for Occupational Safety and Health.

Krause, N., Lynch, J., Kaplan, G. A., Cohen, R. D., Goldberg, D. E., & Salonen, J. T. (1997). Predictors of disability retirement. *Scandinavian Journal of Work, Environment and Health, 23*, 403–413.

Krause, N., & Ragland, D. R. (1994). Occupational disability due to low back pain: A new interdisciplinary classification based on a phase model of disability. *Spine, 19,* 1011–1020.

Kraut, A. (1994). Estimates of the extent of morbidity and mortality due to occupational diseases in Canada. *American Journal of Industrial Medicine, 25,* 267–278.

Kristensen, T. S. (1991). Sickness absence and work strain among Danish slaughterhouse workers: An analysis of absence from work regarded as coping behaviour. *Social Science and Medicine, 32,* 15–27.

Lanier, D. C., & Stockton, P. (1988). Clinical predictors of outcome of acute episodes of low back pain. *Journal of Family Practice,* 27, 483–489.

Lawrence, R. H., & Jette, A. M. (1996). Disentangling the disablement process. *The Journals of Gerontolology. Series B, Psychological Sciences and Social Sciences, 51,* 173–182.

Leigh, J. P., Markowitz, S. B., Fahs, M., Shin, C., & Landrigan, P. J. (1997). Occupational injury and illness in the United States: Estimates of costs, morbidity, and mortality. *Archives of Internal Medicine, 157,* 1557–1568.

Loisel, P., Abenhaim, L., Durand, P., Esdaile, J. M., Suissa, S., Gosselin, L., et al. (1997). A population-based, randomized clinical trial on back pain management. *Spine, 22,* 2911–2918.

Loisel, P., Durand, P., Abenhaim, L., Gosselin, L., Simard, R., Turcotte, J., et al. (1994). Management of occupational back pain: The Sherbrooke model. Results of a pilot and feasibility study. *Occupational and Environmental Medicine, 51,* 597–602.

Loisel, P., Durand, P., Gosselin, L., Simard, R., & Turcotte, J. (1996). *La clinique des maus de dos. Un modele de prise en charge, en prevention de la chronicite.* Montreal, Quebec: Centre hopitalier universitaire de Sherbrooke.

Lund, T., Iversen, L., & Poulsen, K. B. (2001). Work environment factors, health, lifestyle and marital status as predictors of job shift and early retirement in physically heavy occupations. *American Journal of Industrial Medicine, 40,* 161–169.

MacKenzie, E., Morris, J. A., Jurkovich, G., Yasui, Y., Cushing, B., Burgess, A., et al. (1998). RTW following injury: The role of economic, social, and job-related factors. *American Journal of Public Health, 88,* 1630–1637.

Maeland, J. G., & Havik, O. E. (1986). RTW after a myocardial infarction: The influence of background factors, work characteristics and illness severity. *Scandinavian Journal of Social Medicine, 14,* 183–195.

Marklund, S. (Ed.). (1995). *Rehabilitering i ett samhällsperspektiv.* Lund, Sweden: Studentlitteratur.

McIntosh, G., Frank, J., Hogg-Johnson, S., Bombardier, C., & Hall, H. (2000). Prognostic factors for time on workers' compensation benefits in a cohort of low back patients. *Spine, 25,* 147–157.

Melles, T., McIntosh, G., & Hall, H. (1995). Provider, payor, and patient outcome expectations in back pain rehabilitation. *Journal of Occupational Rehabilitation, 5,* 57–69.

Nagi, S. Z. (1965). Some conceptual issues in disability and rehabilitation. In M. B. Sussman (Ed.), *Sociology and Rehabilitation* (pp. 100–113). Washington, DC: American Sociological Association.

Oleinick, A., Gluck, J. V., & Guire, K. (1996). Factors affecting first RTW following a compensable occupational back injury. *American Journal of Industrial Medicine, 30,* 540–555.

Oleinick, A., Guire, K., Hawthorne, V. M., Schork, M. A., Gluck, J. V., Lee, B., et al. (1993). Current methods of estimating severity for occupational injuries and illnesses: Data from the 1986 Michigan Comprehensive Compensable Injury and Illness Database. *American Journal of Industrial Medicine, 23,* 231–252.

Peterson, M. A., Reville, R. T., Kaganoff-Stern, R., & Barth, P. S. (1997). *Compensating permanent workplace injuries: A study of the California system* (RAND, MR-920-ICJ). Santa Monica, CA: Institute for Civil Justice.

Polatin, P. B., Gatchel, R. J., Barnes, D., Mayer, H., Arens, C., & Mayer, T. G. (1989). A psychosociomedical prediction model of response to treatment by chronically disabled workers with low-back pain. *Spine, 14,* 956–961.

Pope, A. M., & Tarlov, A. R. (Eds.). (1991). *Disability in America: Toward a national agenda for prevention.* Washington, DC: National Academy Press.

Pransky, G., Benjamin, K., Hill-Fotouhi, C., Himmelstein, J., Fletcher, K. E., Katz, J. N., et al. (2000). Outcomes in work-related upper extremity and low back injuries: Results of a retrospective study. *American Journal of Industrial Medicine, 37,* 400–409.

Prochaska, J. O., DiClemente, C. C., & Norcross, J. C. (1992). In search of how people change: Applications to addictive behaviors. *American Psychologist, 47,* 1102–1114.

Rael, E. G. S. (1992). *An epidemiological study of the incidence and duration of compensated lost time injury for construction workmen, Ontario, 1989: An assessment and application of workers' compensation board and labor force data.* Toronto, Ontario: University of Toronto Press.

Reville, R. T. (1999). The impact of a permanently disabling workplace injury on labor force participation and earnings. In J. Lane (Ed.), *The creation and analysis of linked employer–employee data* (pp. 147–174). Amsterdam: Elsevier Science.

Ronnevik, P. K. (1988). Predicting RTW after acute myocardial infarction: Significance of clinical data, exercise test variables and beta-blocker therapy. *Cardiology, 75,* 230–236.

Ryden, L. A., Molgaard, C. A., & Bobbitt, S. L. (1988). Benefits of a back care and light duty health promotion program in a hospital setting. *Journal of Community Health, 13,* 222–230.

Schmidt, S. H., Oort-Marburger, D., & Meijman, T. F. (1995). Employment after rehabilitation for musculoskeletal impairments: The impact of vocational rehabilitation and working on a trial basis. *Archives of Physical Medicine and Rehabilitation, 76,* 950–954.

Sinclair, S. J., Hogg-Johnson, S. H., Mondloch, M. V., & Shields, S. A. (1997). The effectiveness of an early active intervention program for workers with soft-tissue injuries: The Early Claimant Cohort Study. *Spine, 22*, 2919–2931.

Smith, J., Tarasuk, V., Ferrier, S., & Shannon, H. (1996). Relationship between workers' reports of problems and legitimacy and vulnerability in the workplace and duration of benefits for lost-time musculoskeletal injuries [Abstract]. *American Journal of Epidemiology, 143*, 17.

Smith, J., Tarasuk, V., Shannon, H. & Ferrier, S. (1998). *Prognosis of musculoskeletal disorders: Effects of legitimacy and job vulnerability* (IWH working paper no. 67). Toronto, Ontario: Institute for Work and Health.

Snook, S. H. (1988). The costs of back pain in industry. *Occupational Medicine, 3*, 1–5.

Snook, S. H., & Webster, B. S. (1987). The cost of disability. *Clinical Orthopaedics and Related Research, 221*, 77–84.

Spitzer, W. O., Le Blanc, F. E., Dupuis, M., Abenhaim, L., Belanger, A. Y., Bloch, R. (1987). Scientific approach to the assessment and management of activity-related spinal disorders: A monograph for clinicians. *Spine, 12*(Suppl. 7), 1–59.

Stone, D. A. (1984). *The disabled state*. Philadelphia: Temple University Press.

Syme, S. L. (1990). Control and health: An epidemiological perspective. In K. W. Schaie, J. Rodin, & C. Schooler (Eds.), *Self-directedness: Cause and effects throughout the life course* (pp. 213–229). Hillsdale, NJ: Erlbaum.

Syme, S. L. (1991). Control and health: A personal perspective. *Advances, 7*, 16–27.

Tarasuk, V., & Eakin, J. (1994). Back problems are for life: Perceived vulnerability and its implications for chronic disability. *Journal of Occupational Rehabilitation, 4*, 55–64.

Tarasuk, V., & Eakin, J. (1995). The problem of legitimacy in the experience of work-related back injury. *Qualitative Health Research, 5*, 204–221.

Tate, D. G. (1992). Workers' disability and RTW. *American Journal of Physical Medicine and Rehabilitation, 71*, 92–96.

Theorell, T., Harms-Ringdahl, K., Ahlberg-Hulten, G., & Westin, B. (1991). Psychosocial job factors and symptoms from the locomotor system: A multicausal analysis. *Scandinavian Journal of Rehabilitation Medicine, 23*, 165–173.

U.S. Department of Labor, Bureau of Labor Statistics. (1997). *Lost-worktime injuries: Characteristics and resulting time away from work, 1995*. Washington, DC: U.S. Government Printing Office.

van der Weide, W. E., Verbeck, J. H., Sallé, H. J. A., & van Dijk, F. J. H. (1999). Prognostic factors for chronic disability from acute low back pain in occupational health care. *Scandinavian Journal of Work, Environment and Health, 25*, 50–56.

Verbrugge, L. M., & Jette, A. M. (1994). The disablement process. *Social Science and Medicine, 38*, 1–14.

Von Korff, M. (1994). Studying the natural history of back pain. *Spine, 19*(Suppl. 18), 2041–2046.

West, M. D. (1995). Aspects of the workplace and RTW for persons with brain injury in supported employment. *Brain Injury*, 9, 301–313.

Wiesel, S. W., Boden, S. D., & Feffer, H. L. (1994). A quality-based protocol for management of musculoskeletal injuries A ten-year prospective outcome study. *Clinical Orthopaedics and Related Research, 301*, 164–176.

Williams, S. A., Feuerstein, M., Durbin, D., & Pezzullo, J. (1998). Health care and indemnity costs across the natural history of disability in occupational low back pain. *Spine*, *23*, 2329–2336.

Yassi, A., Khokhar, J. B., Tate, R., Cooper, J. E., Snow, C., & Vallentyne, S. (1995). The epidemiology of back injuries in nurses at a large Canadian tertiary care hospital: Implications for prevention. *Occupational Medicine*, *45*, 215–221.

Yassi, A., Tate, R., Cooper, J. E., Snow, C., Vallentyne, S., & Khokhar, J. B. (1995). Early intervention for back-injured nurses at a large Canadian tertiary care hospital: An evaluation of the effectiveness and cost benefits of a two-year pilot project. *Occupational Medicine (Oxford)*, *45*, 209–214.

Yelin, E. (1986). The myth of malingering: Why individuals withdraw from work in the presence of illness. *Milbank Quarterly*, *64*, 622–649.

Yelin, E., Meenan, R., Nevitt, M., & Epstein, W. V. (1980). Work disability in rheumatoid arthritis: Effects of disease, social, and work factors. *Annals of Internal Medicine*, 93, 551–556.

Yelin, E., Henke, C. J., & Epstein, W. V. (1986). Work disability among persons with musculoskeletal conditions. *Arthritis and Rheumatism*, *29*, 1322–1333.

IV

CONCLUSIONS

14

COMMON THEMES AND FUTURE DIRECTIONS

MICHAEL R. FRONE AND JULIAN BARLING

We point out in chapter 1 that despite evidence of interest dating back 75 to 100 years, industrial/organizational psychology has not played a major role in the study of unintentional occupational injuries. Nonetheless, the contributions to this volume collectively provide a broad sampling of avenues where industrial/organizational psychology can play an important role in understanding the etiology and prevention of unintentional occupational injuries. We also anticipate that the role of psychology in the study of workplace safety may become more prominent with recent growth in the interdisciplinary specialty known as occupational health psychology. Occupational health psychology "applies psychology in organizational settings for the improvement of work life, the protection and safety of workers, and the promotion of healthy work" (Quick, 1999, p. 123). More information on the history and recent developments in this area can be found in the *Handbook of Occupational Health Psychology* (Quick & Tetrick, 2003); the *Journal of Occupational Health Psychology*; and several introductory articles (Quick, 1999; Quick et al., 1997; Sauter & Hurrell, 1999; Schneider, Camara, Tetrick, & Stenberg, 1999).

In this brief concluding chapter, we first summarize several common themes that emerge throughout this book. We then end with a discussion of some general avenues for future research on the psychology of workplace safety that go beyond the many suggestions already made in the individual chapters.

COMMON THEMES

The chapters in this volume cover a wide range of important issues and topics on the etiology and prevention of unintentional occupational injuries. Many issues are addressed in more than one chapter, often from a slightly different vantage point. Before turning to common themes, we would like to point out that the chapters collectively make clear that occupational injuries and other relevant safety outcomes are affected by a variety of working conditions, climates, and cultures. The management of occupational injuries falls squarely on traditional organizational human resource management functions, such as compensation and benefits, leadership development, and employee training, as well as more recent developments that are part of high-performance work systems. The primary stakeholders in workplace safety include employees, employers, unions, and governments. It is important to pay attention to potentially vulnerable populations of employees, such as young workers, contingent workers, and individuals who use alcohol and other drugs. Finally, not only do we need to understand the dynamic causes of working safely and occupational injuries, but we also need to understand the dynamic process of returning individuals to work after experiencing a debilitating injury at work.

Among the important general themes is the need to consider occupational injuries from multiple levels (e.g., Bleise & Jex, 2002; Klein, Dansereau, & Hall, 1994; Klein & Kozlowski, 2000), which includes individual employees, work groups and teams, and the organization. Exploration of the causes of occupational injuries can occur at each of these three levels. For example, individuals' work demands or on-the-job substance use may lead to injuries among the afflicted individual employees and to the people who work around them. Likewise, differences in team structure may lead to higher injury rates across work teams. The predictors of occupational injuries and other relevant workplace safety outcomes may operate across levels. For example, organizational climates that place a low value on safety performance may cause lower levels of safety compliance and an increased likelihood of being injured among individual employees.

Another related theme is context (Johns, 2001), which broadly represents the mediating and moderating processes that explain how and when a given causal force may enhance or compromise workplace safety. For example, we have learned that when alcohol and drugs are used at levels, times, and places that do not result in direct on-the-job impairment, there is little reason to expect that employees who use alcohol and other drugs will experience increased rates of occupational injury. Whether or not autonomous work groups reduce the risk of injury may depend on the social environment within the work groups, and the external environment in which the work groups need to operate. The success of safety training depends on a number of contextual factors, such as whether employees are given the resources and support to enact the recommended procedures. Job insecurity may lead to employee injuries because it undermines the motivation to follow safety policies and procedures, and the acquisition of knowledge regarding safety policies and procedures. Virtually every chapter discusses the role of context or presents a conceptual model that depicts key contextual factors that help explain how and when a specific causal factor will affect occupational injuries or related workplace safety outcomes.

A final theme is that a precise understanding of the etiologic process underlying unintentional occupational injuries may require research based on a multidimensional view of workplace safety. The outcome of interest in this volume is unintentional occupational injuries, which can range from fatal injuries to major nonfatal injury to minor nonfatal injury. However, a number of other important outcomes can be gleaned from the chapters. Although not addressed in detail in any of the chapters, the issue of near-miss events is mentioned. A near-miss event represents an event that did not result in injury but could have resulted in an injury. Similarly, Zohar (2000, 2002) introduced the notion of what he terms "microaccidents," which represent injuries in the workplace that may require a visit to the first aid station but do not require time off work (and therefore would not appear in official statistics). Clearly, near-miss events represent noninjury events and microaccidents refer to minor injuries. A primary reason to study near-miss events and minor injuries is to avoid problems associated with the analysis of low-base-rate outcomes. In other words, in any single study, the number of individuals experiencing an unintentional occupational injury declines quickly as we begin to focus on more severe and reportable injuries. However, the value of using near-miss events as a proxy for occupational injuries in order to better understand the etiology of the latter hinges on the extent to which near-miss events are predictive of actual occupational injuries. Some recent research suggests that, relative to workers who do not report near-miss events, those who do are also more likely to report an occupational injury (e.g., Hayes, Perander, Smecko, & Trask, 1998; Lilley, Feyer, Kirk, & Gander, 2002; Zacharatos & Barling, 2003). And Zohar (2000, 2002) has begun to demonstrate the value of studying microaccidents (i.e., minor injuries). In addition to near-miss events and microaccidents, several chapters in this volume suggest that measures of safety compliance, safety initiative, and procedural errors represent important process outcomes in their own right and have been found to precede the occurrence of occupational injuries. Finally, returning to work is an important outcome in the debilitating injury cycle.

FUTURE DIRECTIONS

Most chapters in this volume explore factors that can be viewed as antecedents of unintentional occupational injuries. The outcomes of occupational injuries can be classified as economic (e.g., loss of income); clinical (e.g., range of physical motion); and social (e.g., family performance; Keller, 2001). For psychologists, it is important to note that the social outcomes of occupational injuries have received much less research attention than the economic and clinical outcomes. We need a better understanding of the short- and long-term impact of various occupational injuries on employees' work attitudes and performance (Barling, Kelloway, & Iverson, 2003), and more broadly on employees' quality of life and that of their social networks, which includes family and friends. Also, we need to explore the effect of occupational injuries on employees' ability to meet the demands and responsibilities of their non-work-related social roles, such as spouse, parent, and community member.

Most of the research summarized in this volume was based on quantitative cross-sectional survey designs. Therefore, there is a need for quantitative research that will provide a stronger basis for drawing causal inferences. Research on the causal antecedents of occupational injuries and on the effectiveness of preventive intervention efforts aimed at reducing or eliminating known risk factors for occupational injuries needs to make use of more rigorous experimental, quasi-experimental, and longitudinal designs that will allow tests of causal processes as they unfold over time. Although the use of such designs by psychologists in research on occupational injuries is extremely rare, they are well known to psychologists (e.g., Cook, Campbell, & Peracchio, 1990; Reis & Judd, 2000; Sackett & Larson, 1990; Shadish, 2002). However, studying the causes of unintentional occupational injuries can present some special problems for many traditional research designs. For example, as mentioned earlier, more severe injuries represent a low-base-rate outcome and some of the causes of work injuries are transient and have an acute effect. Therefore, in addition to focusing on near misses and microaccidents, psychologists may also benefit from quantitative research designs developed in epidemiology. For example, the causes of low-base-rate outcomes, such as disabling occupational injuries, may be more efficiently studied using a case-control design, whereas the study of risk factors for occupational injuries that are intermittently present and transient in their effect (e.g., alcohol or drug use) can be studied using a case-crossover design. (For more details on these and other epidemiologic study designs, see Mittleman, Maldonado, Gerberich, Smith, & Sorock, 1997; Rothman & Greenland, 1998; Sorock & Courtney, 1997.)

We also should point to opportunities for psychologists to use analog and experimental studies to explore the causes of occupational injuries. Probst (2002) recently conducted an experimental study in which job insecurity was manipulated in a group of students, and its effects on safety process outcomes and job performance were studied. Although such studies will always have to confront the issue of ecological validity, the consideration of experimental designs will encourage and challenge psychologists to think about innovative ways in which the etiologic processes leading to occupational injuries can be studied.

In addition to better quantitative research, there is a need for more qualitative research. As noted earlier, a common theme across the chapters in this volume is that the impact of the causal antecedents of injuries and the success of intervention efforts are likely to depend on a number of contextual factors. However, relevant contextual influences may not be known and may not be obvious. They also may be unique to specific settings. Further, theoretical models of workplace safety processes and occupational injuries may not be developed to the point that all relevant causes and all potential outcomes of occupational injuries are known. Therefore, qualitative or ethnographic research methods offer a unique means to develop better models of workplace safety that can be subsequently subjected to quantitative testing. Among the techniques that may be particularly useful are semistructured interviews, focus groups, and participant observation (Atkinson, Coffey, Delamont, Lofland, & Lofland, 2001; Bernard, 2000; Flick, 2002; MacDougall & Fudge, 2001; Powell, Single, & Lloyd, 1996).

Finally, one cannot bring up research methods without some discussion of data analysis. Psychologists are well acquainted with the general linear model through the use of ordinary least squares (OLS) regression and analysis of variance. However, many outcomes reflecting the occurrence, frequency, severity, or type of occupational injuries do not take on the necessary distributional characteristics required for the appropriate use of OLS regression or analysis of variance: continuous and unbounded. The outcomes of interest may be dichotomous (injured vs. not injured), discrete counts of rare events (number of injuries experienced during the preceding 6 months), or nominal categories (types of injuries). Thus, psychologists exploring the etiology of occupational injuries need to rely on a broader set of regression models developed for categorical and limited dependent variables, such as logistic or probit regression for dichotomous outcomes, Poisson or negative binomial regression for count outcomes, and multinomial logistic regression for nominal outcomes. Introductions to these and other relevant nonlinear regression models can be found in Long (1997) and Hardin and Hilbe (2001). As mentioned earlier, an additional complication is that causes of occupational injuries can occur at multiple levels simultaneously and may need to be assessed at multiple points in time. Extensions of linear and nonlinear regression models to multiple levels of analysis and to longitudinal and clustered data have been developed (e.g., Burton, Gurrin, & Sly, 1998; Raudenbush & Bryk, 2002).

In conclusion, all of the participating chapter authors point to ways in which psychological knowledge and research methods can be valuable in providing a greater understanding of the nature, causes, and consequences of unintentional occupational injuries. With continued research by psychologists specializing in industrial/organizational psychology and occupational health psychology, workplaces can be made safer, thereby enhancing employees' physical health and psychological well-being.

REFERENCES

Atkinson, P. A., Coffey, A. J., Delamont, S., Lofland, J., & Lofland, L. H. (2001). *Handbook of ethnography.* Thousand Oaks, CA: Sage.

Barling, J., Kelloway, E. K., & Iverson, R. (2003). Accidental outcomes: Attitudinal consequences of workplace injuries. *Journal of Occupational Health Psychology, 8,* 74–85.

Bernard, H. R. (2000). *Social research methods: Qualitative and quantitative approaches.* Thousand Oaks, CA: Sage.

Bleise, P. D., & Jex, S. M. (2002). Incorporating a multilevel perspective into occupational stress research: Theoretical, methodological, and practical implications. *Journal of Occupational Health Psychology, 7,* 265–276.

Burton, P., Gurrin, L., & Sly, P. (1998). Extending the simple linear regression model to account for correlated responses: An introduction to generalized estimating equations and multi-level mixed modeling. *Statistics in Medicine, 17,* 1261–1291.

Cook, T. D., Campbell, D. T., & Peracchio, L. (1990). Quasi experimentation. In M. D. Dunnette & L. M. Hough (Eds.), *Handbook of industrial and organizational psychology* (2nd ed., Vol. 1, pp. 491–576). Palo Alto, CA: Consulting Psychologists Press.

Flick, U. (2002). *An introduction to qualitative research.* Thousand Oaks, CA: Sage.

Hardin, J., & Hilbe, J. (2001). *Generalized linear models and extensions.* College Station, TX: Stata Corporation.

Hayes, B. E., Perander, J., Smecko, T., & Trask, J. (1998). Measuring perceptions of workplace safety: Development and validation of the work safety scale. *Journal of Safety Research, 29,* 145–161.

Johns, G. (2001). In praise of context. *Journal of Organizational Behavior, 22,* 31–42.

Keller, S. D. (2001). Quantifying social consequences of occupational injuries and illnesses: State of the art and research agenda. *American Journal of Industrial Medicine, 40*, 438–451.

Klein, K. J., Dansereau, F., & Hall, R. J. (1994). Levels issues in theory development, data collection, and analysis. *Academy of Management Review, 19*, 105–229.

Klein, K. J., & Kozlowski, S. W. J. (Eds.). (2000). *Multilevel theory, research, and methods in organizations: Foundations, extensions, and new directions*. New York: Wiley.

Lilley, R., Feyer, A. M., Kirk, P., & Gander, P. J. (2002). A survey of forest workers in New Zealand: Do hours of work, rest and recovery play a role in accidents and injury? *Journal of Safety Research, 33*, 53–71.

Long, J. S. (1997). *Regression models for categorical and limited dependent variables*. Thousand Oaks, CA: Sage.

MacDougall, C., & Fudge, E. (2001). Planning and recruiting the sample for focus groups and in-depth interviews. *Qualitative Health Research, 11*, 117–126.

Mittleman, M. A., Maldonado, G., Gerberich, S. G., Smith, G. S., & Sorock, G. S. (1997). Alternative approaches to analytic designs in occupational injury epidemiology. *American Journal of Industrial Medicine, 32*, 128–141.

Powell, R. A., Single, H. M., & Lloyd, K. R. (1996). Focus groups in mental health research: Enhancing the validity of user and provider questionnaires. *International Journal of Social Psychiatry, 42*, 193–206.

Probst, T. M. (2002). Layoffs and tradeoffs: Production, quality, and safety demands under the threat of layoff. *Journal of Occupational Health Psychology, 7*, 211–220.

Quick, J. C. (1999). Occupational health psychology: The convergence of health and clinical psychology with public health and preventive medicine. *Professional Psychology: Research and Practice, 30*, 123–128.

Quick, J. C., Camara, W. J., Hurrell, J. J., Jr., Johnson, J. V., Piotrowski, C. S. Sauter, S. L., et al. (1997). Introduction and overview. *Journal of Occupational Health Psychology, 2*, 3–6.

Quick, J. C., & Tetrick, L. E. (Eds.). (2003). *Handbook of occupational health psychology*. Washington, DC: American Psychological Association.

Raudenbush, S. W., & Bryk, A. S. (2002). *Hierarchical linear models: Applications and data analysis methods* (2nd ed.). Thousand Oaks, CA: Sage.

Reis, H. T., & Judd, C. M. (Eds.). (2000). *Handbook of research methods in social and personality psychology*. New York: Cambridge University Press.

Rothman, K. J., & Greenland, S. (1998). *Modern epidemiology* (2nd ed.). New York: Lippincott Williams & Wilkins.

Sackett, P. R., & Larson, J. R., Jr. (1990). Research strategies and tactics in industrial and organizational psychology. In M. D. Dunnette & L. M. Hough (Eds.), *Handbook of industrial and organizational psychology* (2nd ed., Vol. 1, pp. 419–489). Palo Alto, CA: Consulting Psychologists Press.

Sauter, S. L., & Hurrell, J. J., Jr. (1999). Occupational health psychology: Origins, content, and direction. *Professional Psychology: Research and Practice, 30,* 117–122.

Schneider, D. L., Camara, W. J., Tetrick, L. E., & Stenberg, C. R. (1999). Training in occupational health psychology: Initial efforts and alternative models. *Professional Psychology: Research and Practice, 30,* 138–142.

Shadish, W. R. (2002). Revisiting field experimentation: Field notes for the future. *Psychological Methods, 7,* 3–18.

Sorock, G. S., & Courtney, T. K. (1997). Advancing analytic epidemiologic studies of occupational injuries. *Safety Science, 25,* 29–43.

Zacharatos, A., & Barling, J. (2003). *High performance work systems and occupational safety.* Manuscript submitted for publication.

Zohar, D. (2000). A group-level model of safety climate: Testing the effect of group climate on microaccidents in manufacturing jobs. *Journal of Applied Psychology, 85,* 587–596.

Zohar, D. (2002). The effects of leadership dimensions, safety climate and assigned priorities on minor injuries in work groups. *Journal of Organizational Behavior, 23,* 75–92.

AUTHOR INDEX

Numbers in italics refer to the listings in the reference sections.

SUBJECT INDEX

ABOUT THE EDITORS

Julian Barling is the Queen's Research Chair in the School of Business, Queen's University, Kingston, Ontario. He earned his PhD from the University of the Witwatersrand, South Africa. His research focuses on the different aspects of employee well-being and employee safety, leadership, and workplace violence. He is the author and editor of several books. Dr. Barling is the editor of the American Psychological Association's (APA) *Journal of Occupational Health Psychology* and serves on the editorial boards of the *Journal of Applied Psychology*, *Leadership and Organizational Development Journal*, and *Stress and Health*. He previously served as consulting editor of the *Journal of Organizational Behavior* and is chair of the APA's Task Force on Workplace Violence. In 1997, Dr. Barling received the annual award for Excellence in Research from Queen's University; in 2001, he received the National Post's Leaders in Business Education award; and in 2002, he was elected a fellow of the Royal Society of Canada.

Michael R. Frone is a senior research scientist at the Research Institute on Addictions, State University of New York at Buffalo. He earned his PhD from the State University of New York at Buffalo. He has published extensively in leading journals on work–family dynamics and the work-related predictors of employee mental health, physical health, and substance use. Dr. Frone is associate editor of the *Journal of Occupational Health Psychology* and serves or has served on the editorial boards of the *Journal of Applied Psychology*, *Journal of Occupational Health Psychology*, *Journal of Organizational Behavior*, *Organizational Behavior and Human Decision Processes*, and *Organizational Research Methods*. He has been principal investigator or coinvestigator on research grants totaling more than $5.5 million. Dr. Frone is currently the principal investigator on a grant from the National Institutes of Health to conduct a large national telephone survey of workplace health and safety.